D1708691

Developmental Robotics

Intelligent Robotics and Autonomous Agents

Edited by Ronald C. Arkin

Dorigo, Marco, and Marco Colombetti, *Robot Shaping: An Experiment in Behavior Engineering*

Arkin, Ronald C., *Behavior-Based Robotics*

Stone, Peter, *Layered Learning in Multiagent Systems: A Winning Approach to Robotic Soccer*

Wooldridge, Michael, *Reasoning about Rational Agents*

Murphy, Robin R., *An Introduction to AI Robotics*

Mason, Matthew T., *Mechanics of Robotic Manipulation*

Kraus, Sarit, *Strategic Negotiation in Multiagent Environments*

Nolfi, Stefano, and Dario Floreano, *Evolutionary Robotics: The Biology, Intelligence, and Technology of Self-Organizing Machines*

Siegwart, Roland, and Illah R. Nourbakhsh, *Introduction to Autonomous Mobile Robots*

Breazeal, Cynthia L., *Designing Sociable Robots*

Bekey, George A., *Autonomous Robots: From Biological Inspiration to Implementation and Control*

Choset, Howie, Kevin M. Lynch, Seth Hutchinson, George Kantor, Wolfram Burgard, Lydia E. Kavraki, and Sebastian Thrun, *Principles of Robot Motion: Theory, Algorithms, and Implementations*

Thrun, Sebastian, Wolfram Burgard, and Dieter Fox, *Probabilistic Robotics*

Matarić, Maja J., *The Robotics Primer*

Wellman, Michael P., Amy Greenwald, and Peter Stone, *Autonomous Bidding Agents: Strategies and Lessons from the Trading Agent Competition*

Floreano, Dario, and Claudio Mattiussi, *Bio-Inspired Artificial Intelligence: Theories, Methods, and Technologies*

Sterling, Leon S., and Kuldar Taveter, *The Art of Agent-Oriented Modeling*

Stoy, Kasper, David Brandt, and David J. Christensen, *An Introduction to Self-Reconfigurable Robots*

Lin, Patrick, Keith Abney, and George A. Bekey, editors, *Robot Ethics: The Ethical and Social Implications of Robotics*

Weiss, Gerhard, editor, *Multiagent Systems*, second edition

Vargas, Patricia A., Ezequiel A. Di Paolo, Inman Harvey, and Phil Husbands, editors, *The Horizons of Evolutionary Robotics*

Murphy, Robin R., *Disaster Robotics*

Cangelosi, Angelo, and Matthew Schlesinger,

Developmental Robotics: From Babies to Robots

Developmental Robotics

From Babies to Robots

Angelo Cangelosi and Matthew Schlesinger

The MIT Press
Cambridge, Massachusetts
London, England

MIT Press books may be purchased at special quantity discounts for business or sales promotional use. For information, please email special_sales@mitpress.mit.edu.

This book was set in ITC Stone Serif Std by Toppan Best-set Premedia Limited, Hong Kong. Printed and bound in the United States of America.

Library of Congress Cataloging-in-Publication Data

Cangelosi, Angelo, 1967–
Developmental robotics : from babies to robots / Angelo Cangelosi and Matthew Schlesinger ; foreword by Linda B. Smith.
pages cm.—(Intelligent robotics and autonomous agents)
Includes bibliographical references and index.
ISBN 978-0-262-02801-1 (hardcover : alk. paper) 1. Autonomous robots. 2. Machine learning. 3. Robotics. 4. Self-organizing systems. I. Schlesinger, Matthew, 1967– II. Title.
TJ211.495.C36 2015
629.8'92—dc23
2014012489

10 9 8 7 6 5 4 3 2 1

To my parents Vita and Salvatore (AC)
To Angie, Nick, and Natalie (MS)

Contents

Foreword

Linda B. Smith

The dominant method of science is analysis and simplification. This was clearly articulated by Descartes in 1628: In studying any phenomenon, simplify it to its essential components, dissecting away everything else. This approach is motivated by the belief that complicated systems will be best understood at the lowest possible level. By reducing explanations to the smallest possible entities, the hope is that we will find entities that are simple enough to fully analyze and explain. The spectacular success of this methodology in modern science is undeniable. Unfortunately, it has not given us an understanding of how systems made up of simple elements can operate with sufficient complexity to be autonomous agents. Building artificial agents who can act and adapt in complex and varying environments requires a different kind of science, one that is principally about integration and complexity rather than analysis and simplification. The theoretical task of understanding of developmental process in biological systems also requires a science of integration.

Developmental robotics is based on the premise that principles of developmental process are the key to engineering adaptive and fluid intelligence. Although the promise of this idea is not yet fully realized, remarkable progress has been made over the last decade and half. This book presents the current state of the art. In so doing, the authors also make a case for deeper collaborations between developmental roboticists and developmental psychologists. At present the ties are weak. We are working on related problems, reading the same literatures, sometimes participating in joint conferences, but only rarely actually collaborating in a sustained way. I firmly believe that remarkable gains could be made in both fields through programmatic research by teams of researchers in human development and robotics. For developmental psychology, the promise is both better theory and new ways to test theories by manipulating the pathways and experiences using artificial developing intelligent systems. Accordingly, in this foreword, I highlight seven fundamental aspects of the human developmental process that might be better understood through developmental robotics.

1. *Extended immaturity*. Development, like evolution and culture, is a process that creates complexity by accumulating change. At any moment, the developing agent is

a product of all previous developments, and any new change begins with and must build on those previous developments. Biological systems that are flexibly smart have relatively long periods of immaturity. Why is this? Why and how does "slow accumulative" intelligence yield higher and more abstract forms of cognition? One possibility is that a slow accumulative system—one that does not settle too fast—can acquire the massive amounts of experience that yield multiple layers of knowledge at multiple granularities. A second related possibility concerns what developmentalists sometimes call "readiness" and what recent research in robotics has called "learning progression."[1] As learning progresses, new structures and new ways of learning emerge so that that the same experiences later in development have different effects on the learning system than those experiences earlier in development. If these ideas are correct, then the developmental pathway itself may be part of the explanation as to why human intelligence has the properties that it does. It simply may not be possible to shortcut development—to try to build just the adult system—and achieve fluid and adaptive intelligence that characterizes biologically developing systems.

2. *Activity*. Learning experiences do not passively "happen" to infants. Piaget[2] described a pattern of infant activity that is highly illustrative of this point. He placed a rattle in a four-month-old infant's hands. As the infant moved the rattle, it would both come into sight and also make a noise, arousing and agitating the infant and causing more body motions, and thus causing the rattle to move into and out of sight and to make more noise. The infant has no prior knowledge of the rattle but discovers—through activity—the task and goal of rattle shaking. As the infant accidentally moves the rattle, and sees and hears the consequences, the infant will become captured by the activity—moving and shaking, looking and listening—and incrementally through this repeated action gain intentional control over the shaking of the rattle and the goal of making noise. Action and exploration creates opportunities for learning and new tasks to be conquered. This role of action is well covered in this book and is an area in which developmental robotics is clearly demonstrating its relevance to theories of development.

3. *Overlapping tasks*. Developing organisms do not solve just one task; they solve many overlapping tasks. Consider again the rattle example. The infant's shaking of the rattle couples auditory, motor, and visual systems creating and changing the specialized regions in the brain and the connections between them.[4] But these same systems and functional connections enter into many other behaviors and so achievements in shaking rattles may extend to influence means-end reasoning and or the processing of multimodal synchronicities. Developmental theory deeply needs a way to explore how multimodal and multitask experiences create an abstract, general purpose, and inventive intelligence. This is also an area in which developmental robotics is ready to make big contributions.

4. *Degeneracy*. Degeneracy as it is used in computational neuroscience[3] refers to complex systems in which individual components may contribute to many different functions and in which there is more than one route to the same functional end. Degeneracy is believed to promote robustness in developmental outcomes. Because functionally redundant pathways can compensate for one another, they provide a kind of insurance against pathway failure. Robotic models may exploit these principles to build systems that are robust and that can succeed—over the long term and in multiple tasks—even given breakdowns in some components. Such robotic models also offer a rigorous way to test the implications of multicausality and complex systems of causes may constrain developmental outcome.

5. *Cascades*. Developmental theorists often refer to the far reach of early developments on later ones in terms of the "developmental cascade." These cascades often evident in the perturbed patterns of atypical development also characterize typical development and such seemingly distinct domains of intelligence as sitting and visual object representation and walking and language input.[4] Here is the deeper theoretical question: Are the facts of these cascades—the way earlier developments start the pathway for quite different later developments—relevant to how and why human intelligence has the properties that it does? Developmental robotics may not only advance the engineering of robots by taking on this question, but also provide a platform for understanding how the integrative nature the complex pathways that characterize human cognitive development are essential to human intelligence.

6. *Ordered tasks*. Biologically developing systems typically confront classes of experiences and tasks in a particular sequence and there is a large theoretical and experimental literature on the cascading developmental consequences of altering that natural order of sensorimotor development in animals.[5] Human infants travel through a systematic of set of changing environments in the first two years of life as they proceed to rolling over, reaching, sitting steadily, crawling, and walking. The series of changes in motor skills in the first two years of human life provide strong and most likely evolutionarily selected gates on experience. The consequences and importance of ordered experiences and the significance of perturbations in that ordering have not been theoretically well specified in humans nor systematically pursued in developmental robotics; this is an important next frontier.

7. *Individualism*. It is the individual that develops. The history of the species may be in the intrinsic biology and environment may contain conspecifics who scaffold development but each developing organism has to travel the path. Because developmental pathways are degenerate, because development builds on itself, because intrinsic biologies and environments are inherently unique, different developing agents may come to comparable functional skills through different paths. This is a theoretically important idea to understanding both the robustness and variability in human intelligence and

perhaps also a foundational idea for building multifunctional adaptive robots that can be intelligent in whatever environment they find themselves in.

This book is an excellent steppingstone to future advances in developmental science.

Notes

1. Gottlieb, J., P. Y. Oudeyer, M. Lopes, and A. Baranes, "Information-Seeking, Curiosity, and Attention: Computational and Neural Mechanisms," *Trends in Cognitive Science* 17 (11) (2013): 585–593.

2. J. Piaget, *The Origins of Intelligence in the Child,* trans. M. Cook (New York: International Universities Press, 1952. (Original work published in 1936.)

3. O. Sporns, *Networks of the Brain* (Cambridge, MA: MIT Press, 2011).

4. L. Byrge, O. Sporns, and L. B. Smith, "Developmental Process Emerges from Extended Brain-Body-Behavior Networks," *Trends in Cognitive Science* (in press); L. B. Smith, "It's All Connected: Pathways in Visual Object Recognition and Early Noun Learning," *American Psychologist* 68 (8) (2014): 618.

5. G. Turkewitz, and P. A. Kenny, "Limitations on Input as a Basis for Neural Organization and Perceptual Development: A Preliminary Theoretical Statement," *Developmental Psychobiology* 15 (4) (1982): 357–368.

Preface

Instead of trying to produce a programme to simulate the adult mind, why not rather try to produce one which simulates the child's? If this were then subjected to an appropriate course of education one would obtain the adult brain.

—Alan Turing, "Computing Machinery and Intelligence"

The idea that the human child can be used as a template for designing an intelligent machine is rooted in the early days of modern artificial intelligence (AI). Alan Turing was part of a large community of researchers in the interdisciplinary field of cognitive science, which included Marvin Minsky, Jean Piaget, Noam Chomsky, and Herbert Simon, who collectively argued that the same principles could be used to study both biological organisms and "artificial" or man-made systems. Nevertheless, over the next fifty years the concept of *child-inspired* AI failed to gain widespread appeal, and instead made only sporadic progress. By 2000, however, a critical mass of researchers had formed in psychology, computer science, linguistics, robotics, neuroscience, and a number of other related disciplines, and as we highlight in chapter 1, two new scientific communities were established (autonomous mental development; epigenetic robotics) and two conference series (IEEE ICDL: IEEE International Conference on Development and Learning; EpiRob: International Workshop on Epigenetic Robotics) and an international IEEE journal (*IEEE Transactions in Autonomous Mental Development*) were subsequently launched, all devoted to the study of developmental robotics.

It is now just over a decade later, and the two groups have merged into a unified research community (see icdl-epirob.org). The time is right for a comprehensive volume that not only surveys the previous dozen years of work in the interdisciplinary field of developmental robotics, but more important, also articulates the core principles that shape and guide the discipline.

There are three key goals that we pursued while writing our book. First, much of the decision making about what to include (as well as the technical level at which it was presented) followed the premise that the book should be *broadly accessible*. In particular, whether the reader is an engineer or a philosopher, an anthropologist or a neuroscientist, a developmental psychologist or a roboticist, our objective was to ensure that a wide audience could read and comfortably digest what is often relatively complex material. On this note, we also envisioned that our text would be a good fit for both advanced undergraduate students and graduate students in the engineering, biological, and social sciences, as well as the humanities.

Our second goal was to deliberately take a *behavior-centered approach*, which means we focused on robotics research that could be mapped in a relatively direct way to comparable studies with human infants and children. In other words, we highlight here robotics work (or more broadly speaking, computational models) that either seeks to directly simulate and replicate a specific developmental study, or more generally, to capture a well-defined developmental phenomenon (e.g., the emergence of crawling, first words, face perception, etc.).

This lays the foundation for our third goal, which was to demonstrate the *collaborative, interdisciplinary nature of developmental robotics*. Thus, an important benefit gained by focusing on embodied, perceiving, acting, autonomous agents is that we can then illustrate a variety of examples in which ongoing work in the developmental sciences is informed by parallel efforts in robotics, engineering, and computer science, and vice versa. As part of this goal, in each chapter we strategically selected and profiled a specific human-developmental study, and where possible, also presented a comparable robotics study that was intentionally designed to simulate the same task, behavior, or developmental phenomenon. We hope that by juxtaposing these analogous studies of natural and artificial organisms, we can make a clear and convincing case that humans and machines indeed have much to learn from each other!

Acknowledgments

This volume is the result not only of the effort of the two authors, but also the contribution from the wider community of collaborators in our own labs, and within the broader international community of developmental robotics.

Many colleagues kindly and patiently offered to go through some sections of the draft manuscript, especially to make sure that our description of their models and experiments was correct and clear. In particular, we would like to thank the following colleagues for their review and feedback on specific sections of chapter 2 (and for providing images of their own baby robots): Gordon Cheng, Paul Baxter, Minoru Asada, Yasuo Kuniyoshi, Hiroshi Ishiguro, Hisashi Ishihara, Giorgio Metta, Vadim Tikhanoff, Hideki Kozima, Kerstin Dautenhahn, William De Braekeleer (Honda Motor Europe), Oliver Michel (Cyberobotics), Jean-Christophe Baillie and Aurea Sequeira (Aldebaran Robotics), and Masahiro Fujita (SONY Corporation). Lisa Meeden provided feedback on chapter 3, and Daniele Caligiore reviewed parts of chapter 5. Verena Hafner, Peter Dominey, Yukie Nagai, and Yiannis Demiris reviewed sections of chapter 6. Chapter 7 was reviewed by Anthony Morse (who contributed box 7.2), Caroline Lyon, Joe Saunders, Holger Brandl, Christian Goerick, Vadim Tikhanoff, and Pierre-Yves Oudeyer (extra thanks to Pierre-Yves for kindly providing feedback on many other chapters). Marek Rucinki (who contributed box 8.2) and Stephen Gordon provided feedback on chapter 8. Kerstin Dautenhahn and Tony Belpaeme reviewed the section on assistive robotics in chapter 9. Moreover, the recommendation and feedback from the three referees was invaluable to improve the final version of the monograph. We are also grateful to many colleagues who gave us the original image files of many of the figures in the book (their names are acknowledged in the figure captions).

We are particularly grateful to the PhD students and postdocs at the Plymouth Centre for Robotics and Neural Systems for help on the book formatting, figures, and references, in particular: Robin Read (who also produced some of the figures), Ricardo de Azambuja, Giovanni Sirio Carmantini (again, for producing some of the drawings and figures), Giulia Dellaria (for her hard work on the name and subject indexes), Matt Rule (who had to check hundreds of references!), and Elena Dell'Aquila.

We also thank the staff at MIT Press, Ada Brunstein for her original enthusiastic support of this book proposal, Marie L. Lee and Marc Lowenthal for their continuous support in the later stages of the manuscript preparation, and Kathleen Caruso and Julia Collins for the help in the editing of the final manuscript.

Cangelosi would also like to thank Rolf Pfeifer who, without realizing it, gave the decisive motivational push to embark on the book project by inspiring Cangelosi with his influential books on embodied intelligence.

This volume received essential and generous support from research grants from the European Union Framework 7 Programme (through the projects ITALK and POETICON++ and the Marie Curie ITN ROBOT-DOC), the UK Engineering and Physical Sciences Research Council (BABEL project), and the U.S. Air Force Office of Science and Research (through the EOARD grant on distributed communication).

And finally, we extend a big and heartfelt thank you to our families for their patience when we had to steal time from them to work on this book. We hope they will agree that this was, after all, "time well spent" and that they might even enjoy reading all about baby robots.

1 Growing Babies and Robots

Human development is one of the most fascinating phenomena in nature. Babies are born as helpless individuals, with simple motor and cognitive skills not even sufficient to allow them to survive and fend for themselves without the support of their parents and caregivers. However, within a few years, they reach a sophisticated level of mental development. A ten-year-old child can play chess and computer games, solve increasingly complex math problems, master one or more languages, build a theory of mind of self and others, cooperate altruistically with peers and adults, excel at gym exercises, and use complex tools and machines. These slow but impressive developmental changes pose a series of key questions on the understanding of human development: What are the mechanisms that allow the child to develop autonomously such mental capabilities? How does the social and physical environment, with which the child interacts, shape and scaffold the child's developing cognitive skills and knowledge? What is the relative contribution of *nature* (i.e., genes) and *nurture* (i.e., environment) in the development of human intelligence? What do qualitative stages during development, and body and brain maturational changes tell us about the mechanisms and principles supporting development?

Developmental psychology is the discipline that aims at understanding the child's autonomous mental development, through field and laboratory experiments with children of different ages and varying cultural backgrounds, and through comparative psychology studies. These empirical investigations lead to the definition of theories and hypotheses of motor, cognitive, and social development and to the identification of general developmental principles underlying the acquisition of mental capabilities.

Such a growing set of empirical data and theoretical knowledge on human development, in addition to benefiting human sciences such as psychology, philosophy, and cognitive science, can have tremendous technological implications. If we understand the underlying principles and mechanisms of the development of natural cognition in human babies through social interaction, we can use this knowledge to inform the design of cognitive capabilities in artificial agents such as robots. Such principles and mechanisms can be implemented in the cognitive architecture of robots and tested through developmental experiments with robots. This is the aim of developmental robotics, and this volume will explore the current achievements and challenges in the design of autonomous mental development via social interaction in robots and the benefit of a mutual interaction between developmental psychologists and developmental roboticists.

1.1 Developmental Theories of Nature and Nurture

One of the oldest, and endless, debates in psychology, as well as in philosophy, is the contribution of nature and nurture in the development of human intelligence. The baby's prolonged interaction with its physical and social environment is essential to, and significantly influences, its full mental development. At the same time, the baby's genome plays a fundamental role both in the physical and cognitive development of the child. Some traits, especially physical body characteristics, but also cognitive skills such as color perception, can be strongly determined by the baby's own genes, with little influence of environmental phenomena.

This debate has led to various developmental psychology theories on the role of nature versus nurture (Croker 2012). Nativist theories tend to stress the fact that children are born with innate, domain-specific knowledge, which is the result of direct influence of the genes on mental development, with little or no influence from the environment. One of the best-known nativist theories is Chomsky's hypothesis on the language acquisition device and universal grammar (Chomsky 1957; Pinker 1994; see also Pinker and Bloom 1990). This nativist theory proposes that children are born with innate knowledge of linguistic and syntactic principles, whose parameters are then fine-tuned through experience of the language of their parents. In other fields, Leslie (1994) hypothesized that children are born with a theory of mind, and Wynn (1998) that they have innate knowledge of math concepts. On the opposite end, empiricist theories stress the importance of the social and cultural environment in cognitive development. This is the case of Vygotsky's (1978) sociocultural theory, where the role of adults and peers is essential to guide the child to exploit her "zone of proximal development," meaning, the space of the infant's potential capabilities. Similarly, Bruner's socio-cognitive theory of development (Bruner and Haste 1987) stresses the importance of social interaction and interpersonal communication in the various stages of learning. Tomasello (2003) proposes an empiricist theory of language development based on the principle of constructivist and emergent development, whereby the child constructs her own language competence through interaction with other language-speaking agents.

Within these extremes, Piaget (1971) has proposed one of the most influential theories in developmental psychology that combines the contribution of nature and nurture mechanisms. The key tenet of Piaget's theory is that a child goes through different *stages* of development, where at each stage the infant develops qualitatively different and increasingly complex *schemas*, the building block of intelligence. These stages are influenced by maturational constraints, determined by genetic influence, and called "epigenetic" in Piaget's theory (ibid.). However, the child goes through a process of *adaptation*, where the contribution of the external environment is important in the adaptation of existing schemas to new knowledge (assimilation) and the modification and creation of new schemas (accommodation). Piaget proposed four key stages of development of mental capabilities, with a particular focus on the development of thinking capabilities and the origin of abstract thought schemas in sensorimotor knowledge. In the Sensorimotor Stage (Stage 1, 0–2 years old), the child starts with the acquisition of sensorimotor schemas, which initially consist of motor reflexes. In the Preoperational Stage (Stage 2, 2–7 years old), children acquire egocentric symbolic representations of objects and actions, which allow them to represent objects even when these are not visible (object permanence task, when the child understands that a moving object reappears after hiding behind an obstacle). In the subsequent Concrete Operational Stage (Stage 3, 7–11 years old) children can adopt other people's perspectives on object representation and perform mental transformation operations on concrete objects (e.g., liquid conservation task). This finally leads to the Formal Operational Stage (Stage 4, 11+ years old) with the acquisition of full abstract thinking capabilities and complex problem-solving skills. Piaget's theory and stages will be further described in chapter 8, on the models of abstract knowledge.

Another theory that considers the simultaneous contribution of biological and environmental factors is Thelen and Smith's (1994) dynamic systems theory of development. This considers the complex dynamic interaction of various neural, embodiment, and environmental factors in the self-organization of cognitive strategies (see section 1.3.1 for more details).

The nature/nurture debate and nativist/empiricist theories have significantly influenced other fields interested in intelligence, specifically in artificial intelligence and robotics. When building artificial cognitive systems, as with adaptive agents in artificial intelligence and with cognitive robots in robotics, it is possible to use a nativist approach. This implies that the agent's cognitive architecture is fully predefined by the researcher, and does not change significantly during the agent's interaction with the environment. On the other end, the utilization of a more empiricist approach in artificial intelligence and robotics requires the definition of a series of adaptation and learning mechanisms that allow the agent to gradually develop its own knowledge and cognitive system through interaction with other agents and human users. The developmental robotics approach presented in this volume mostly follows a balanced nativist/empiricist approach to robot design as it puts a great emphasis on the development of the robot's capability during interaction with the environment, as well as on the maturational and embodiment factors that constrain development. In particular, Piaget's theory, in addition to being the most influential theory in developmental psychology, has strongly influenced the field of developmental robotics, including the use of the term "epigenetic" in the "Epigenetic Robotics" conference title series. This is because Piaget's theory emphasizes the sensorimotor bases of mental development and the balanced biological and environmental approach.

Together with Piaget, another well-known developmental psychologist, Lev Vygotsky, has also significantly influenced the field of developmental robotics. Vygotsky's theory puts much emphasis on the role of social environment on mental development and on the effects that the social and physical environment have on the *scaffolding* of the child's cognitive system during development (Vygotsky 1978). His insights have therefore contributed to social learning and human-robot imitation studies, and to the developmental robotics theory of scaffolding (Asada et al. 2009; Otero et al. 2008; Nagai and Rohlfing 2009).

In the following sections, after defining developmental robotics and presenting a brief historical overview, we will discuss the main defining characteristics and principles of this approach, which combines the dynamic interaction of biological and cultural phenomena in the autonomous mental development of robots.

1.2 Definition and Origins of Developmental Robotics

Developmental robotics is the *interdisciplinary approach to the autonomous design of behavioral and cognitive capabilities in artificial agents (robots) that takes direct inspiration from the developmental principles and mechanisms observed in the natural cognitive systems of children.* In particular, the main idea is that the robot, using a set of intrinsic developmental principles regulating the real-time interaction between its body and brain and its environment, can autonomously acquire an increasingly complex set of sensorimotor and mental capabilities.

Developmental robotics relies on a highly interdisciplinary effort of empirical developmental sciences such as developmental psychology, neuroscience, and comparative psychology; and computational and engineering disciplines such as robotics and artificial intelligence. Developmental sciences provide the empirical bases and data to identify the general developmental principles, mechanisms, models, and phenomena guiding the incremental acquisition of cognitive skills. The implementation of these principles and mechanisms into a robot's control architecture and the testing through experiments where the robot interacts with its physical and social environment simultaneously permits the validation of such principles and the actual design of complex behavioral and mental capabilities in robots. Developmental psychology and developmental robotics mutually benefit from such a combined effort.

Historically, developmental robotics traces its origins to the years 2000–2001, in particular in coincidence with two scientific workshops that, for the first time, gathered together scientists interested in developmental psychology principles in both humans and robots. These workshops had been preceded by some work and publications advocating an explicit link between human development and robotics, such as in Sandini, Metta, and Konczak (1997); Brooks et al. (1998); Scassellatti (1998); and Asada et al. (2001).

The first event was the Workshop on Development and Learning (WDL) organized by James McClelland, Alex Pentland, Juyang (John) Weng, and Ida Stockman and held on April 5–7, 2000, at Michigan State University, in East Lansing, Illinois. This workshop subsequently led to the establishment of the annual International Conference on Development and Learning (ICDL). At the WDL the term "developmental robotics" was publicly used for the first time. In addition, the workshop contributed to the coinage of the term "autonomous mental development," to stress the fact that robots develop mental (cognitive) capabilities in an autonomous way (Weng et al. 2001). Autonomous mental development has in fact become a synonym for developmental robotics, and is the name of the main scientific journal in this field, *IEEE Transactions on Autonomous Mental Development*.

The second event to contribute to the birth of developmental psychology as a scientific discipline was the First International Workshop on Epigenetic Robotics: Modeling Cognitive Development in Robotic Systems, which again led to the establishment of the subsequent Epigenetic Robotics (EpiRob) conference series. This workshop was organized by Christian Balkenius and Jordan Zlatev, and was held at Lund University (Sweden) September 17–19, 2001. The workshops borrowed the term "epigenetic" from Piaget. As noted earlier, in Piaget's Epigenetic Theory of human development, the child's cognitive system develops as a result of the interaction between genetic predispositions and the organism's interaction with the environment. As such the choice of the term "epigenetic robotics" was justified by Piaget's stress on the importance of the role of interaction with the environment, and in particular on the sensorimotor bases of higher-order cognitive capabilities. Moreover, this early definition of epigenetic robotics also complemented Piaget's sensorimotor bases of intelligence with Lev Vygotsky's emphasis on social interaction (Zlatev and Balkenius 2001).

In addition to the term "developmental robotics" used in this volume and in other review publications (e.g., Metta et al. 2001; Lungarella et al. 2003; Vernon, von Hofsten, and Fadiga 2010; Oudeyer 2012), and the related term "cognitive developmental robotics" used in Asada et al. (2001, 2009), in the literature other names have been proposed to refer to the same approach and interdisciplinary field. Some authors prefer the term "autonomous mental development" (Weng et al. 2001), while others use the term "epigenetic robotics" (Balkenius et al. 2001; Berthouze and Ziemke 2003).

The use of these different terms mostly reflects historical factors, as discussed, rather than real semantic differences. As a matter of fact, in 2011 the two communities of the ICDL conference series (preferring the term "autonomous mental development") and of the EpiRob series (preferring the term "epigenetic robotics") joined forces to organize the first joint International Conference on Developmental and Learning and on Epigenetic Robotics (IEEE ICDL-EpiRob). This joint conference, continued since 2011, has become the common home for developmental robotics research, with a web presence on http://www.icdl-epirob.org, through the activities of the IEEE Technical Committee on Autonomous Mental Development, which coordinate such joint efforts.

1.3 Principles of Developmental Robotics

The field of developmental robotics has been strongly influenced by developmental psychology theories, as seen in section 1.1. As discussed, developmental robotics models follow an approach based on the coupled interaction of both nativist and empiricist phenomena, though with a stronger emphasis on environmental and social factors. The consideration of the influence of biological and genetic factors includes the effects of maturational phenomena in both the agent's body and brain, the exploitation of embodiment constraints for the acquisition of sensorimotor and mental capabilities, and the role of intrinsic motivation and the instinct to imitate and learn from others. Empiricist and constructivist phenomena considered in developmental robotics research include a focus on situated learning and the contribution of both the social and physical environment in shaping development, and of an online, open-ended and cumulative acquisition of cognitive skills. Moreover, both biological and environmental factors are coupled in an intricate and dynamic way resulting in stage-like qualitative changes of cognitive strategies dependent on a nonlinear dynamical system interaction of genetic, embodiment, and learning phenomena.

A series of general principles can be identified that reflect the numerous factors and processes implicated in the design of autonomous mental development in robots and that have guided developmental robotics practice. These principles can be grouped as shown in table 1.1, and will then be briefly analyzed in the following subsections.

1.3.1 Dynamical Systems Development

An important concept taken from mathematics and physics, and which has significantly influenced general theories of human development, is that of *dynamical systems*. In mathematics, a dynamical system is characterized by complex changes, over time, in the phase state, and which are the result of the self-organization of multifaceted interactions between the system's variables. The complex interaction of nonlinear phenomena results in the production of unpredictable states of the system, often referred to as *emergent* states. This concept has been borrowed by developmental psychologists, and in particular by Thelen and Smith (1994; Smith and Thelen 2003), to explain child development as the emergent product of the intricate and dynamic interaction of many decentralized and local interactions related to the child's growing body and brain

Table 1.1
Principles and characteristics of developmental robotics

	Principles	Characteristics
1	Development as a dynamical system	Decentralized system Self-organization and emergence Multicausality Nested timescales
2	Phylogenetic and ontogenetic interaction	Maturation Critical period Learning
3	Embodied and situated development	Embodiment Situatedness Enaction Morphological computation Grounding
4	Intrinsic motivation and social learning	Intrinsic motivation Value systems Imitation
5	Nonlinear, stage-like development	Qualitative stages U-shaped phenomena
6	Online, open-ended, cumulative learning	Online learning Cumulative Cross-modality Cognitive bootstrapping

and her environment. Thus Thelen and Smith have proposed that the development of a child should be viewed as change within a complex dynamic system, where the growing child can generate novel behaviors through her interaction with the environment, and these behavioral states vary in their stability within the complex system.

One key concept in this theory is that of *multicausality*, for example, in the case when one behavior, such as crawling and walking, is determined by the simultaneous and dynamic consequences of various phenomena at the level of the brain, body, and environment. Thelen and Smith use the example of the dynamic changes in crawling and walking motor behaviors as an example of multicausality changes in the child's adaptation to the environment, in response to body growth changes. When the child's body configuration produces sufficient strength and coordination to support its body through the hands and knee posture, but not to support upright walking, the child settles for a crawling strategy to locomote in the environment. But when the infant's body growth results in stronger and more stable legs, the standing and walking behavior emerges as the stable developmental state, which as a consequence destabilizes, and gradually replaces, the pattern of crawling. This demonstrates that rather than following a predetermined, top-down genetic-controlled developmental trajectory that first controls crawling and then walking, the locomotion behavior is the result of self-organizing dynamics of decentralized factors such as the child's changing body (stronger legs and better balance) and its adaptation to the environment. This illustrates the principle of multicausality, as there are many parallel factors causing varying behavioral strategies.

Another key concept in Thelen and Smith's dynamic systems view of development is that of *nested timescales*, in other words, neural and embodiment phenomena acting at different timescales, and all affecting development in an intricate, dynamic way. For example the dynamics of the very fast timescale of neural activity (milliseconds) is nested within the dynamics of the other slower timescales such as reaction time during action (seconds or hundreds of milliseconds), learning (after hours or days), and physical body growth (months).

One of the best-known developmental psychology examples used by Thelen and Smith to demonstrate the combined effects of the concepts of multicausality and nested timescales is that of the A-not-B error. This example is inspired by Piaget's object permanence experiment, when one toy is repeatedly hidden under a lid at a location A (right) during the first part of the experiment. Toward the end of the task, the experimenter hides the same toy in the location B (left) for a single trial, and then asks the child to reach for the object. While infants older than twelve months have no problem in reaching for the toy in its correct location B, unexpectedly most eight-to-ten-month-old infants produce the curious error of looking for the object in location A. This error is only produced when there is a short delay between hiding and reaching. While psychologists such as Piaget have used explanations based on age (stage) differences linked to qualitative changes in the capability to represent objects and space, a computational simulation of the dynamic system model (Thelen et al. 2001) has demonstrated that there are many decentralized factors (multicausality) and timing manipulations (nested timing) affecting such a situation. These for example depend on the time delay between hiding and reaching, the properties of the lids on the table, the saliency of the hiding event, the past activity of the infant, and her body posture. The systematic manipulation of these factors results in the appearance, stopping, and modulation of the A-not-B errors.

The use of a dynamical systems approach as a theory of development, and the general dynamic linking of body, neural, and environmental factors, have had significant influence in developmental robotics research, as well in other fields of robotics and cognitive systems (Beer 2000; Nolfi and Floreano 2000). This theory has been applied for example to developmental robotics models of early motor development, as in Mori and Kuniyoshi's (2010) simulation on the self-organization of body representation and general movements in the fetus and newborn (section 2.5.3). Also a developmental robotics model of early word learning (Morse, Belpaeme, et al. 2010) uses a setup similar to the A-not-B error to investigate dynamics interactions between embodiment factors and higher-order language development phenomena (section 7.3).

1.3.2 Phylogenetic and Ontogenetic Interaction

Discussion of the dynamical systems approach has already stressed the importance of different timescales during development, including the ontogenetic phenomena of *learning*, over a timescale of hours or days, and *maturational* changes, occurring for periods of months or years. An additional, slower, timescale to consider when studying development is that of the phylogenetic time dimension, that is, the effect of evolutionary changes in development. Therefore the additional implication of the interaction between ontogenetic and phylogenetic phenomena should be considered in robotics models of development.

In this section we will discuss the importance of maturational changes, as these more closely relate to phylogenetic changes. The effect of cumulative changes due to learning new behaviors and skills will be discussed in sections 1.3.5 and 1.3.6.

Maturation refers to changes in the anatomy and physiology of the child's brain and body, especially during the first years of life. Maturational phenomena related to the brain include the decrease of brain plasticity during early development, and phenomena like the gradual hemispheric specialization and the pruning of neurons and connections (Abitz et al. 2007). Brain maturation changes have also been evoked to explain the critical periods in learning. Critical periods are stages (window of time) of an organism's lifespan during which the individual is more sensitive to external stimulation and more efficient at learning. Moreover, after a critical period has ended, learning becomes difficult or impossible to achieve. The best know example of critical period (also known as the sensitive period) in ethology is Konrad Lorenz's study on imprinting, that is, the attachment of ducklings to their mother (or to Lorenz!), which is only possible within the first few hours of life and has a long-lasting effect. In vision research, Hubel and Wiesel (1970) demonstrated that the cat's visual cortex can only develop its receptive fields if the animal is exposed to visual stimuli in the first few months of life, and not when there is total visual deprivation by covering the kitten's eyes. In developmental psychology, the best-studied critical period is that for language learning. Lenneberg (1967) was one of the first to propose the critical period hypothesis for language development that claims that the brain changes occurring between the age of two and seven years, specifically for the hemispheric specialization gradually leading to lateralization of the linguistic function in the left hemisphere, are responsible for the problems in learning language after this age. The critical period hypothesis has also been proposed to explain the limitation in the acquisition of a second language after puberty (Johnson and Newport 1989). Although this hypothesis is still debated in the literature, there is general agreement that brain maturation changes significantly affect language learning beyond the period of puberty.

Maturation in the body of the child is more evident given the significant morphological changes a child goes through from birth to adolescence. These changes naturally affect the motor development of the child, as in Thelen and Smith's analysis of crawling and walking. Morphological changes occurring during development also have implication for the exploitation of embodiment factors, as discussed in section 1.3.3, on the morphological computation effects of embodiment.

Some developmental robotics models have explicitly addressed the issue of brain and body maturation changes. For example, the study by Schlesinger, Amso, and Johnson (2007) models the effects of neural plasticity in the development of object perception skills (section 4.5). The modeling of body morphology development is also extensively discussed in chapter 4 on motor development.

The ontogenetic changes due to maturation and learning have important implications for the interaction of development with phylogenetic changes due to evolution. Body morphology and brain plasticity variations can in fact be explained as evolutionary adaptations of the species to changing environmental context. These phenomena have been analyzed, for example, in terms of genetic changes affecting the timing of ontogenetic phenomena, known as heterochronic changes (McKinney and McNamara 1991). Heterochronic classifications are based on the comparison of ontogenies that differ for the onset of growth, the offset of growth, and the rate of growth of an organ or a biological trait. Namely, the terms "predisplacement" and "postdisplacement" refer respectively to an anticipated and a postponed onset of morphological growth, "hypermorphosis" and "progenesis" refer respectively to a late and an early offset of growth, and "acceleration" and "neoteny" refer respectively to a faster and a slower rate of growth. Heterochronic changes have been used to explain the complex interaction between nature and nurture in models of development, an in Elman et al.'s (1996) proposal that the role of genetic factors in development is to determine the architectural constraints, which subsequently control learning. Such constraints can be explained in terms of brain adaptation and neurodevelopmental and maturational events.

The interaction between ontogenetic and phylogenetic factors has been investigated through computational modeling. For example, Hinton and Nowlan (1987) and Nolfi, Parisi, and Elman (1994) have developed simulation models explaining the effects of learning in evolution, as for the Baldwin effect. Cangelosi (1999) has tested the effects of heterochronic changes in the evolution of neural network architectures for simulated agents. Furthermore, the modeling of the evolution of varying body and brain morphologies in response to phylogenetic and ontogenetic requirements is also the goal of the "evo-devo" computational approach. This aims at simulating the simultaneous effects of developmental and evolutionary adaptation in body and brain morphologies (e.g., Stanley and Miikkulainen 2003; Kumar and Bentley 2003; Pfeifer and Bongard 2007). Developmental robotics models normally are based on robots with fixed morphologies and cannot directly address the simultaneous modeling of phylogenetic changes and their interaction with ontogenetic morphological changes. However, various epigenetic robotics models take into consideration the evolutionary origins of the ontogenetic changes of learning and maturation, especially for studies including changes in brain morphology.

1.3.3 Embodied, Situated, and Enactive Development

Growing empirical and theoretical evidence exists on the fundamental role of the body in cognition and intelligence (*embodiment*), the role of interaction between the body and its environment (*situatedness*), and the organism's autonomous generation of a model of the world through sensorimotor interactions (*enaction*). This embodied, situated, and enactive view stresses the fact that the body of the child (or of the robot, with its sensors and actuators), and its interaction with the environmental context determines the type of representations, internal models, and cognitive strategies learned. As Pfeifer and Scheier (1999, 649) claim, "intelligence cannot merely exist in the form of an abstract algorithm but requires a physical instantiation, a body."

In psychology and cognitive science, the field of embodied cognition (aka grounded cognition) has investigated the behavioral and neural bases of embodiment, specifically for the roles of action, perception, and emotions in the grounding of cognitive functions such as memory and language (Pecher and Zwaan 2005; Wilson 2002; Barsalou 2008). In neuroscience, brain-imaging studies have shown that higher-order functions such as language share neural substrates normally associated with action processing (Pulvermüller 2003). This is consistent with philosophical proposals on the embodied mind (Varela, Thompson, and Rosch 1991; Lakoff and Johnson 1999) and situated and embodied cognition (Clark 1997).

In robotics and artificial intelligence, embodied and situated cognition has also received great emphasis through the approach of embodied intelligence (Pfeifer and Scheier 1999; Brooks 1990; Pfeifer and Bongard 2007; Pezzulo et al. 2011). Ziemke (2001) and Wilson (2002) analyze different views of embodiment and their consideration in computational models and psychology experiments. These different views range from considering embodiment as the phenomenon of the "structural coupling" between the body and the environment, to the more restrictive "organismic" embodiment view, based on the autopoiesis of living systems, that is, that cognition actually is what living systems *do* in interaction with their world (Varela, Thompson, and Rosch 1991). Along the same lines, the paradigm of enaction highlights the fact that an autonomous cognitive system interacting in its environment is capable of developing its own understanding of the world and can generate its own models of how the world works (Vernon 2010; Stewart, Gapenne, and Di Paolo 2010).

Embodied and situated intelligence has significantly influenced developmental robotics, and practically any developmental model places great emphasis on the relation between the robot's body (and brain) and the environment. Embodiment effects concern pure motor capabilities (morphological computation) as well as higher-order cognitive skills such as language (grounding). *Morphological computation* (Bongard and Pfeifer 2007) refers to the fact that the organism can exploit the body's morphological properties (e.g., type of joint, length of limbs, passive/active actuators), and the dynamics of the interaction with the physical environment (e.g., gravity) to produce intelligent behavior. One of the best-known examples of this is the passive dynamic walker, that is, bipedal robots that can walk on a slope without any actuator, thus not requiring any explicit control, or bipedal robots only requiring minimal actuation to start movement (McGeer 1990; Collins et al. 2005). The exploitation of morphological computation has important implications for energy consumption optimization in robotics, and for the increasing use of compliant actuators and soft robotics material (Pfeifer, Lungarella, and Iida 2012).

On the other end, an example of the role of embodiment in higher-order cognitive functions can be seen in the models of the grounding of words in action and perception (Cangelosi 2010; Morse, Belpaeme, et al. 2010, see section 7.3) and the relationship between spatial representation and numerical cognition in psychology and developmental robotics (Rucinski, Cangelosi, and Belpaeme 2011, see section 8.2).

1.3.4 Intrinsic Motivation and Social Learning Instinct

Conventional approaches to designing intelligent agents typically suffer from two limitations. First, the objectives or goals (i.e., the value system) are normally imposed by the model-builder, rather than determined by the agent themselves. Second, learning is often narrowly restricted to performance on a specific, predefined task. In response to these limitations, developmental robotics explores methods for designing *intrinsically motivated* agents and robots. An intrinsically motivated robot explores its environment in a completely autonomous manner, by deciding for itself what it wants to learn and what goals it wants to achieve. In other words, intrinsic motivation enables the agent to construct its own value system.

The concept of intrinsic motivation is inspired by a variety of behaviors and skills that begin to develop in infancy and early childhood, including diverse phenomena such as curiosity, surprise, novelty seeking, and the "drive" to achieve mastery. Oudeyer and Kaplan (2007) propose a framework for organizing research on models of intrinsic motivation, including two major categories: (1) knowledge-based approaches (which are subdivided into novelty-based and prediction-based approaches), and (2) competence-based approaches. Within this framework, a large number of algorithms can be defined and systematically compared.

Novelty-based approaches to intrinsic motivation often utilize mobile robots, which learn about their environments by exploring and discovering unusual or unexpected features. A useful mechanism for detecting novelty is habituation: the robot compares its current sensory state to past experiences, devoting its attention toward situations that are unique or different from those that have already been experienced (e.g., Neto and Nehmzow 2007).

Prediction-based approaches are a second type of knowledge-based intrinsic motivation, as they also rely on accumulated knowledge. However, in this case prediction-based models explicitly attempt to predict future states of the world. A simple example could be a robot that pushes an object toward the edge of the table, and predicts that it will make a sound when it drops on the floor. The rationale of this approach is that incorrect or inaccurate predictions provide a learning signal, that is, they indicate events that are poorly understood, and require further analysis and attention. As an example of this approach, Oudeyer et al. (2005) describe the Playground Experiment, in which the Sony AIBO robot learns to explore and interact with a set of toys in its environment.

The third approach to modeling intrinsic motivation is competence based. According to this view, the robot is motivated to explore and develop skills that effectively produce reliable consequences. A key element of the competence-based approach is *contingency detection*: this is the capacity to detect when one's actions have an effect on the environment. While the knowledge-based approach motivates the agent toward discovering properties of the world, the competence-based approach, in contrast, motivates the agent to discover *what it can do* with the world.

Child development research has shown the presence of social learning capabilities (instincts). This is evidenced for example by observations that newborn babies have an instinct to imitate the behavior of others from the day they are born and can imitate complex facial expressions (Meltzoff and Moore 1983). Moreover, comparative psychology studies have demonstrated that 18- to 24-month-old children have a tendency to cooperate altruistically, a capacity not observed in chimpanzees (Warneken, Chen, and Tomasello 2006).

As we highlight in chapter 3, the development of intrinsic motivation has direct implications for how infants perceive and interact with others. For example, young infants quickly learn that people in their environment respond contingently to their movements and sounds. Thus, babies may be intrinsically motivated to orient toward and interact with other people.

Developmental robotics places a heavy emphasis on social learning, and as demonstrated in the numerous studies discussed in chapter 6, various robotics models of joint attention, imitation, and cooperation have been tested.

1.3.5 Nonlinear, Stage-Like Development

The literature on child psychology has plenty of theories and models proposing a sequence of developmental *stages*. Each stage is characterized by the acquisition of specific behavioral and mental strategies, which become more complex and articulated as the child progresses through these stages. Stages are also linked to specific ages of the child, except for individual differences. Piaget's four stages of development of thought are the prototypical example of a theory of development centered on stages (chapter 8). Numerous other examples of stage-based development exist, and a few will be described in the chapters that follow, as Courage and Howe's (2002) timescale of self-perception (chapter 4), Butterworth's (1991) four stages of joint attention and Leslie's (1994) and Baron-Cohen's (1995) stages of the theory of mind (chapter 6), the sequential acquisition of lexical and syntactic skills (chapter 8), and the stages of numerical cognition and of rejection behavior (chapter 9).

In most theories, the transition between stages follows nonlinear, qualitative shifts. Again, in the example of Piaget's four stages, the mental schemas used in each stage are qualitatively different, as they are the results of accommodation processes that change and adapt the schema to new knowledge representations and operations. Another well-known developmental theory based on qualitative changes during development is the Representational-Redescription Model of Karmiloff-Smith (1995). Although Karmiloff-Smith explicitly avoids the definition of age-determined stage models, as in Piaget, her model assumes four levels of development going from the use of implicit representation to different levels of explicit knowledge-representation strategies. When a child learns new facts and knowledge about specific domains, she develops new representations, which are gradually "redescribed" and increase the child's explicit understanding of the world. This has been applied to a variety of knowledge domains such as physics, math, and language.

The nonlinearity of the developmental process and the qualitative shifts in the mental strategies and knowledge representations employed by the child at different stages of development have been extensively investigated through "U-shaped" learning error patterns and with the vocabulary spurt phenomenon. The prototypical case study of the U-shaped phenomenon in child development is in the patterns of errors produced by children during the acquisition of the verb morphology for the English past tense. The (inverted) U-shaped phenomenon consists of the low production of errors at the beginning of learning, which is then followed by an unexpected increase in errors, subsequently followed by an improved performance and low error production again. In the case of the English past tense, initially children produce few errors as they can say the correct past tense for high-frequency irregular verbs, such as "went," and the correct "ed" suffix form for regular verbs. At a later stage, they pass through a stage of "over-regularization," and start producing morphological errors for irregular verbs, as with "goed." Eventually, children can again distinguish the multiple forms of irregular past tenses. This phenomenon has been extensively studied in psychology, and has caused heated debate between the proponents of a rule-based strategy for syntax processing (Pinker and Prince 1988), and the advocates of a distributed representation strategy, which is supported by demonstration that connectionist networks can produce a U-shaped performance by using distributed representations (e.g., Plunkett and Marchman 1996). U-shaped learning phenomena have also been reported in other domains, such as in phonetic perception (Eimas et al. 1971; Sebastián-Gallés and Bosch 2009), face imitation (Fontaine 1984), and in Karmiloff-Smith's (1995) explanation of a child's performance and errors due to the changing representational strategies.

The vocabulary spurt phenomenon in lexical acquisition is another case of nonlinear and qualitative shifts during development. The vocabulary spurt (also called the "naming explosion") occurs around the eighteen- to twenty-four-month period, when the child goes from an initial pattern of slow lexical learning, with the acquisition of few words per month, to the *fast mapping* strategy, whereby a child can quickly learn tens of words per week by single exposure to the lexical item (e.g., Bloom 1973; Bates et al. 1979; Berk 2003). The vocabulary spurt typically happens when a child has learned around 50–100 words. This qualitative change of strategy in word learning has been attributed to a variety of underlying cognitive strategies, including the mastering of word segmentation or improvements in lexical retrieval (Ganger and Brent 2004).

Many developmental robotics studies aim to model the progression of stages during the robot's development, with some directly addressing the issue of nonlinear phenomena in developmental stages as a result of learning dynamics. For example Nagai et al. (2003) explicitly modeled the joint attention stages proposed by Butterworth (1991). However, the model shows that qualitative changes between these stages are the result of gradual changes in the robot's neural and learning architecture, rather than ad hoc manipulations of the robot's attention strategies (see section 6.2). Some models have also directly addressed the modeling of U-shaped phenomena, such as in the Morse et al. (2011) model of error patterns in phonetic processing.

1.3.6 Online, Open-Ended, Cumulative Learning

Human development is characterized by online, cross-modal, continuous, open-ended learning. *Online* refers to the fact that learning happens while the child interacts with its environment, and not in an offline mode. *Cross-modal* refers to the fact that different modalities and cognitive domains are acquired in parallel by the child, and interact with each other. This is for example evidenced in the interaction of sensorimotor and linguistics skills, as discussed in the embodiment section 1.3.3. *Continuous* and *open-ended* refers to the fact that learning and development do not start and stop at specific stages, but rather constitute a lifelong learning experience. In fact, developmental psychology is often framed within the wider field of the psychology of life cycles, ranging from birth to aging.

Lifelong learning implies that the child *accumulates* knowledge, and thus learning never stops. As seen in the previous sections, such continuous learning and accumulation of knowledge can result in qualitative changes of cognitive strategies, as in the language vocabulary spurt phenomenon, and in Karmiloff-Smith's theory on the transition from implicit to explicit knowledge through the Representational-Redescription model.

One consequence of cumulative, open-ended learning is *cognitive bootstrapping*. In developmental psychology, cognitive bootstrapping has been mostly applied to numerical cognition (Carey 2009; Piantadosi, Tenenbaum, and Goodman 2012). According to this concept, a child acquires knowledge and representation from learned concepts (e.g., numerical quantities and counting routines) and then inductively uses this knowledge to define the meaning of new number words learned subsequently, and with a greater level of efficiency. The same idea can be applied to the vocabulary spurt, in which the knowledge and experience from the slow learning of the first 50–100 words causes a redefinition of the word learning strategy, and to syntactic bootstrapping, by which children rely on syntactic cues and word context in verb learning to determine the meaning of new verbs (Gleitman 1990). Gentner (2010) has also proposed that general cognitive bootstrapping is achieved through the use of analogical reasoning and the acquisition of symbolic relationship knowledge.

Online learning is implemented in developmental robotics, as will be demonstrated in most of the studies presented in the next chapters. However, the application of cross-modal, cumulative, open-ended learning, which can lead to cognitive bootstrapping phenomena, has been investigated less frequently. Most of the current models typically focus on the acquisition of only one task or modality (perception, or phonetics, or semantics, etc.), and few consider the parallel development, and interaction, between modalities and cognitive functions. Thus a truly online, cross-modal, cumulative, open-ended developmental robotics model remains a fundamental challenge to the field.

The presentations of various examples of developmental robotics models and experiments will show how most of the preceding principles guide and inform the design of the cognitive architecture and the experimental setups of developmental robots.

1.4 Book Overview

In this introductory chapter we have defined developmental robotics and discussed its grounding in developmental psychology theories. The discussion of the main principles at the basis of developmental robotics has also highlighted the common defining characteristics of such an approach.

In chapter 2 we provide further introductory material, in particular a working definition of robots and a look at the different types of android and humanoid robots. Chapter 2 also includes an overview of sensor and actuator technology for humanoid robotics and the primary baby robot platforms used in developmental robotics, as well as the simulation tools.

In the experiment-focused chapters 3–8 we look in detail at how developmental robotics models and experiments have explored the realization of various behavioral and cognitive capabilities (motivation, perception, action, social, linguistic, and abstract knowledge). Then we consider the achievements and challenges in these areas. Each of these chapters begins with a concise overview of the main empirical findings and theoretical positions in developmental psychology. Although each overview is aimed at readers who are not familiar with the child psychology literature, at the same time it provides the specific empirical grounding and reference work pertaining to the individual developmental issues modeled in the robotics studies described in the rest of each chapter. Each experiment-focused chapter then discusses seminal experiments demonstrating the achievement of developmental robotics in modeling autonomous mental development. These examples explicitly address key issues in child psychology research. Moreover, most experimental chapters include boxes that provide technical and methodological details on exemplar psychology and robotics experiments, and highlight methodological implications in developmental robotics and their direct correspondence to child psychology studies.

Of these six experimental chapters, chapter 3 specifically concerns the developmental robotics models of intrinsic motivation and curiosity, looking in particular at the neural, conceptual, and computational bases of novelty, prediction, and competence. Chapter 4 discusses the models of perceptual development, concentrating in detail on the models of face recognition, perception of space, robot's self-perception, and recognition of objects and of motor affordances. Chapter 5 analyzes motor development models contemplating both manipulation (i.e., reaching and grasping) and locomotion capabilities (i.e., crawling and walking). In chapter 6 we look at the developmental work of social learning, with emphasis on the models of joint attention, imitation learning, cooperation and shared plans, and robot's theory of mind. Chapter 7 focuses on language and analyzes the development of phonetic babbling, the grounded acquisition of words and meaning, and the development of syntactic processing skills. Chapter 8 focuses on developmental robotics models of abstract knowledge, with a discussion of number learning models, abstracts concepts, and reasoning strategies.

Finally, the concluding chapter 9 reflects on the achievements in developmental robotics common to the different cognitive areas, and looks ahead to consider future research directions and developments in the field.

Additional Reading

Thelen, E., and L. B. Smith. *A Dynamic Systems Approach to the Development of Cognition and Action*. Cambridge, MA: MIT Press, 1994.

This is a seminal book on theoretical developmental psychology that, in addition to its impact in child psychology, has greatly inspired developmental robotics, and in general dynamical systems approaches to cognitive modeling. The volume provides the original and detailed account of the dynamical systems approach to development briefly introduced in section 1.3.1.

Pfeifer, R., and J. Bongard. *How the Body Shapes the Way We Think: A New View of Intelligence*. Cambridge, MA: MIT Press, 2007.

This book presents an inspiring analysis of the concept of embodied intelligence in natural and artificial cognitive systems. Its aim is to demonstrate that cognition and thought are not independent of the body, but rather are tightly constrained by embodiment factors, and at the same time the body enables and enriches cognitive capabilities. The book is centered on the idea of "understanding by building," meaning, the fact that we understand and can build intelligent agents and robots gives us a better understanding of intelligence in general. As such the book discusses many examples from robotics, biology, neuroscience, and psychology, and covers specific applications in ubiquitous computing and interface technology, in business and management for building intelligent companies, in the psychology of human memory, and in everyday robotics.

Nolfi, S., and D. Floreano. *Evolutionary Robotics: The Biology, Intelligence, and Technology of Self-Organizing Machines*. Cambridge, MA: MIT Press, 2000.

This book offers detailed discussion of the evolutionary robotics principles and of the pioneering evolutionary models and experiments by Nolfi and Floreano. It can be considered a companion of *Developmental Robotics: From Babies to Robots* because it introduces the complementary area of evolutionary robotics. *Evolutionary Robotics* comes with a free software simulator for evolutionary robotics experiments with mobile robots (gral.istc.cnr.it/evorobot) as well as the more recent evolutionary robotics simulator for the iCub humanoid (laral.istc.cnr.it/farsa).

2 Baby Robots

In this chapter we will introduce the main robotic platforms, and associated robot simulation software, used in developmental robotics. Most of these platforms will be referred to when presenting specific studies on developmental cognitive modeling in the chapters that follow. As the book is also aimed at readers not familiar with robotics concepts, section 2.1 that follows introduces the basic concepts of robotics, including humanoid and android robots, and gives an overview of sensor and actuator technology for humanoid robotics.

2.1 What Is a Robot?

The definition of a robot and its core characteristics is the obvious starting point for a book on developmental robotics.

Historically, the etymology of the word "robot" derives from the Slavic word *robota*, which is used to refer to slave or forced labor. This word first appeared in the play *R.U.R.* (Rossum's Universal Robots), written by the Czech writer Karel Čapek (1920). This etymology indicates that robots are built to support humans in everyday and work tasks, and in some cases to fully or partially substitute for humans (like slaves) by performing tasks in their place, as in industrial robotics.

What is a robot? The *Oxford English Dictionary* defines a robot as "a machine capable of carrying out a complex series of actions automatically, especially one programmable by a computer." This definition contains four key concepts that are important for our interest in developmental robotics: (1) *machine*, (2) *complex . . . actions*, (3) *automatically*, and (4) *programmable by a computer*. The use of the first concept, *machine*, is important because it encompasses a variety of platforms currently considered to be robots. We could instinctively think of a robot as a human-like (humanoid or android—the definitions and distinction of humanoid and android robots follow) machine, as we are used to thinking about robots through famous movies such as *A.I. Artificial Intelligence*. In fact, the great majority of robots currently used in the world are nonhuman-looking industrial manufacturing and packaging machines that perform repetitive tasks in factories. These industrial robots have the shape of a mechanical, multijoint arm that performs precision tasks such as welding metal parts (as in a car factory plant) or lifting and moving objects (as in a food packaging factory). Other common shapes a robotic platform has are the wheeled mobile cart used to transport boxes and parts in a factory and the round-wheeled vacuum-cleaning robot present in an increasing number of households.

The second concept included in the *Oxford English Dictionary* definition of "robot" is that a robot performs "a complex series of actions." This is in part true; for example, many manufacturing robots can perform fine-tool manipulation tasks. However in most cases what robots in use today actually do is to perform simple, repetitive tasks, in a context primarily oriented toward saving time and increasing safety.

The third concept of *automatically* is the key defining characteristic of a robot. What truly distinguishes a robot from a machine or piece of equipment is the fact that the robot operates automatically, without the direct and continuous control of a person. In particular, as a robot has sensors to perceive the state of the world (including its own internal state) and actuators to operate in the world, the automaticity (and autonomy—to be discussed) refers to the robot having the capability to integrate such input information and select an action to achieve a specific goal. This is better reflected by Matarić's definition of robot as "an autonomous system which exists in the physical world, can sense its environment, and act on it to achieve a goal" (Matarić 2007, 2).

A term similar to automaticity, but which is more often used in cognitive and developmental robotics, as in Matarić's definition, is that of "autonomy," or "autonomous" robots. This term still refers to the fact that the robot performs tasks automatically, though autonomy underlies a more general, high-level decision-making capability. This stresses the fact that the robot can autonomously choose the action to perform in response to sensory feedback and to its own internal cognitive system. It should however be noted that in robotics there is a branch of teleoperation robots that still fall under the category of robots, but rather than performing automatically, instead require the presence of a human teleoperator. This is the case of medical surgical robots, whose task is not to be fully autonomous in performing surgery, but rather to support the expert human surgeon during an operation, in a safe and less intrusive way, and only to perform some surgical tasks semi-autonomously.

Finally, the fourth key concept in the definition of a robot is *programmable by a computer*, that is, the machine is controlled by computer software programmed by a human expert. But beyond a pure software implementation point of view, the issue of writing a program that implements a cognitive controller (architecture) is highly relevant to developmental robotics, as the challenging task in this field is to design and implement, through computer programs and artificial intelligence algorithms, cognitive theories of behavior control.

This analysis of the definition of "robot" and explanation of its key characteristics illustrates that "robot" is not a univocal, unambiguous term. It rather implies a continuum between different extremes. From a physical appearance and mechanical point of view, the continuum goes from a humanoid-looking robot at one extreme, to an industrial robot or a vacuum cleaner at the other extreme. In terms of control, robots range from fully autonomous systems to semi-automatic or teleoperated machines, and with regard to behavioral capacity they range from operating simple, repetitive actions to performing complex tasks. For detailed analyses of all aspects of robotics, their varying physical, control, and behavioral characteristics, and the diverse application areas, we refer the interested reader to the comprehensive handbook of robotics edited by Siciliano and Khatib (2008).

2.1.1 Humanoid Robots

Given the wide use of human- and childlike robot platforms in developmental robotics, we discuss here the definition of "humanoid robots" and the differences between humanoid and android robots. A *humanoid robot* has an anthropomorphic body plan (i.e., with a head, torso, two arms, and two legs) and human-like senses (cameras for vision, microphones for audio perception, touch sensors for tactile perception) (de Pina Filho 2007; Behnke 2008). Their appearance varies from a complex electronic machine with visible cables, motors, and electronic components (figure 2.1a) to a robot body [with plastic covers resembling space suits or toys] (figure 2.1b).

A special type of humanoid platform is the *android robot*, which in addition to an anthropomorphic body has a human-like "skin" surfacing (figure 2.1c). The prototypical examples of android robots are the Geminoid adult robot (Nishio, Ishiguro, and Hagita 2007; Sakamoto et al. 2007) developed by Hiroshi Ishiguro and colleagues at the ATR Intelligent Robotics and Communication Laboratories and at Osaka University in Japan. Infant and child versions of android robots also exist for developmental robotics research, as with the Repliee R1 platform of a four-year-old girl (Minato et al. 2004; see also section 2.3).

An android robot is intentionally designed with the goal of being indistinguishable from humans in its external appearance and behavior (MacDorman and Ishiguro 2006a). Therefore it is important to understand the implications of using androids in human-robot interaction because of the "uncanny valley" phenomenon. Mori (1970/2012) initially proposed this as, with the development of increasingly human-like robots, he hypothesized a shift of the human's response to a humanlike android robot from empathy to revulsion. This sense of revulsion and eeriness, in other words, the uncanny valley, is caused by the failure to attain full human-like qualities in both behavior and appearance. Figure 2.2 visualizes Mori's representation of the uncanny valley as a drop in the curve between the robot's human-like appearance and people's

a) Humanoid COG b) Humanoid iCub c) Android GEMINOID

Figure 2.1
Humanoid robot COG with visible electronic equipment, courtesy of Brian Scassellatti (a); humanoid robot iCub with plastic cover, courtesy of Giorgio Metta, Italian Institute of Technology (b); android robot Geminoid H1–4 with his creator Hiroshi Ishiguro, courtesy of Hiroshi Ishiguro, Osaka University (Geminoid is a registered trademark of Advanced Telecommunications Research Institute International, ATR) (c).

affinity for such a robotic entity. The shift from industrial robots to toy robots only causes a partial increase of similarity to the human body, and is expected to cause a positive increase in people's affinity response to the robot. However, the use of a robotic prosthetic limb that is almost indistinguishable from the human arm but has subtle flaws in appearance and behavior will cause a significant drop in the curve of affinity (hence the valley concept) with a discomforting sense of eeriness. As such, the uncanny valley is a key issue in robotics and human-robot interaction. Numerous studies have started to investigate both the biological and social factors behind Mori's uncanny valley hypothesis and the implications of such a phenomenon in human-robot interaction with humanoid and android robots, as well as with computer-graphics animation and virtual agents (MacDorman and Ishiguro 2006b).

Developmental robotics has mostly been based on humanoid robots. However, some researchers, especially in Japan, have investigated the use of humanoid-android baby robot platforms (Guizzo 2010). In section 2.3 we will review in detail the baby robots used in developmental robotics research with both humanoid and android robots, as well as some studies with mobile robots such as the SONY AIBO dog cub. But first we

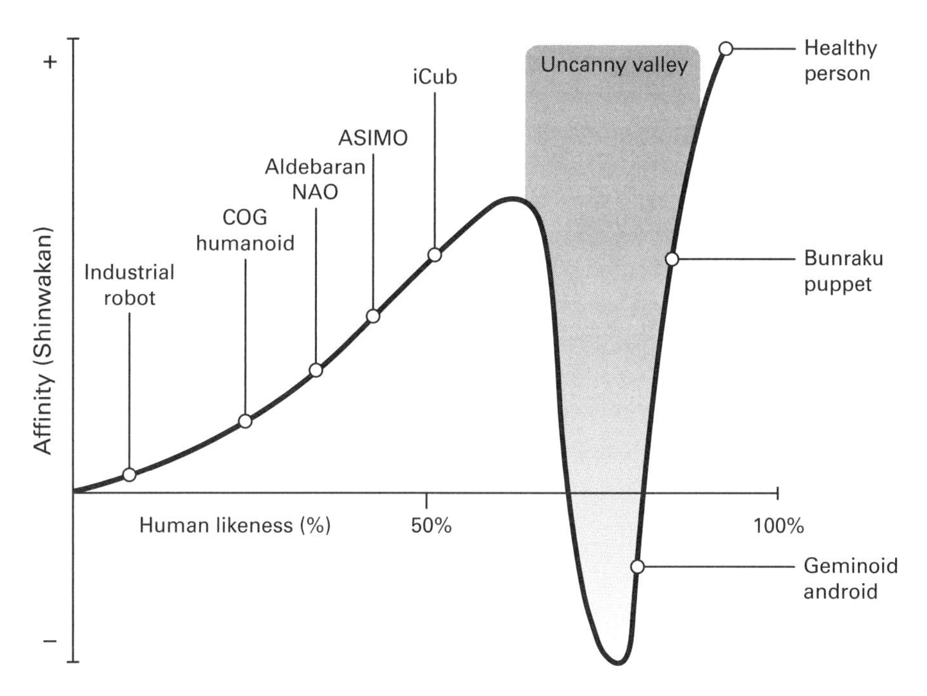

Figure 2.2
Uncanny valley. Figure adapted from Mori 2012.

need a crash course on the basic concepts and technologies of robot platform construction with actuators and sensors.

2.2 Introduction to Robotics

This section gives an introduction to the main terminology and hardware technologies used in robotics, and is aimed at readers who are not familiar with such concepts. Terms such as degrees of freedom (DOFs), infrared sensors, and electric and pneumatic actuators will be used extensively in this book, so our readers from the cognitive and human sciences with no direct experience in robotics will benefit from such a brief primer on robot technology. This technical overview is in part based on Maja Matarić's concise volume *The Robotics Primer* (Matarić 2007). Readers should refer to this book for a more detailed description of robotics concepts. For a further, in-depth, and complete understanding of robotics, we recommend the Siciliano and Khatib (2008) handbook mentioned earlier.

This section will first cover the DOF concept and its implementation through a variety of actuators used in humanoid robotics. It will then present the types of sensors used to handle visual, auditory, tactile, and proprioceptive/force sensors and some basic signal processing concepts.

2.2.1 Degrees of Freedom, Effectors, and Actuators

A robot has a body with a set of sensors to perceive the environment (e.g., camera for vision, microphones for audio perception, and infrared sensors for distance estimation) and a set of actuators to produce actions (e.g., move wheels or legs for locomotion, or move an articulated arm to manipulate objects).

To identify the possible dimensions of the actions that a robot can perform, the *degree of freedom* concept (DOF henceforth) is used. DOF is the dimension in which a movement can be produced in the 3D x, y, z space coordinates. For example, our shoulder has three DOFs because it can move the upper arm in all three horizontal (x), vertical (y) and rotational (z) dimensions, through the ball socket articulation structure. The articulation between the upper arm and the forearm only has one DOF in the elbow, to open or close the forearm. The eyes typically have three DOFs, for movements in the vertical, horizontal, and rotational directions.

Robots use effectors and actuators to perform actions along one or more DOFs. An *effector* is a mechanical device that allows a robot to produce an effect on the environment. In the human body, the effectors are the arms, fingers, eyes, and legs. In robots, the effectors are wheels and legs, fingers and grippers, eyes, arms, wings, and flippers. An *actuator* is the mechanism that enables the robot's effector to perform an action in at least one DOF by converting energy into a movement. Primary examples of the human body actuators are the muscles and tendons controlling the joints. In robots, the most common actuators are the motors (electric, hydraulic, or pneumatic) or other materials that can change their shape and properties to produce a movement. Table 2.1 lists the most common actuators and their main features.

An actuator connecting two body parts is called a *joint*. Given a joint can have more than one DOF, in general a separate actuator is needed for each DOF. More recently, motors allowing more than one DOF have been developed.

A robot can either use a passive effector or, more commonly, an (active) actuator. *Active actuators* are active joints that use energy to provide power to move the effector to a desired position. These actuators can also use a compliant mechanism. *Passive effectors*, in contrast, are based on passive actuation mechanisms and exploit the energy produced by the actuator configuration (e.g., the shape of aircraft wings supporting aerodynamic lift) and its interaction with the environment (e.g., gravity, wind) to move the robot. The best-known robotic example of a passive effector is the passive walker (Collins et al. 2005), which exploits the gravity in a sloped track and the structure of two legs with flexible knees in order to move in a downward direction without the need of applying energy. Many animals and part of the human motor system exploit the properties of passive and compliant effectors, and there is a branch of robotics called *embodied intelligence* that takes inspiration from

Table 2.1
Main types of actuators

Actuator	Description	Notes
Electric motor	Rotating motors producing movement due to electric current	Pros: simple, affordable, common Cons: produce heat, not very energy efficient Main types: DC motors, geared DC motors, servo motors, stepper motors
Hydraulics	Actuators producing movement due to changes in fluid pressure	Pros: for powerful, precise movements Cons: large, dangerous, risk of leaking Main types: pistons
Pneumatics	Actuators producing movement due to changes in air pressure	Pros: for powerful movements, quick and accurate response, passive dynamics Cons: dangerous, noisy, risk of leaking Main types: McKibben muscles
Reactive materials	Materials producing small movement (shrinking/ elongation) due to reaction to light, chemical substances, or temperature	Pros: suitable for microrobots, linear actuators Cons: only for small/weak movements; chemical risks Main types: photoreactive materials, chemically reactive materials, thermally reactive materials

biological systems to develop low-energy-consumption robots based on passive or hybrid passive-active actuators (Pfeifer and Bongard 2007). An actuator can be *compliant* if it has a mechanism to respond to external force stimulation by stopping its motor. A humanoid robot with a compliant arm, and engaged in an object-handling task, will stop the motor closing its fingers and hand when its force sensors feel the opposing force applied by the object surface. Without compliance the motor will continue to apply a force until the target shaft position of the motor has been achieved, thus with the risk of crushing the object. Compliant actuators are also very important for safe human-robot interaction because a robot must be able to stop moving its effector if it is hitting a surface or object, especially if its movement risks hurting a human user.

The most common type of actuator is an electric motor. These are used extensively in robot platforms as motors are affordable, require relatively low energy consumption (though they are not energy efficient), and are based on the standard, simple technology of electric motors available in many other engineering fields. There are four main types of electric motors currently used in robotics: (1) DC motors, (2) geared DC motors, (3) servo motors, and (4) stepper motors. These are rotational motors as they transform energy into the motion of a shaft. The first three are variations of the same main type of electric motor, while the stepper motor is based on different motor design principles and significantly differs in its use and cost in robotic platforms.

The two main parameters of rotational motors are the *torque*, the rotational force produced at a given distance, and the *speed*, the rotational velocity (measured in revolutions per minutes, rpm, or revolutions per second, rps). Most motors work with a *position control* mechanism, which is the controller that drives the motor to reach a desired position. Even if an obstacle and a force are applied, the controller will react to the obstacle's stiffness and produce a counterforce in the motor to keep the desired position. An alternative is to use *torque control*, by which the motor aims to keep a target torque, ignoring the actual position of the shaft. A compliant actuator will have a feedback mechanism to feel the opposing force and stop the motor from reaching its target position or torque.

DC motors (direct current motors) are by far the most commonly used type of actuator in robotics. They are devices that convert electric energy into rotating movements. DC motors are based on the electromechanical principles of current passing through electric wire loops that generate magnetic fields that turn the motor shaft. The amplitude of the electric current can be modulated to cause rotations of different speeds (rotational velocity) and torques (rotational force), with the actuator's power proportional to the torque and the speed. However, most DC motors typically have a high speed of 50–150 revolutions per second (3,000–9,000 revolutions per minute) and low torque, which can be a problem for robots that need high rotational force and low speed. *Geared DC motors* address this problem by a gear mechanism that converts the speed and torque of the rotating shaft into slower and stronger actuator movements. This is based on the principles that two interconnected gears with different diameters produce a change in speed and torque. When a small gear is linked to a motor shaft and drives a larger gear, this causes an increase in rotational force and a decrease in rotational velocity in the large gear. When instead a large gear is linked to a motor shaft and drives a smaller gear, this causes a decrease in rotational force and an increase in rotational velocity in the smaller gear. So the connection between a shaft-led small gear and a larger interconnected gear allows the robot to achieve its aim of using DC motors to apply high forces at low rotational speeds. Combinations of more than two gears (ganged gears) further allow a motor to be able to achieve various multiplications of torque and rotation effects.

The third type of motor is a servo motor (often referred to simply as a servo), that is, an electric motor actuator that rotates the shaft to a specific angular position. It combines a standard DC motor mechanism, an electronic component to tell the motor how much to turn the shaft and to which position, and a potentiometer to sense the current position of the shaft. Servo motors use feedback signals to calculate error and adjust final position, for example, compensating when a payload is applied. One limitation is that the shaft rotation is typically restricted to 180 degrees. However, because of their precision, servo motors increasingly are being used in robotics to define and reach a target angular position.

The final type of electric actuator is the *stepper motor*, an electric device that rotates in specified "steps," or degrees of rotation, and can move the actuator to a desired angular position. Stepper motors can also be combined with gears to increase the torque and decrease the rotational velocity.

Hydraulic and pneumatic motors (pistons) are linear actuators where liquid (the former) or compressed air (the latter) is inserted in a tube causes a linear shrinking or elongation of the actuator. They can apply a strong force, and are therefore common in industrial robotics. Hydraulic actuators can require heavy equipment, with difficult maintenance, so they are more common in automated manufacturing industry. Hydraulic motors also offer a high level of precision for the desired target position. The disadvantages of hydraulic and pneumatic actuators in smaller research platforms, such as humanoid robots, are that they can be very noisy, are subject to liquid/air leakage risk, and may require large compressors. However, there are some advantages for developmental robotic platforms in the use of pneumatic actuators, as they can model the dynamic properties of muscles. This is the case, for example, for the use of the McKibben pneumatic muscles in the Pneuborn-13 baby robot (Narioka et al. 2009, see section 2.3). Such a pneumatic actuator consists of an inner rubber tube and an outer nylon sleeve, and the pumping of compressed air within the inner tube causes the muscle to contract, up to a 25 percent, and the stiffness to increase and be modulated as needed.

Finally, *reactive material actuators* provide an alternative to electric- and pressure-based systems. These consist of a variety of special materials, such as fabric or polymers and chemical compounds, which produce small movement (shrinking/elongation of a tissue) as a reaction to light (photoreactive), or to chemical substances and acidic/alkaline solutions (chemically reactive), or to temperature (thermally reactive).

An actuator, or a combination of these, can control one effector to perform a specific motor function. In robotics, we can differentiate between two main classes of effectors depending on their main function, that is, to control locomotion or manipulation. *Locomotion effectors* can implement a variety of movements (e.g., bipedal walking, four-leg walking, swinging, jumping, crawling, climbing, flying, swimming) to move the robot to different locations. Locomotion effectors include legs, wheels, wings, and flippers. Arms too can sometimes be used for locomotion, as for crawling or climbing. *Manipulation effectors* are used to act on objects and typically include mono- or bi-manual manipulators using arms and hands similar to the human upper limbs. Alternative manipulation effectors are inspired by animal manipulators. An example of animal-inspired robots is the soft material effectors of the arms of an octopus (Laschi et al. 2012) or the trunk of an elephant (Hannan and Walker 2001; Martinez et al. 2013).

The arm and hand manipulator effectors are important in developmental robotics as they are common in the robotic arms of the humanoid robots used in this field, and as the development of manipulation skills has been linked to the evolutionary and ontogenetic development of various higher-level cognitive functions (Cangelosi et al. 2010). A manipulator in a robotic arm would typically involve shoulder joints, elbow joints, wrist joints, and finger joints, with a simplified DOF structure, similar to that of the human body. The human arm, excluding the hand, has seven DOFs, with three in the shoulder (up-down, left-right, rotation of the arm axis), one in the elbow (open-close) and three in the wrist (up-down, left-right, rotation of the hand axis). Humanoid robots created for cognitive and developmental models will tend to replicate a more precise organization of the human body's arm/hand manipulation system, although typically with limitations in the number of DOFs in the hand and fingers.

Finally, one important issue to consider in manipulation is how to find the position of the end effector (i.e., the effector containing the gripper or fingers that has to handle the object) using such a high number of DOFs. This can be based on kinematics and the inverse kinematic problem, or on machine learning methods (Pattacini et al. 2010) and bio-inspired models (e.g., Massera, Cangelosi, and Nolfi 2007; see also the discussion of motor control in chapter 5 of this book).

2.2.3 Sensors

Robots must use information from the external environment and from their own internal states to be able to act appropriately in the environment. This is where sensors play a crucial role. Robot sensors are electromechanical devices that measure quantitatively the physical properties of the (internal and external) environment.

To be able to sense both internal and external states, the robot needs proprioceptive and exteroceptive sensors. *Proprioceptive (internal) sensors* can perceive the position of the robot's own wheels and the values of the joint angles in the different actuators (e.g., force and torque sensors). This is similar to the human body's proprioceptive sensors as those measuring the stretching of the muscles. *Exteroceptive (external) sensors* permit the sensing of the external world as the distance from obstacles and walls, forces applied by other robots/humans, and the presence and characteristics of objects and entities in the world through vision, sounds, and smell.

Table 2.2 lists the main types of sensors used in robotics. The first four types are exteroceptive sensors that measure signals from the external environment (light, sound, distance, and position), while the other two are proprioceptive sensors that measure the states of internal parts and states (motor force torque and acceleration and inclination). The table also lists some of the common properties and issues related to the use of each specific sensor.

Sensors can use either passive or active sensing technologies. Passive sensors consist of a detector to sense changes in the environment, as with a bump sensor that detects pressure or a light switcher that detects light signals. Active sensors need energy and an emitter to send their own signal to the environment as well as a detector to sense the signal's return and determine the quantitative effects (e.g., return delays) caused by the environment. The most common active sensors are the "reflective optosensors," and use either reflectance sensors or break beam sensors. In reflectance sensors the emitter and the detector are on the same side of the device and the light signal must

Table 2.2
Main types of sensors

Sensors	Devices	Notes
Vision (light)	Photocell	For intensity of light
	1D camera	For perception of horizontal direction
	2D B&W or color camera	Full visual processing sensing; Computation intensive but information rich
Sound	Microphone	Full audio processing sensing; Computation intensive but information rich
Distance and proximity	Ultrasonic (sonar, radar)	Time-of-flight for the return of emitted ultrasonic sound waves; Limits of specular reflection for non-smooth surfaces
	Infrared (IR)	Use of reflective optosensor for infrared light waves; Modulated IR to reduce interference
	Camera	Binocular disparity or visual perspective
	Laser	Time of flight for the return of emitted laser light; No specular reflection issue
	Hall effect	Ferromagnetic materials
Contact (tactile)	Bump switch	Binary on/off contact
	Analogical touch sensors	Spring coupled with a shaft; Soft conductive material that change its resistance according to compression
	Skin	Sensors distributed over body
Position and localization	GPS	Global Positioning System Accuracy from 1.5m (GPS) to 2cm (DGPS)
	SLAM (optical, sonar, vision)	Simultaneous localization and mapping; Use of light, sound, or vision sensors
Force and torque	Shaft encoder	For number of rotations of the motor's shaft; Use of break-beam optosensor speedometer for speed of rotation; Odometer for number of rotations
	Quadratic shaft encoder	For the direction of rotation of the motor shaft
	Potentiometer	For motor's shaft position; In servo motors, to detect shaft position
Inclination and acceleration	Gyroscope	For inclination and acceleration
	Accelerometer	For acceleration

be reflected back to the detector due to the presence of an obstacle. These reflective sensors use a triangulation method to compute the time it takes between a light signal being actively emitted and its return to the detector. Active reflectance sensors are used with light signals in infrared (IR) sensors and with ultrasound signals in sonar (SOund NAvigation and Ranging) and with laser sensors. The second type of active reflective optosensor is the break beam sensor, where the detector is at the opposite side of the emitter. The presence of the obstacle breaks the light beam sent by the emitter, with the time delay and lack of signal detection indicating the presence of an obstacle.

An important issue to consider when working with robot sensors is the *uncertainty* of the quantitative information being provided by the sensors themselves. Such uncertainty is inherent in any physical measurement system, and comes from a variety of sources such as the sensor noise and error, its measurement range limitations, noise and error from the robot's effectors, and unknown and unforeseeable properties and changes in the environment. Although there are methods available to deal with some of the noise and limitations of the sensors, these can only partially address the uncertainty issue. For example, calibration can be used for light sensors to deal with ambient light when using sensors such as LEDs, IR sensors, and cameras. Ambient light is the default level of light in the environment that a reflectance light sensor should ignore to recognize and process the light emitted by the sensor itself. This can be done through sensor calibration, in other words, by taking a reading when the emitted light is turned off and then subtracting this from the light sensor reading when the emitted light is on.

The information provided by most sensors normally is fed to the robot's controller to contribute to the selection of an action. For simple sensors, such as IR and bump switches, the controller can use raw data directly. For more complex sensors, instead, advanced signal processing techniques are needed before the robot uses the information. This is the case, for example, of vision and sound sensors, and from combinations of active sonar/light sensors for localization and mapping. We will describe here briefly the main signal processing methodologies related for 2D camera vision sensors, to audio microphone sensors and to simultaneous localization and mapping (SLAM) sensors (Thrun and Leonard 2008), as these methodologies will be referred to in various experiments described in the book. Vision and audio perception are crucial features in most developmental robotics studies, and SLAM can also be used for navigation control in human-robot interaction experiments.

The robot's *visual perception* uses digital images generated by its (color) cameras to provide a rich set of information. Once the visual signal is processed, the agent can use it to build a representation of the external word. This consists for instance in the segmentation and identification of objects and their properties for manipulation tasks, of the environment layout for navigation and object avoidance, and of other human and robotic agents for social interaction.

Research in artificial vision (also called machine vision) has developed a series of standard image processing routines and algorithms widely used in robotics. Although vision remains a big challenge in artificial intelligence and robotics, these routines provide useful tools for object segmentation, identification, and tracking in robots (Shapiro and Stockman 2002; Szeliski 2011).

In robotics vision, a set of feature extraction methods typically can be applied to derive a *saliency map* (Itti and Koch 2001), that is, the identification of the parts of image that are important for the robot behavior. Specifically, feature extraction is used to detect and isolate the various features of a digitized image such as color, shape, edges, and motion. For example, in Schlesinger, Amso, and Johnson's (2007) developmental model of object completion (see chapter 4), a saliency map is created by combining a set of feature extraction methods for color, motion, orientation, and brightness. These features are combined to generate the whole salience map used by the model to fixate the objects (figure 2.3).

The processing of a digital camera image typically requires two main stages of visual analysis: (1) image processing, for example smoothing to remove noise and identification of edges and regions to segment an object of interest from the background and from other objects; and (2) scene analysis to create a description of objects, as integrating parts of an object (e.g., body parts like limbs, head, torso) into a model of the whole object (e.g., human body).

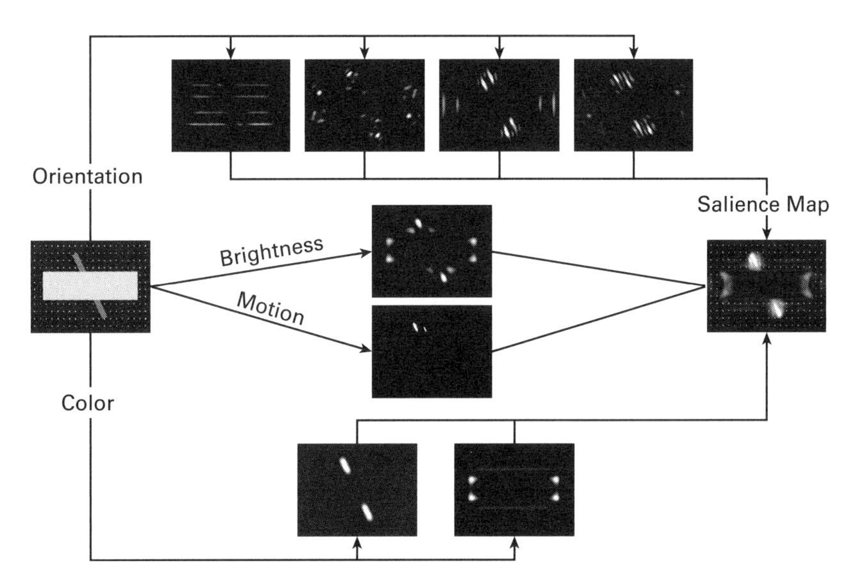

Figure 2.3

Combination of feature extraction for salience map generation in Schlesinger, Amso, and Johnson 2007. See chapter 4 for more details on this model.

The smoothing process removes noise from the original image, as spurious pixels and irregularities (e.g., elimination of thin lines and thickening of thick lines). Edge detection requires the identification of the points in an image at which the image brightness changes sharply. Smoothing is typically done with the method of *convolution*, in other words, the sliding of an averaging window and summing up of the pixel intensity numbers within the center of the sliding window. A sliding window can typically use a Gaussian function. More advanced methods use averaging window functions that combine smoothing and edge detection. For example, the Sobel operator combines Gaussian smoothing and differentiation for edge detection. The Sobel filter is based on the convolution of the image with a small, separable, and integer-valued filter in horizontal and vertical directions. Alternatively, a Gabor filter uses linear filters for edge detection, and is inspired by the functioning of neurons in the primary visual cortex. A 2D Gabor filter is based on a Gaussian function modulated by a sinusoidal plane wave, and can use different frequencies and orientation settings.

For region identification, methods like histogram and split-and-merge can be used. The histogram-based methods produce a histogram of the pixels brightness in the image, and the peaks and valleys in this distribution are used to locate the regions (clusters) in the image. This is a very efficient method because these techniques typically require only one pass through the pixels. The split-and-merge method uses a "quadtree" partition of an image. It starts by analyzing the pixel intensities of the whole image. If it finds a nonhomogeneous intensity of the pixels in the whole image, it splits the image into four quadrants, to check again the homogeneity of the pixel in each new, smaller region. This process is done iteratively with ever-smaller quadrants. After the splitting method, the adjacent homogeneous areas are merged into larger areas, which constitute the segmented regions.

Other filters can be applied to generate additional features of the object, such as color filtering, typically using the red-green-blue (RGB) primary color components, and motion filtering to identify parts of the image where pixels (or regions/objects) have moved. Once a fully segmented image has been extracted, various scene analysis methods can be used to create description of objects. For example, a model-based approach can be used where, say, in the case of recognition of a human/robot body, the different parts of a body must be segmented and identified (e.g., body parts like limbs, head, torso) and matched to an expected model of the whole human body. In face recognition models, this would require the identification of crucial points and features of a face, such as the position and structure of the eyes and mouth.

Software libraries provide off-the-shelf tools to process a robot's camera images with a variety of feature extraction filters and algorithms. OpenCV (opencv.org; Bradski and Kaehler 2008) is one of the most commonly used open source libraries for visual processing in developmental robotics. Such libraries are commonly integrated along with robotic software, such as the iCub and YARP middleware (see section 2.3.1).

Other artificial vision methods are based on the neurophysiology and functioning of the brain's visual system. For example, log-polar vision methods model the neurophysiology of the distribution of visual sensors (retina cones) in the visual system of mammals (Sandini and Tagliasco 1980). In these animals, as in humans, the distribution of the receptors (rods and cones) in the retina is denser in the central fovea and sparser in the periphery. This results in a nonlinear resolution of the image, with higher resolution at the center and lower resolution at the sides. This distribution follows a radial symmetry, which can be approximated by a polar distribution. Moreover, the projection of the array of cones from the retina to the primary visual cortex can also be approximated by a logarithmic-polar (log-polar) distribution mapped onto a rectangular surface (the cortex). Thus the cortical representation of the fovea has more neurons dedicated to it, while the periphery uses fewer neurons with a coarser resolution representation. This log-polar method converts a standard rectangular image into a space-variant image. This topological representation can be easily employed for real-time object segmentation and tracking tasks, as it utilizes a resource-economic image representation and processing methodology. Log-polar mappings have been extensively used in cognitive robotics research (Traver and Bernardino 2010), for example, for color object tracking with the Babybot platform, a precursor of the iCub humanoid robot (Metta, Gasteratos, and Sandini 2004).

Some vision processing tools are based on artificial neural networks methodologies (e.g., Riesenhuber and Poggio 1999; Serre and Poggio 2010; see Borisyuk et al. 2009, for a robotic application). These do not perform hand-defined analyses of the image features with predefined filtering methods. Rather, they use neural computation algorithms inspired by the physiology of the visual system. Neural networks process an image using a hierarchical processing architecture inspired by the way the visual cortex hierarchically processes visual stimuli from the primary cortex V1-V4 (with selectivity for stimulus orientation, size, depth, and motion direction) to the anterior inferotemporal (AIT) cortex (for object representation). For example, Riesenhuber and Poggio (1999) propose a hierarchical computational model of visual object recognition in the brain.

Software libraries based on neural computational models of vision include the Spikenet Vision System (spikenet-technology.com), which permits template matching of segmented objects with a set of pre-trained target images. Dominey and Warneken (2011) used this library in developmental robotics modeling to identify the location of the target object and landmark in their human-robot cooperation experiments (see section 6.4 in chapter 6).

Sound and speech perception in robots can be realized through signal processing methods of raw audio signal, or through more advanced pattern recognition systems for sound and speech perception (Jurafsky et al. 2000). Classical methods to analyze raw sound signals include Fourier analysis for the extraction of the first formants of speech sounds (as in the case of vowel representation), hidden Markov models (HMM), and artificial neural networks. The ad-hoc processing of sound and speech signals in robots is typically applied in cases when no predefined sound systems exist, as in robotics models of the evolution of phonetic systems (see Oudeyer's study in chapter 7), or when focusing on the phonetic, prelexical capabilities of a robot or on the neural modeling of sound perception (e.g., Westermann and Miranda 2004).

For speech perception, in developmental robotics it is more common to utilize automatic speech recognition (ASR) systems that can extract a string of recognized words, and an automatic parser that performs grammatical analysis of the words. Some speech processing software packages include both ASR and parser recognition systems. The ASR systems most commonly used in developmental robotics are the open source libraries SPHINX, ESMERALDA, and JULIUS, commercial software such as DRAGON DICTATE, as well as the standard ASRs included in the computer operating systems of Windows and Apple. SPHINX, developed at Carnegie Mellon University, is one of the most commonly used ASR applications. SPHINX actually includes a group of speech recognition systems, with a primary reliance on N-gram and HMM methods. It permits the recognition of continuous speech with large vocabularies, after preliminary training with a speaker to create an acoustic model of the individual's phonetic systems. ESMERALDA, developed at Bielefeld University, provides a software framework for the development of speech recognition systems using a variety of ASR methods. JULIUS, developed by the Japanese Continuous Speech Recognition Consortium, is a real-time, high-speed continuous speech recognition system for large vocabularies, based on the N-gram and context-dependent HMM methods. It also includes a grammar-based recognition parser named "Julian." Julius was originally developed for the Japanese language, though acoustic models now exist for other languages including English. On the commercial side, Dragon Dictate (Nuance Communications) is one of the most popular choices, as it also includes a free version with reduced functionalities.

Developmental experiments on language learning have used various ASR systems such as SPHINX (Tikhanoff, Cangelosi, and Metta 2011), ESMERALDA (Saunders et al. 2009) and the Apple and Dragon Dictate software (Morse et al. 2011) (see also chapter 7).

For the robot to perceive its location on a map, the signals from a combination of position, localization, and distance sensors can be integrated to allow the robot to construct a map of the environment and identify its position within the map. Through the method of simultaneous localization and mapping (also known as concurrent mapping and localization), the robot can perform the two tasks of building a map and localizing itself at the same time. SLAM does not require the human programmer to provide the robot with a predefined plan of a room/building or the external environment, and does not require the absolute location coordinates, as the Global Positioning System (GPS) does. SLAM is in fact essential in scenarios where the robot has to explore an unknown environment for which a map is not available, or when GPS-like sensors cannot function, as in an indoor environment. Simultaneous localization and mapping is hard to achieve because it is a chicken-and-egg situation: how can the robot determine its position if it doesn't have a map, yet how can it create a map if the robot does not know where it is. Moreover, this problem is made harder by the fact that the sensor data collected in different parts of the environment might look the same, or data from the same/near locations might appear different due to sensor noise or different orientation. Identification of landmarks can help address this issue, as well as odometry information, though SLAM remains a hard problem for roboticists to solve due to the inherent complexity of the problem and the sensor data.

Sensor data to perform SLAM can be collected in a variety of ways, including from optical sensors (e.g., 1D or 2D sweeping laser rangefinders), ultrasonic sensors (2D or 3D sonar sensors), and visual camera sensors and robot wheel odometry sensors. Often, combinations of these sensors are used, as with sonar signals complemented by camera information for landmark detection. Various signal processing and information fusion solutions have been developed for SLAM that address the localization and mapping problem, with varying degrees of success. A good SLAM algorithm was a key element in the success of Stanford University's team and the Stanley robot car at the 2005 DARPA Grand Challenge. For a review of SLAM methods, see for example chapter 37, by Thrun and Leonard, in Siciliano and Kathib's *Springer Handbook of Robotics* (2008).

2.3 Baby Humanoid Robots

The birth of developmental robotics research in the early 2000s has led to the design and production of a variety of platforms for research on baby humanoid robots. This has been accompanied by the parallel development of new, more standard adult-like humanoid platforms, also used in some developmental studies.

The following sections briefly describe the main characteristics of these robots. Table 2.3 gives a comparative overview of these characteristics. It lists the twelve most commonly used robot platforms in developmental robotics, with details on their designer or manufacturer, total number of degrees of freedom, type and position of actuators, skin sensors, appearance and size, and main literature reference and year of delivery. A brief analysis of this comparison table shows that the majority of robots are full-body humanoid robots, with actuators in both the upper torso and legs, and most of the time based on electric actuators. A notable exception is COG, the Massachusetts Institute of Technology platform, one of the first robots used in cognitive and developmental robotics. The latest baby robot Affetto currently only has upper-torso actuators, though this platform is being extended to a full-body humanoid. As for skin sensors, only a few (iCub, CB^2, and Robovie) have tactile skin sensors distributed in all or most of the body. Most robots have a mechanical humanoid structure, with the exception of the typical baby-girl android Repliee R1, and in part for the silicon-type full-body

Table 2.3
Humanoid robots used in developmental robotics research

	Manufacturer	DOFs	Actuator type	Actuators position	Soft sensitive skin	Human-like appearance	Child size	Height/ weight	History (Model)	Main research references
iCub	IIT (Italy)	53	Electric	Whole-body	Yes	No	Yes	105 cm 22 kg	2008	Metta et al., 2008 Parmiggiani et al., 2012
NAO	Aldebaran (France)	25	Electric	Whole-body	No	No	Yes	58 cm 4.8 kg	2005 (AL-01) 2009 (Academic)	Gouaillier et al. 2008
ASIMO	Honda (Japan)	57	Electric	Whole-body	No	No	Yes	130 cm 48 kg	2011 (All New ASIMO)	Sakagami et al., 2002 Hirose & Ogawa, 2007
QRIO	Sony (Japan)	38	Electric	Whole body	No	No	Yes	58 cm 7.3 kg	2003 (SDR-4X-II)	Kuroki et al., 2003
CB	SARCOS (USA)	50	Hydraulic	Whole body	No	No	No	157 cm 92 kg	2006	Cheng et al., 2007b
CB^2	JST ERATO (Japan)	56	Pneumatic	Whole body	Yes	Yes	Yes	130 cm 33 kg	2007	Minato et al., 2007
Pneuborn-13 (Pneuborn-7II)	JST ERATO (Japan)	21	Pneumatic	Whole body	No	No	Yes	75 cm 3.9 kg	2009	Narioka et al., 2009
Repliee R1 (Geminoid)	ATR, Osaka, Kokoro (Japan)	9 (50)	Electric (Pneumatic)	Head (upper body)	No (Yes)	Yes (Yes)	Yes (No)	(150 cm)	2004 (2007)	Minato et al., 2004 (Sakamoto et al., 2007)
Infanoid	NICT (Japan)	29	Electric	Upper body	No	No	Yes		2001	Kozima, 2002
Affetto	Osaka (Japan)	31	Pneumatic and electric	Upper body	No	Yes	Yes	43 cm 3 kg	2011	Ishihara et al., 2011
KASPAR	Hertfordshire (UK)	17	Electric	Upper body	No	Yes	Yes	50 cm 15 kg	2008	Dautenhahn et al., 2009
COG	MIT (USA)	21	Electric	Upper body	No	No	No		1999	Brooks et al., 1999

cover of CB^2. Not surprising for developmental robotics is the fact that the majority of these platforms (ten out of twelve) have a childlike appearance or size, and two have a standard adult humanoid size. Most of these platforms are noncommercial robots with restricted availability for academic and industry research laboratories, and typically fewer than ten copies exist for most of these humanoid robots, or single copies in the case of robots such as Repliee R1 and CB^2. The main exception to this is the Aldebaran NAO robot, which is commercially available for research and robot competitions, and as of 2013 has sold over 2,500 copies. The iCub robot is another exception. Although it is not commercially available, twenty-eight copies exist in laboratories worldwide (as of 2013). This is the result of significant public research investment by the European Union's research Framework Programmes 6 and 7 (FP6 and FP7) in Cognitive Systems and Robotics, and the open source approach adapted by the RobotCub-iCub consortium led by the Italian Institute of Technology.

The twelve humanoid robot platforms included in table 2.3 were chosen because they have been primarily developed and used for developmental robotics research. Likewise a growing number of baby and adult humanoid robot platforms now exist and are used in part for cognitive research. For example, in July 2010 *IEEE Spectrum* (Guizzo 2010) published a comparison matrix of thirteen baby robots along the two dimensions of robot appearance and behavioral complexity. This classification includes four of the robots listed in table 2.1 (iCub, NAO, CB^2, Repliee R1), and nine additional baby robots: NEXI (MIT, United States), SIMON (Georgia Tech, United States), M3 NEONY (JST ERATO Asada Project, Osaka University, Japan), DIEGO-SAN (University of California at San Diego, United States, and Kokoro Corporation, Japan), ZENO (Hanson Robotics, United States), KOJIRO (Tokyo University, Japan), YOTARO (Tsukuba University, Japan), ROBOTINHO (Bonn University, Germany), and REALCARE BABY (Realityworks, United States). Of these, some are toy-like entertainment products such as the REALCARE BABY and YOTARO.

Other adult-size humanoid robots that have been in part used for general research in cognitive robotics include PR2 (Willow Garage Inc., United States), the suite of HRP-2, HRP-3, and HRP-4 humanoids (National Institute of Advanced Industrial Science and Technology, and Kawada Industries, Japan), LOLA (Technical University of Munich, Germany), HUBO (Korea Advanced Institute of Science and Technology), BARTHOC (Bielefeld University, Germany), Robovie (ATR Japan), Toyota Partner Robot (Toyota, Japan), and ROMEO (Aldebaran Robotics, France). Many of these platforms have been developed primarily for research on bipedal walking (e.g., HRP), for entertainment systems (e.g., Toyota music-playing robot), and for general research on manipulation and service robotics.

2.3.1 iCub

The *iCub* humanoid baby robot (Metta et al. 2008; Metta et al. 2010; Parmiggiani et al. 2012; www.icub.org) is one of the most widely used platforms for developmental robotics research. This robot was built with the explicit purpose of supporting cross-laboratory collaboration through the open-source licensing model. Its open model has made it one of the key developmental robotics benchmarking platforms, allowing laboratories working on the iCub to replicate and validate results and integrate existing software modules and cognitive models for more complex cognitive capabilities.

The iCub was the result of collaborative effort from different European laboratories, led by the Italian Institute of Technology, through the European Union-funded research consortium robotcub.org (Sandini et al. 2004; Metta, Vernon, and Sandini 2005). Two precursors of the iCub robot were developed by researchers at the LIRA-Lab of the University of Genoa and the Italian Institute of Technology. Babybot, whose design started in 1986, is an upper-torso humanoid robot, which in its final configuration had eighteen DOFs for the head, arm, torso, and hand (Metta et al. 2000). Both the head and hand were custom designed at the LIRA-Lab, while the Babybot's arm was realized through a commercial PUMA robot arm. Experience working on Babybot led to the subsequent development of James, a more advanced upper-torso humanoid robot with twenty-three DOFs (Jamone et al. 2006). The testing of mechatronics and electronic solutions for both Babybot and James had a big impact on the subsequent design of the iCub robot, with the EU project robotcub.org started in 1996.

The iCub is 105 cm tall and weighs approximately twenty-two kilograms, and its body was designed to model that of a three-and-a-half-year-old child (figure 2.4). The robot has a total of fifty-three DOFs, which is a high number of actuators in comparison with

Figure 2.4
iCub robot: head (top left), hand (bottom left), and human-robot interaction setup (right). Figures courtesy of Giorgio Metta, Italian Institute of Technology and Plymouth University.

related humanoid platforms of similar size. As the robot was designed with the primary aim of studying manipulation and mobility, it has a high number of DOFs in the hands and upper torso. The fifty-three DOFs consist of six for the head, fourteen for the two arms, eighteen for the two hands, three for the torso, and twelve for the two legs.

In particular, the six DOFs of the head include three for the neck to provide full head movement, and three for the eyes for vertical/horizontal tracking and vergence. Each hand has nine DOFs, with three independently controlled fingers, and the fourth and fifth finger controlled by a single DOF. The hand actuators are tendon driven, with most of the motors located in the forearm. The overall size of the hand is thirty-four millimeters wide at the wrist, sixty millimeters at the fingers, and only twenty-five millimeters thick (figure 2.4). The legs have six DOFs each and are strong enough to allow bipedal locomotion and crawling. The six DOFs of each leg consist of three at the hip, one at the knee, and two at the ankle (flexion/extension and abduction/adduction).

The suite of sensors in the iCub includes two digital cameras (640 × 480 pixels, 30 fps high resolution) and microphones. For inertia sensors there are three gyroscopes, three linear accelerometers, and one compass. It also has four custom-made force/torque sensors at the shoulders and hips. Tactile sensors in the palm and fingertips are based on capacitive sensing. A distributed sensorized skin has also been developed using capacitive sensor technology, and is usually located on the arms, fingertips, palms, and legs (Cannata et al. 2008). Each joint is equipped with positional sensors, in most cases using absolute position encoders.

A Pentium-based PC104 card, in the head, handles synchronization and reformatting of the various data streams integrating all sensory and motor-state information. However, the computationally intensive and time-consuming computation is carried out externally on a cluster of machines. The communication with the robot occurs via a Gbit Ethernet connection, with an umbilical cord providing network connectivity and external power. A set of DSP-based control cards, interconnected via CAN bus, and designed to fit the iCub's limited space requirements, are responsible locally for the low-level control loop in real-time. Ten CAN bus lines connect the various segments of the robot.

YARP (Yet Another Robot Platform; Metta, Fitzpatrick, and Natale 2006) is used as the iCub software middleware architecture. This is a general-purpose open-source software tool for applications like robots, since these platforms are real-time, computation-intensive, and involve interfacing with diverse and evolving hardware. YARP consists of a set of libraries that support modularity by abstracting two common issues: modularity in the algorithms and in the interfacing with the hardware. The first abstractions are defined in terms of protocols for interprocess communication through "ports." These can deliver messages across a network using IP-TCP protocol and can be commanded at run time to connect and disconnect. In the specific case of iCub, the robot is organized into five independent ports (head, right arm, left arm/hand, right leg, left leg). The second abstraction deals with hardware "devices." YARP uses the definition of interfaces for classes of devices to wrap native code APIs (application programming interfaces), typically provided by the hardware manufacturers, or which can be developed and modified when the hardware devices are created/updated.

The development of the iCub started in 2006, with the delivery of the first full-body iCub prototype in autumn 2008, and with twenty-eight iCubs delivered to labs worldwide in 2013. Ongoing developments on the iCub include the delivery of an upgraded 2.0 head, work on more robust legs for bipedal locomotion, and distributed high-efficiency batteries.

The iCub has been extensively used in developmental robotics, given it has been purposely designed for cognitive development modeling through a child robot. This book discusses many examples of experiments on the iCub for motor learning (chapter 4), social cooperation (chapter 6), language (chapter 7), and abstract symbols and numbers (chapter 8). Other developmental-related studies on the iCub not covered in this volume include experiments on grasping (Sauser et al. 2012), object manipulation for the discovery of object motor affordances (Macura et al. 2009; Caligiore et al. 2013; Yürüten et al. 2012), drawing (Mohan et al. 2011), object recognition through oscillatory neural model (Browatzki et al. 2012; Borisyuk et al. 2009), tool use with the passive motion paradigm (Mohan et al. 2009; Gori et al. 2012), and self-body identification based on visuomotor correlation (Saegusa, Metta, and Sandini 2012). See also Metta et al. (2010) and Nosengo (2009) for an overview of applications of the robot to developmental cognitive modeling.

2.3.2 NAO

The humanoid robot *NAO* (Gouaillier et al. 2008; aldebaran-robotics.com), produced by the French company Aldebaran Robotics, is another humanoid platform increasingly used in developmental robotics. The widespread use of the NAO platform not only is because of its affordable price for research purposes (approximately €6,000 in 2014), but also for its selection since 2008 as the "Standard Platform" for the RoboCup robot soccer competition (robotcup.org), thus making NAO available in numerous university labs worldwide. The first NAO robot (AL-01) was produced in 2005, with more recent academic editions available for research purposes since 2009. As of 2012, over 2,500 units of the NAO robot are used on a regular basis in more than 450 research and education institutes in over sixty countries (Aldebaran, personal communication).

NAO is a small humanoid robot fifty-eight centimeters tall, weighing 4.8 kilograms (see figure 2.5). The academic edition, which is most commonly used in developmental robotics, has a total of twenty-five DOFs. These consist of two DOFs for the head, ten for the arms (five each), one for the pelvis, ten for the legs (five each) and two for the hands (one in each hand). In the version currently used for the RoboCup standard competition, the robot only has twenty-three DOFs, as the hands are not actuated.

Figure 2.5
NAO robot by Aldebaran Robotics. © Courtesy of Aldebaran Robotics; photo by Ed Aldcock.

NAO uses two types of motors, patented by Aldebaran, each combining two rotary joints together to produce a universal joint module.

The suite of sensors includes four microphones and two CMOS digital cameras (960 p @30 fps, or 640 × 480 @ 2.5 fps over Wi-Fi). The cameras are not located in the two eyes depicted in the face (which are IR transmitters/receivers), but rather one is in the center of the forehead and one in the chin. Camera location choice is in part explained by NAO's original, primary use for robot soccer, where it is important to see the whole playing pitch (camera in the forehead looking at the whole playing field) and the ball being kicked (camera in the chin, looking downward toward the area in front of the feet). Other sensors include thirty-two Hall Effect sensors for the motor's state, two single-axis gyrometers, a three-axis accelerometer, two foot bumpers, two channel sonars, two infrared sensors, eight force sensitive resistors (FRS) sensors (four per foot), and three tactile sensors in the head.

The NAO robot has two loudspeakers, located in the robot ears, and four microphones around the head (front, back, left, and right). It also uses various LED lights to facilitate human robot interaction, such as twelve LEDs (sixteen blue levels) for the head tactile sensors, and other LEDs for eyes, ears, torso, and feet.

The other hardware specifications include a network connection for Wi-Fi (IEE 802.11 b/g) and a cable Ethernet connection. The 2012 release motherboard consists of a processor with an ATOM Z530 1.6GHz CPU, 1GB RAM, and 2GB flash memory.

The operating system for the on-board CPU uses embedded Linux (32 bit x 86 ELF), and Aldebaran Choregraphe and Aldebaran SDK are the proprietary software. The robot can be controlled via Choregraphe, a user-friendly behavior editor, by programming C++ modules, or by interacting with a rich API from scripting languages. Other programming languages primarily used for the NAO platform are C++, Urbi script, Python, and .Net.

Simulation NAO models exist in various robot simulation software products, such as the Cogmation NAO Sim, Webots, and Microsoft Robotics Studio (see section 2.5 for the Webots NAO model). The proprietary embedded software also features modules for text to speech and voice recognition, through the onboard speakers and microphone system, programs for face and shape detection, obstacle detection through the sonar system, and visual effects through the LEDs.

The NAO platform has been used for various developmental robotics experiments, including studies focused on locomotion, with Li, Lowe, Duran and Ziemke (Li et al. 2011; Li, Lowe, and Ziemke 2013) extending the iCub model of crawling with central pattern generators to compare the gait performance of the iCub and the NAO (see section 5.5). This work was further adapted to modeling early bipedal walking (Lee, Lowe, and Ziemke 2011). Yucel et al. (2009) developed a novel developmentally inspired joint attention fixation mechanism that can estimate the head pose and the gaze direction, also using bottom-up visual saliency.

This robot platform has also been extensively employed in human-robot interaction research, especially for studies with children, given its affordability and commercial safety standards. For example, Sarabia, Ros, and Demiris (2011) used NAO for imitation learning of dancing routines during human-children interaction experiments (see also chapter 7); Andry, Blanchard, and Gaussier (2011) for nonverbal communication to support learning in the robot; Pierris and Dahl (2010) for gesture recognition; Shamsuddin et al. (2012) for interaction with children with autism; and Baxter, Wood, Morse, and Belpaeme (Baxter et al. 2011; Belpaeme et al. 2012) for NAO as a long-term robot companion for hospitalized children. A further area of NAO's use in research is brain-computer interfaces for control of humanoid robots (Wei, Jaramillo, and Yunyi 2012).

2.3.3 ASIMO and QRIO

ASIMO (Honda) and QRIO (Sony) are humanoid robots, ASIMO produced by a major international company in the automotive industry, and QRIO by the electronics entertainment industry giant Sony. They have been used primarily for in-house R&D, and some research with the ASIMO and the QRIO platforms has been specifically based on developmental robotics.

ASIMO (Advanced Step in Innovative MObility) is one of the world's best-known humanoid robots (world.honda.com/ASIMO; Sakagami et al. 2002; Hirose and Ogawa 2007). The research on the ASIMO robot started in 1986, with Honda's first two-legged robot E0, and further prototypes (E1 to E6 series) of legged robots developed between 1987 and 1993. Work on the P1 to P3 series of large full-body humanoid robots from 1993 to 1997 led to the production in 1997 of P3, the first completely independent, two-legged humanoid (1.6 m, 130 kg). This resulted in November 2000 in the production of the first ASIMO robot, a smaller humanoid with a height of 1.2 meters to allow it to operate in the human living space (Sakagami et al. 2002). One of the main characteristics of the ASIMO robot is its efficient bipedal mobility behavior. This is based on the Honda's i-WALK technology, which uses predictive movement control. This was built on the earlier prototypes' walking control technology and extended to produce more smooth and natural locomotion. In addition, since 2002, ASIMO has been endowed with software modules to perform recognition of its environment (objects, moving people), voice, face, posture, and gestures.

The "New ASIMO" (figure 2.6, left) was developed in 2005 and has been used extensively for cognitive robotics research. It is 1.3 meters tall and weighs fifty-four kilograms. The new ASIMO has a total of thirty-four DOFs, with three for head neck joint, fourteen for the arms (seven per arm: three in the shoulder, one in the elbow, and three in the wrist), four in the hand (two fingers for each hand), one for the hips and twelve for the legs (six in each leg: three for the crotch joints, one in the knee joint, and two in the ankle). The humanoid has been used for cognitive robotics research, in scenarios of acting in synchrony with people, such as walking with a person while holding hands, and greeting with a handshake. ASIMO can walk as fast as 2.7 km/hour and run at six km/hour.

In 2011, a further version of the robot was released, called the "All New ASIMO," which has an increased number of DOFs (fifty-seven in total) and lighter weight (forty-seven kilograms), better locomotion performance with a running speed of nine km/hour, and more software functionalities for handling objects.

QRIO is the second R&D humanoid platform developed by a large corporation. Sony Corporation developed QRIO (model name SRD-4X, for Sony Dream Robot) in 2002 (Fujita et al. 2003; Kuroki et al. 2003). In its configuration known as SRD-4X II (Kuroki et al. 2003), the robot is fifty centimeters tall and weighs approximately seven

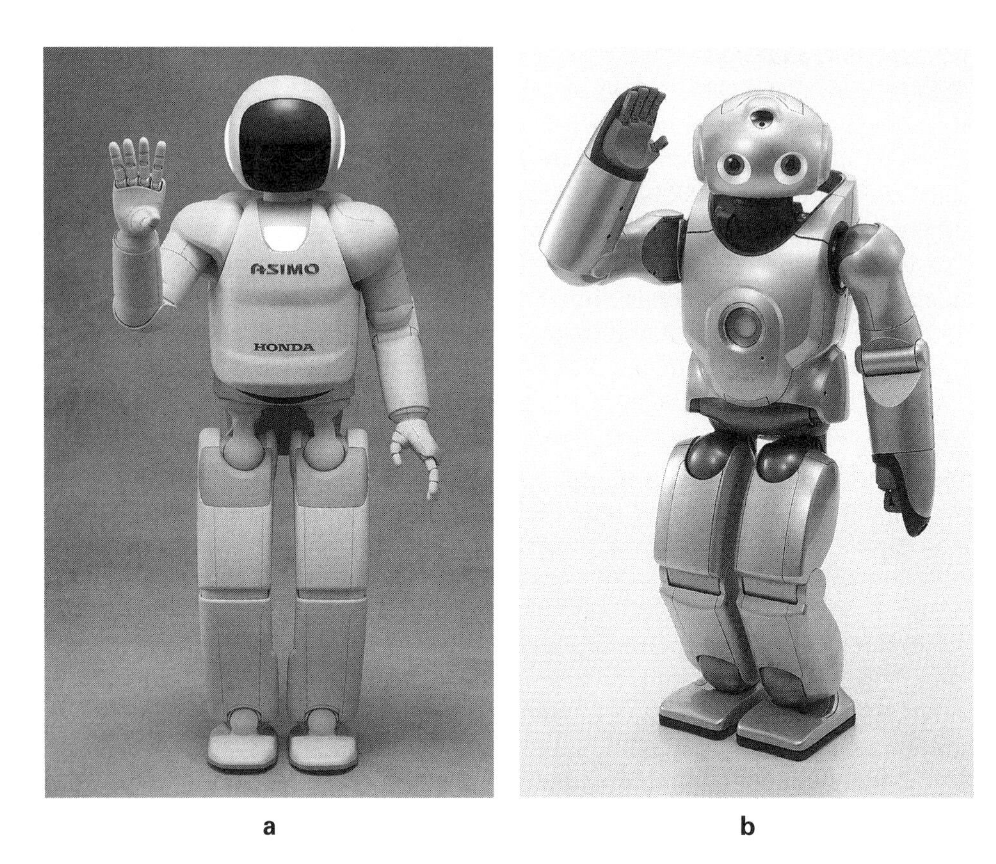

a b

Figures 2.6
Honda ASIMO (a), and Sony QRIO (b). Figures courtesy of Honda Motor Europe and © Sony Corporation.

kilograms (figure 2.6, right). The robot has a total of thirty-eight DOFs, with four in the head, two in the torso, ten for the arms (five in each), ten in the hands (one for each finger), and twelve in the legs (six in each).

The sensors include two small CCD color cameras (110,000 pixels) and microphones. It also has a three-axis accelerometer and gyro in the torso, a two-axis accelerometer, and two force sensors in each foot, touch/pressure sensors in the head, shoulders, and feet, and IR distance sensors in the head and each hand. The robot uses seven microphones located inside its head to detect the direction of sound and reduce the motor noise.

The QRIO has a set of motor capabilities integrated with the proprietary "Real-time Integrated Adaptive Motion Control System." This includes modules responsible for real-time whole body stabilization; terrain adaptive control; integrated, adaptive fall-over and recovery control; and lift-up motion control. It can reach a maximum walking speed of twenty meters per minute on flat surfaces, and up to six meters per minute on uneven terrains. As the robot has been primarily developed for entertainment purposes, given the focus of Sony Corporation, it has been optimized for dance and music performances. In 2004 a QRIO conducted the Tokyo Philharmonic Orchestra in a rehearsal of Beethoven's Fifth Symphony, as part of a children's concert (Geppert 2004). QRIO never reached a commercial production stage, however, in contrast to AIBO (see section 2.4), and its development was stopped in 2006.

Developmental robotics research on the ASIMO platform has been applied to object recognition (Kirstein, Wersing, and Körner 2008), category learning and language for concepts such as "left," "right," "up," "down," "large," "small" (Goerick et al. 2009), for imitation learning of motor skills (Mühlig et al. 2009), and for gaze detection and feedback in human-robot tutoring interaction (Vollmer et al. 2010). Most of the developmental work with the Honda robot uses a neuroscience-inspired developmental architecture called ALIS (autonomous learning and interaction system) (Goerick et al. 2007, 2009). This is a hierarchical and incrementally integrated system that on the humanoid robot ASIMO combines visual, auditory, and tactile saliency; proto-object visual recognition; object classification and naming; and whole-body motion and self-collision avoidance. The ALIS architecture allows interactive learning through human-robot experiments, following a developmental strategy for the incremental acquisition and integration of cognitive skills. This architecture has also been extended and adapted from humanoid robot tasks to automotive-related visual scene exploration and detection of moving objects for driver assistance (Dittes et al. 2009; Michalke, Fritsch, and Goerick 2010).

The QRIO robot has been used for cognitive experiments on imitation and the mirror neuron system (Ito and Tani 2004), compositional motor representation through multiple-scale recurrent neural networks (Yamashita and Tani 2008), communication and construction grammar (Steels 2012), and child education and entertainment support (Tanaka, Cicourel, and Movellan 2007).

2.3.4 CB

CB (computational brain) is an adult-size humanoid robot developed by SARCOS as part of the JST Computational Brain project at the Japan ATR Computational Neuroscience Laboratories in Kyoto (Cheng et al. 2007b). It is 1.57 meters tall and weighs ninety-two kilograms (figure 2.7). The robot has a total of fifty DOFs, with seven for the head (three for the neck, and two each for the two eyes), three for the torso, fourteen for the arms (seven per arm), twelve for the hands (six per hand), and fourteen for the legs (seven in each leg). The fingers and head use passive compliance motors, with active compliance for the rest of the actuators. These actuators were designed to model the human body's physical performance such as fast eye saccadic movement, pointing, and manipulation through grasping and pinching. The suite of sensors includes two

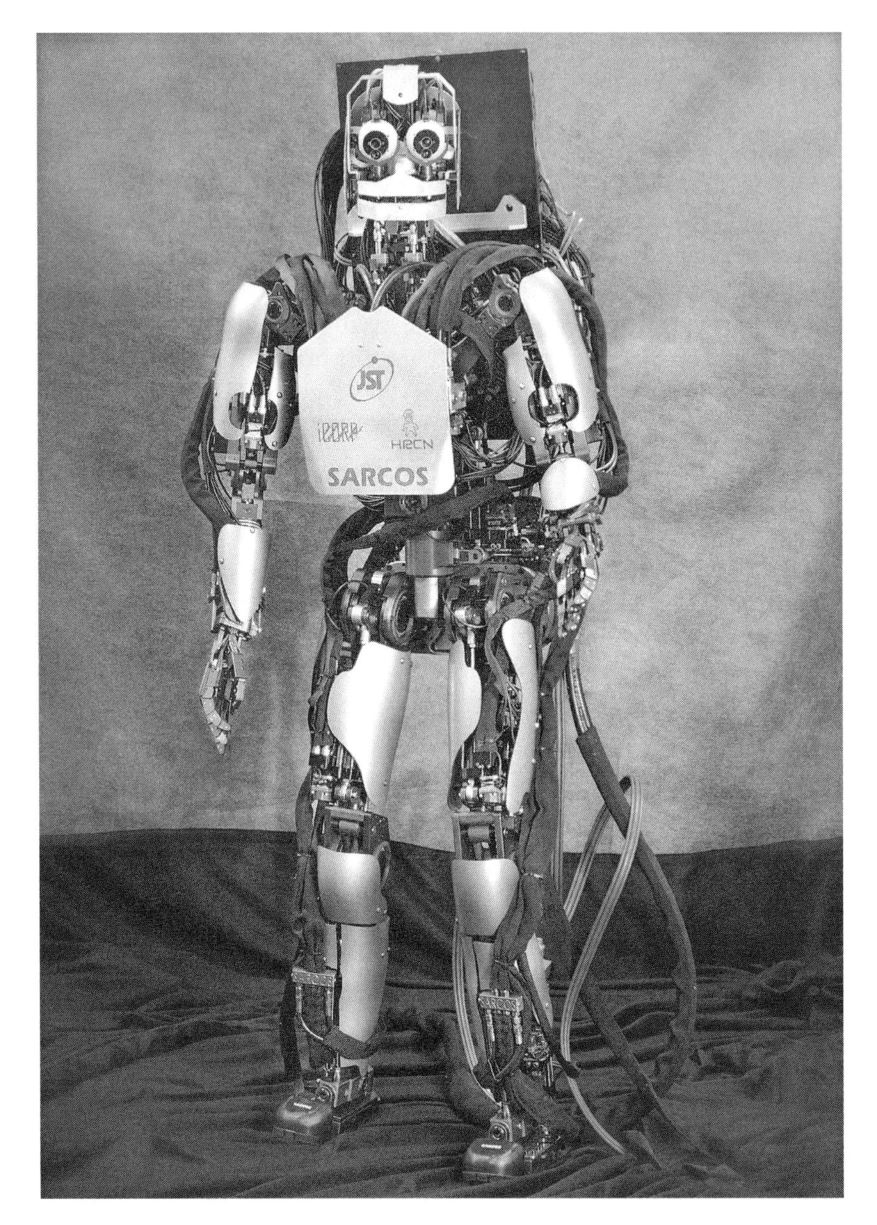

Figure 2.7
The CB robot developed by ATR Japan and SARCO. Figure courtesy of Gordon Cheng.

cameras per eye, with one wide-angle view camera for peripheral vision and one for narrow-view fovea vision. CB also has two microphones for sound perception. Inertia sensors are in the head (three-axis rotational gyro and three-axis translational accelerometer) for head orientation and gaze stabilization, and one in the hip for balancing and orientation of the whole body. Proprioceptive information is provided by various sensors for position, velocity, and torques, which sense the active compliance in the arms, legs, torso, and neck, and foot force sensors for walking and balancing control.

Given the CB's design emphasis on human-like behavior and social neuroscience, this robot has been used for exploring various motor and social learning tasks for a better understanding of human-robot interaction (Chaminade and Cheng 2009). For example, in motor control the robot has been used for the design of biologically inspired walking algorithms using central pattern generators (Morimoto et al. 2006). For social skills, a neuroscience-plausible model of distributed visual attention has also been developed. This is based on the brain processing of several streams (for the visual features of color, intensity, orientation, motion, and disparity), which are integrated to form a global saliency map for the selection of the focus of attention. This also provides relative feedback connection, once the robot achieves selective attention (Ude et al. 2005). CB was also used for brain-machine interfaces for a monkey's motor control of the robot (Kawato 2008; Cheng et al. 2007a).

2.3.5 CB^2 and Pneuborn-13

The JST ERATO Asada Project on Synergistic Intelligence (Asada et al. 2009) has contributed to the design of two child robot platforms specifically produced for developmental robotics research: CB^2 and Pneuborn-13. The project also developed a simulation model of the fetus and newborn (see section 2.5.3 on robot simulator).

CB² (Child-robot with Biomimetic Body) is a child-size humanoid robot with a biomimetic whole-body, and soft silicone skin with distributed tactile sensors. It uses flexible joints through pneumatic actuators (Minato et al. 2007). The robot is about 1.30 meters high and weighs thirty-three kilograms (figure 2.8, left). It has a total of fifty-six DOFs. The actuators are pneumatic, except for the electrical motors for the fast eyeball and eyelid movements. The pneumatic actuators produce flexible joints with mechanical energy through the highly compressed air, and these can also be passively moved by releasing this air. This ensures safe interaction with the human partner. The robot also has an artificial vocal tract, which can produce baby-like vowel sounds. The sensor suite includes two cameras, one per eye, and two microphones. The body has a total of 197 tactile sensors, with pressure sensors based on PVDF films beneath the silicone skin. As the touch sensors are embedded beneath the skin, the robot's own movements can cause self-touches, and these movements and sensor responses can help the self-organization of motor representation.

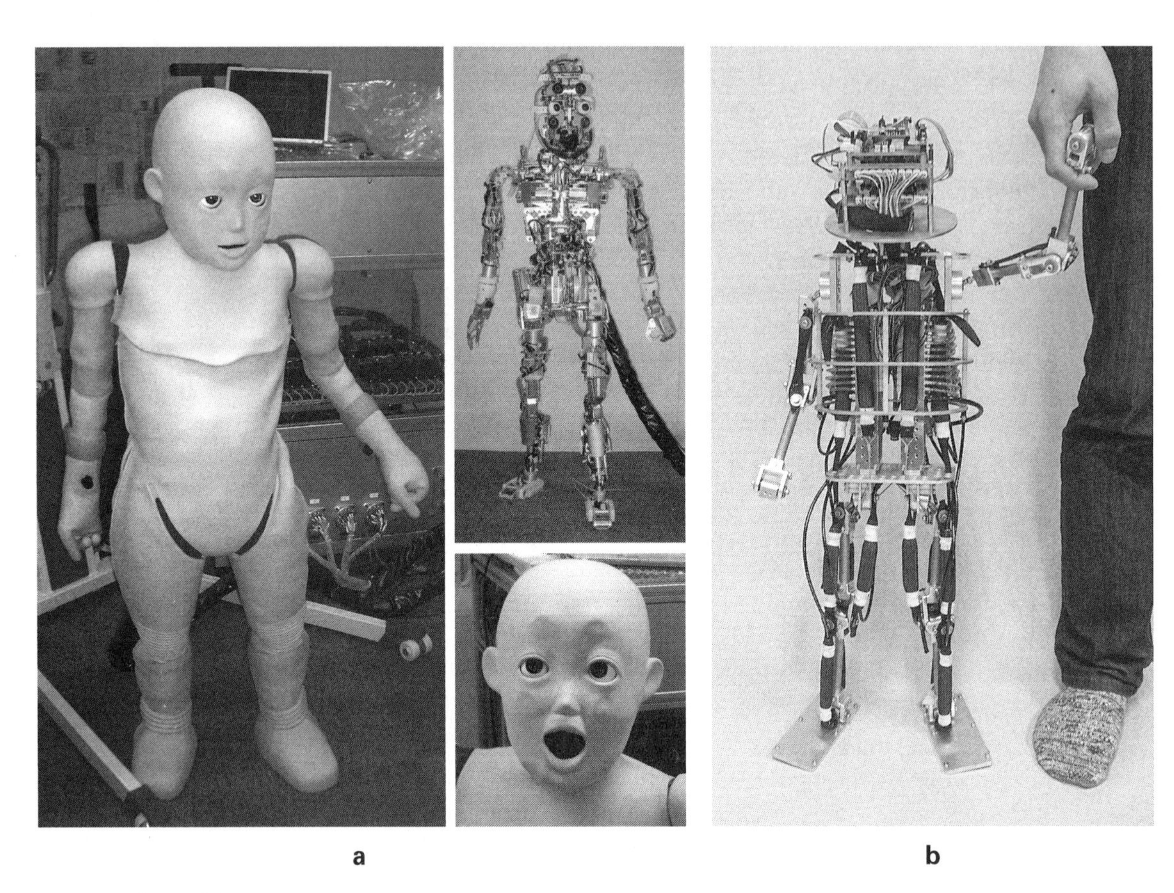

Figure 2.8
The JST ERATO Asada baby robots: CB[2] (left) and Pneuborn-13 (right). Courtesy of JST ERATO Asada Project.

Pneuborn-13 is the second baby robot platform developed in the JST ERATO Asada project (Narioka et al. 2009; Narioka and Hosoda 2008). This is a pneumatic musculoskeletal baby robot 0.75 meters tall and weighing 3.9 kilograms (figure 2.8, right), to reflect the weight and height of a thirteen-month-old infant. It has a total of twenty-one DOFs, with one DOF in the neck, ten in the arms (five per arm, with three controlling the shoulder, and one each for elbow and wrist), ten for the legs (five per leg, with three in the hip, and one each for the knee and ankle). Another prototype of this robot is the Pneuborn-7II, resembling a seven-month-old infant, and capable of rolling over and crawling behavior.

The Pneuborn-13's actuators are inspired by the human infant's musculoskeletal body, with each actuator consisting of a pair of one agonistic and one antagonistic mono-articular muscle that control the joint angle movement and its stiffness. As the Pneuborn-13 was built specifically to study early infant bipedal locomotion, eighteen pneumatic muscles are concentrated at the ankle, knee, and hip joints (nine per leg). Thus a leg has flexion, extension-adduction, abduction, and external rotation in the hip, flexion and extension in the knee, and flexion and extension in the ankle.

Both baby robots have been tested for motor control and learning in human-robot interaction experiments. As the CB^2 was designed to support long-term social interaction between humans and robots, based on infant development principles, a series of experiments has looked at sensorimotor development and the role of humans in scaffolding the robot's motor learning. One study explores the role of human participants in helping the robot stand up (Ikemoto, Minato, and Ishiguro 2009; Ikemoto et al. 2012). The robot can be in three postures: (1) initial sitting posture, (2) intermediate rising with bent knees, (3) final standing posture. To switch between (1) and (2), the robot's hands are pulled up by the human, resulting in a lifting of the CB^2's body with bent knees. For the posture change from (2) to (3), the legs are straightened by the human partner's continued pulling and lifting. The experiment analyzes the robot's timing and motor strategy to switch between postures, and shows that this depends on the robot's anatomy and the novice/expert skills of the human. During training, the robot-human pair goes through improved behavioral coordination, which results in increased synchrony between robot and human during learning. In a second experiment, the CB^2 develops body representations based on the cross-modal integration of visual, tactile, and proprioceptive input (Hikita et al. 2008). For example, when the robot interacts with an object, the tactile information affects the construction of the visual receptive fields for body parts. And if a tool is used to reach distant objects, the tool is integrated into an extended body schema, as observed in Iriki's experiments with monkeys (Iriki, Tanaka, and Iwamura 1996).

The Pneuborn-7II and Pneuborn-13 robots have also been used for preliminary studies on crawling, the standing posture, and stepping motions. Experiments on the Pneuborn-7II focus on rolling over and crawling behavior, including trials with the use of soft skin and tactile sensors in the arm during crawling (Narioka, Moriyama, and Hosoda 2011). The Pneuborn-13 has an autonomous power supply and air valves, and it has been tested for several hours of walking activity without breakdown or overheating (Narioka et al. 2009).

2.3.6 Repliee and Geminoid

Repliee and *Geminoid* are series of android robots developed by Hiroshi Ishiguro and colleagues at the ATR Labs Japan and Osaka University. In particular, Repliee R1 has the appearance of a five-year-old Japanese girl (Minato et al. 2004) (figure 2.9). The prototype has nine DOFs in the head (five for the eyes, one for the mouth, and three for the neck) using electric motors. The other joints in the lower body are passive, free joints to

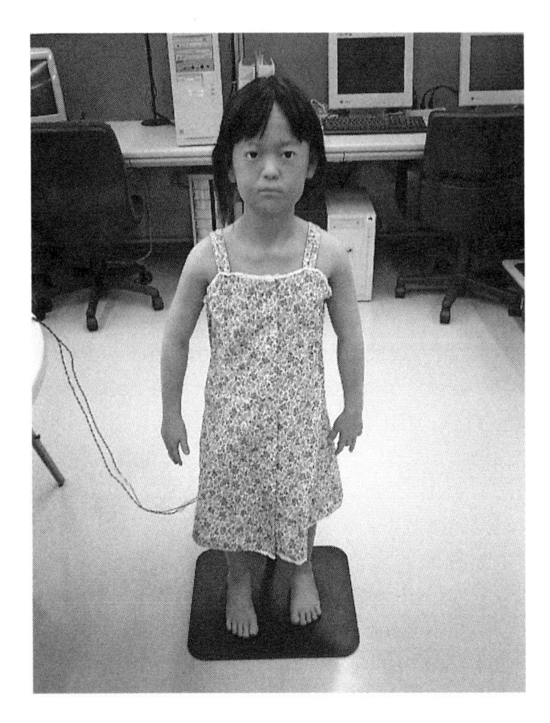

Figure 2.9
Repliee R1 robot. Figures courtesy of Hiroshi Ishiguro, Osaka University.

allow the robot to take on different postures when moved by the human experimenter. The face is covered with a silicone-type molding of a Japanese girl's face. Repliee R1 has four touch sensors under the skin of the left arm, which use a strain rate force sensor that responds in a similar way to human's skin stretching.

An adult female version of this robot, called Repliee Q1, controls the whole upper torso of an adult female, and uses pneumatic actuators (Sakamoto et al. 2007). The later Geminoid android robots also model adults (Nishio, Ishiguro, and Hagita 2007), such as the Geminoid HI-1 copy of Ishiguro himself (see figure 2.1, right). Geminoid HI-1 is 140 centimeters tall, when sitting on a chair (it cannot stand). It has a total of fifty DOFs, with thirteen in the face to mimic realistic human face movements.

The child robot Repliee R1 has been utilized for studies of gaze behavior with androids. Minato et al. (2004) investigated the eye motion of people interacting during a scripted conversation with the child android robot. The comparison of eye fixation times and target location, between a conversation with the robot and that of an actor girl, shows that participants look at the android's eyes more frequently than the real girl's face. Moreover, analyses of the participants' fixation point show that the way participants gaze at the android's eyes differs from how they gaze at a human's eyes.

Moreover, the Repliee and Geminoid androids have been extensively used to investigate social and cognitive mechanisms in the uncanny valley phenomenon of humans' perception of robots, as discussed in section 2.1.

2.3.7 Infanoid

The *Infanoid* robot was developed by Kozima (2002) to study human social development, and as a robot platform supporting child development and education in human-child interaction studies. This is an upper-torso humanoid robot, resembling a three- to four-year-old human child (figure 2.10). The robot has a total of twenty-nine electric motor actuators. For sensing it has four color CCD cameras, two per eye for peripheral and fovea vision, and two microphones in the ear position. The motors have encoders and force sensors.

The hands allow actions such as pointing, grasping, and hand gestures. The motors in the lips and eyebrows are used to produce various facial expressions, including surprise and anger. The eyes can perform saccadic movements and smooth pursuit of

Figure 2.10
The Infanoid robot. Figures courtesy of Hideki Kozima, Miyagi University.

a visual target. The robot's software provides modules for human face detection and eye-gaze direction localization to gaze at objects. The robot has also been equipped with an algorithm to hear and analyze human voices and perform vocal imitation and babbling.

This baby robot has been extensively used for exploring human social development, with a special emphasis on the acquisition of interpersonal communication skills. It has therefore been used in numerous child-robot interaction experiments, including children with disabilities and autism spectrum disorders (Kozima, Nakagawa, and Yano 2005). For example, in a study with five- and six-year-old children, including both normally developing and children with autism, the children's perception of the robot was investigated. The robot was programmed in an autonomous mode, to alternate between eye contact and joint attention with pointing, during a forty-five-minute session with the children. Analyses of the child-robot interaction show the children undergo three stages of "ontological understanding" of the Infanoid robot: (1) *neophobia phase*: in the first three to four minutes, the children show embarrassment and stare at the robot; (2) *exploration phase*: the children explore the perception and response capabilities of the robot by poking it and showing toys; (3) *interaction phase*: the children gradually engage in reciprocal social exchanges, and attribute to the robot mental agencies and feeling like desire and likes/dislikes. The children with autism showed similar responses, with the only difference that they did not get bored after the long interaction (Kozima et al. 2004).

2.3.8 Affetto

Affetto is an upper-torso android robot with an infant's appearance, developed at Asada's Developmental Robotics lab at Osaka University, with the specific aim to study affective interaction between human caregivers and baby robots (Ishihara, Yoshikawa, and Asada 2011) (figure 2.11). To maximize the emotional and attachment quality of the childlike interaction between Affetto and the human caregivers, the design concept followed these three principles: (1) *realist appearance* based on a soft skin (urethane elastomer gel) for the face, and with the use of baby clothes covering the mechanical parts of the upper torso; (2) *actual child size*, as the robot is modeled after a one- to two-year-old child (actual measures taken from a database of Japanese toddler sizes); and (3) *smile-based facial expressions*, to engage human participants in positive emotional interaction, with the complementary use of rhythmic body movements and hand gestures to reinforce the emotional interaction. Moreover, to permit safe, unconstrained physical interaction between human participants and Affetto, these additional technological criteria were followed: (a) use of compliant passive actuators through the use of pneumatic actuators in the joints subjected to external forces from touch interaction; (b) use of tender skin, as the face and arms are covered with soft skin to minimize physical injury risks, thus increasing the participant's willingness to touch

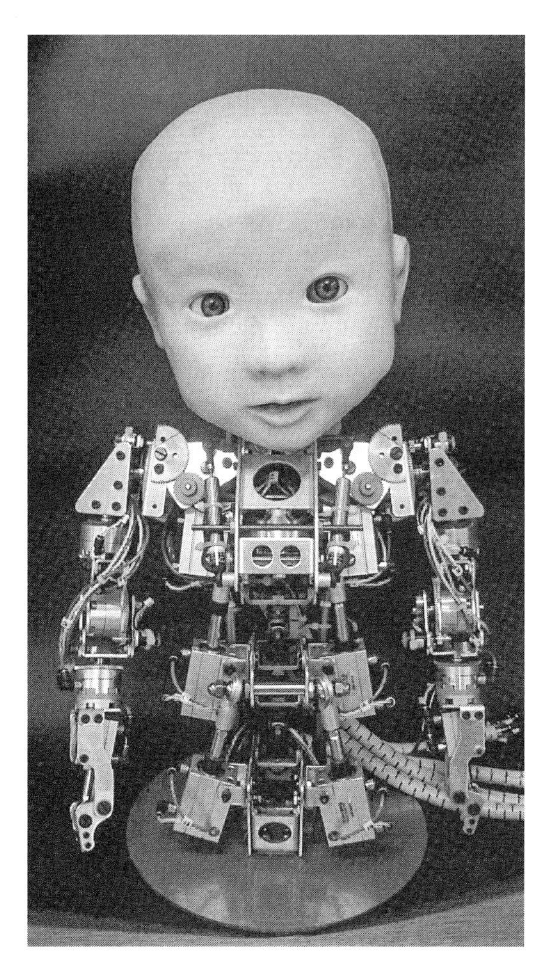

Figure 2.11
The Affetto android infant robot. Figures courtesy of Hisashi Ishihara and Minoru Asada, Osaka University.

the robot; (c) reduction in weight, through the use of the separated packed motors and controllers for additional safe physical interaction and to increase the robot's motion performance, with high power-weight/power-size ratio of the pneumatic actuators; and (d) use of variable deforming points in the face for easy change of facial expressions.

The robot's head is seventeen centimeters high, fourteen centimeters wide, and fifteen centimeters deep. Affetto has twelve DOFs in the head: five DOFs for eyelids and lips, two at the jaw and eyes for up and down movements, two for the eyes for left to right movement, and one yaw axis in the neck. The head's roll and pitch axes are actuated by two pneumatic actuator cylinders. To control the facial movements and emotional expressions, the rotary shafts of DC motors are attached to some of the shafts to pull the wires connected to the inner side of the skin.

The upper torso is twenty-six centimeters high; the entire baby robot is 43 centimeters tall. The overall weight of the head and torso is less than three kilograms, excluding the external controller for the pneumatic actuators. The upper torso to which the head is connected has a total of nineteen DOFs: five in the body's trunk and seven in each arm.

Ishihara, Yoshikawa, and Asada (2011) have proposed four areas of research with the Affetto platform. The first looks at the role of a realistic child appearance and facial expressions in the caregiver's scaffolding strategies during human-robot interaction. Other areas of research include real-world simulation of the development in the attachment relationship, and experiments investigating the robot's and the caregiver's effect on the dynamics of the emotional/affective interaction. In addition, Affetto permits the investigation of multimodal childlike features that in addition to the face/skin might include the auditory sense (high-pitch voice, immature but energetic voice) and the tactile modality.

The robot has been used for preliminary experiments on human-robot interaction (Ishihara and Asada 2013). The interaction scenario concerns a human caregiver who can take both hands of the baby robot and shake them, while the robot is trying to keep an upright posture. Affetto's body parts can follow the caregiver's manipulation smoothly thanks to the intrinsic compliant nature of the mechatronics and without the need for any active computation. Moreover, to facilitate the emergence of rhythmic behaviors in response to the user's interaction, ongoing studies are focusing on the implementation of a rhythm generator using CPG controllers.

2.3.9 KASPAR

KASPAR (www.kaspar.herts.ac.uk) is a child-sized small humanoid robot originally proposed as part of a project on human-robot interaction design (Dautenhahn et al. 2009) (figure 2.12). The design philosophy behind this platform is to use low-cost off-the-shelf components for wider research affordability, giving it a minimally expressive appearance with inspiration from comic design and the Japanese Noh theater

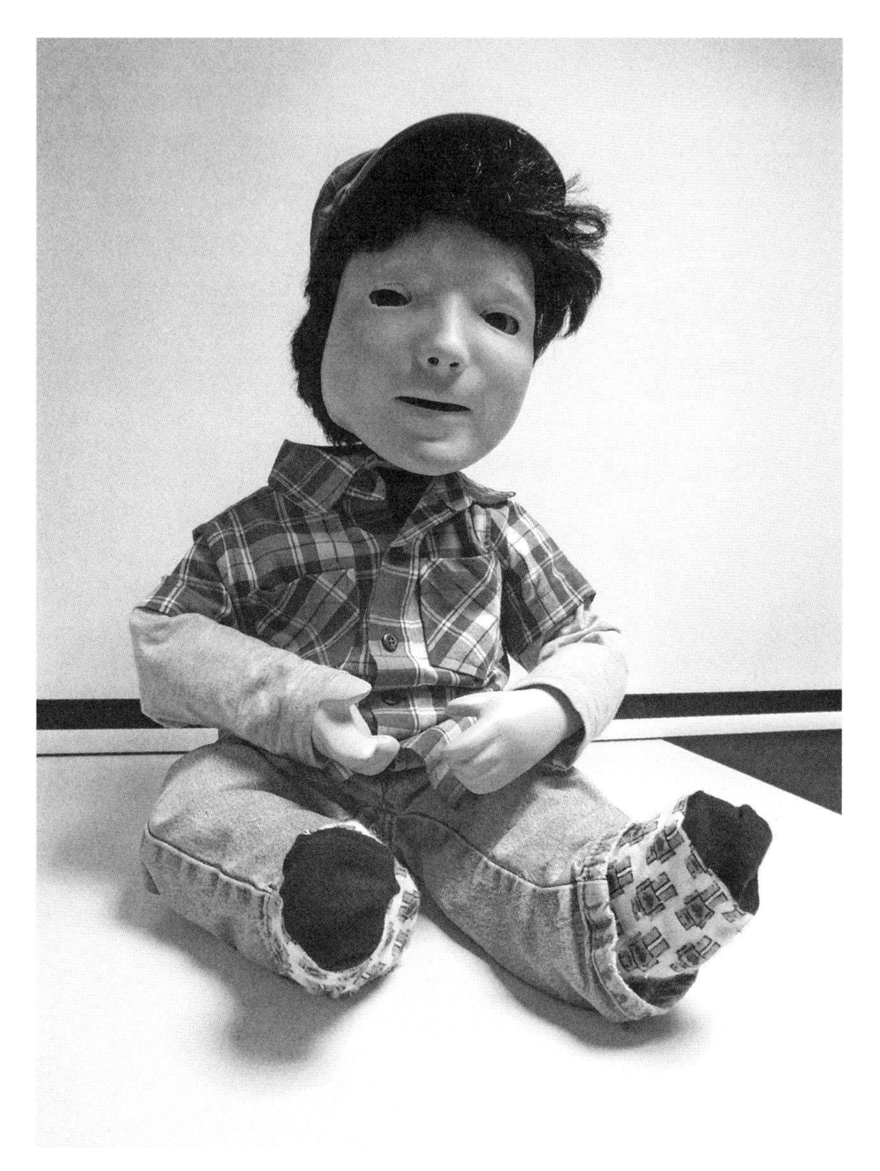

Figure 2.12
The KASPAR robot. Figures courtesy of Kerstin Dautenhahn, Hertfordshire University.

style. Specifically, Dautenhahn et al. propose three principles for the general design of minimally expressive robots such as KASPAR: (1) balanced design, in other words, the consistent use of aesthetic and physical design choices appropriate for the expected human-robot interaction scenarios; (2) expressive features for creating the impression of autonomy, including the robot's attentional capabilities (through head turns and gazing), emotional states (facial expression), and contingency between the human participants' and the robot's behavior; and (3) minimal facial expressive features, through the use of Noh-like elements of design and the control of a limited number of degrees of freedom for expressive behavior like smiling, blinking, and frowning.

The original KASPAR is fifty-five centimeters tall and forty-five centimeters wide. It weighs fifteen kilograms. The most recent version of KASPAR has a total of thirteen DOFs. For the head, the neck has three DOFs (pan, tilt and roll), the eyes have three DOFs (up/down, left/right, eyelids open/close) simultaneously controlling both eyes, and the mouth has two DOFs (open/close, smile/sad). The arms have four DOFs each. The most current version of the robot also has one DOF in the torso so that the robot can turn to the side, as well as touch sensors distributed across its body (Robins et al. 2012a, 2012b). Each eye has a miniature camera with a quarter-inch black-and-white CMOS image sensor producing a PAL output of 288(Horizontal) × 352(Vertical). The head is covered with a rubber mask that is widely used for resuscitation dummies.

A follow-up version, KASPAR II, has improved sensor and actuator technology, still within the low-cost approach of minimally expressive platforms. KASPAR II has the same head mechanisms, except for colored eyes and improved wiring, and a larger body resembling a six-year-old child. The robot also has an extra DOF for each wrist (twist movement), joint position sensors, and a Swiss Ranger 3000 sonar sensor in the chest for depth maps. A new and significantly improved version of KASPAR will be completed in 2015 that is more robust and easier for nonexperts to use, but still low in cost.

Dautenhahn et al. (2009) report three main studies with the KASPAR robot. The first study investigates its application in robot-assisted play and therapy for children with autism. This is part of a wider set of experiments on assistive robotics with children with autism spectrum disorders (ASDs) (Wainer et al. 2010, 2013; Robins, Dautenhahn, and Dickerson 2009) that recently has been extended to children with Down syndrome (Lehmann et al. 2014) (see section 9.2 on assistive robotics for more details). The second study investigates the role of interaction kinesics and gestures, within a musical drumming game scenario, for general human-robot interaction experiments with adults (Kose-Bagci, Dautenhahn, Nehaniv 2008). The third study focuses on the use of KASPAR for developmental robotics research, specifically to investigate the design of cognitive architecture for interaction history suitable for play interaction scenarios such as the Peekaboo game (Mirza et al. 2008). KASPAR can be operated fully autonomously, by using a remote control, or in a hybrid mode combining remote control (in the hands of the teacher or the child) with autonomous behavior.

2.3.10 COG

COG is an upper-torso humanoid robot developed at MIT in the late 1990s (Brooks et al. 1999), and was one of the first humanoid robots specifically developed for cognitive robotics research. The robot has a total of twenty-two DOFs, with twelve for the arms (six for each arm], three for the torso, and seven for the head and neck (see figure 2.1a). Its visual system consists of four color CCD cameras, with two cameras per eye (one with a wide field of view, and the second with a narrower field of 21 degrees). It uses two microphones for sound perception. The COG also has a vestibular system consisting of a three-axis inertial package, and encoders, potentiometers, and strain gauges as kinesthetic sensor.

The COG has been used for developmental robotics, in particular for social learning, active vision, and affordance learning. Scassellati (2002; see also section 6.5 in chapter 6) implements various cognitive and social capabilities in the COG for the developmental model of a robot's theory of mind. Fitzpatrick and Metta (2002) carried out experiments on COG to demonstrate that active exploration of an object through experimental manipulation improves visual recognition. This sensorimotor strategy is based on the observed correlations between arm motion and optic flow, which helps the robot determine both the arm itself and the boundaries of the object.

2.4 Mobile Robots for Developmental Robotics

Most developmental robotics experiments employ the humanoid platforms discussed previously for human-robot experiments. However, a few of the early developmental models used mobile robots, such as the four-legged pet-like platform AIBO, and wheeled mobile robots such as Khepera and PeopleBot. In this section we describe in detail the AIBO robot, as this has been used on a variety of models including intrinsic motivation, and joint attention and language, some of which are described in chapter 3. The other wheeled platforms will only be briefly referred to, with links to the main literature.

The *AIBO* robot (Artificial Intelligence roBOt) was developed in the late 1990s by Sony's Digital Creatures Laboratory led by Toshitada Doi (Fujita 2001) (figure 2.13). The first AIBO was made commercially available in the summer of 1999 (ERS-110 model series), with subsequent upgraded versions in 2001 (ERS-210/220 series), the third generation (ERS-310 series) and the final fourth generation in 2003 (ERS-7 series). It is estimated that over 150,000 copies of the AIBO were sold before Sony stopped its production and commercialization in 2006. The reason for such a success in terms of the high number of robot products sold was its affordability (the first version was for sale

Figure 2.13
The Sony AIBO robot. Figure courtesy of © Sony Corporation.

at $2,500 in 1999, with the last model in 2003 sold for about $1,600) and Sony's aim to make this an entertainment platform, thus targeting the general public. Moreover, its circulation was supported by the fact that the RoboCup event had a specific standard four-legged league based on the AIBO from 1999 to 2008 (subsequently substituted by the NAO standard leagues).

The specifications of the latest ERS-7M3 robot include a total of twenty DOFs: one for mouth, two for tail, three for head, two for ears (one per ear), and twelve for legs (three per leg). The AIBO is 180 millimeters wide, 278 millimeters tall, and 319 millimeters deep, with an approximate weight of 1.65 kilograms (figure 2.13).

The set of sensors include a CMOS color microcamera with 350,000 pixel capacity, and infrared distance sensors in the head and body. It also has a temperature sensor, accelerometer, electric static sensor in the head and back, a vibration sensor, and five pressure sensors in the chin and each paw to respond to the human's patting. Moreover, to facilitate user engagement, the AIBO has numerous status LED lights, with twenty-eight lights in the face to express emotions, and many others distributed all over the head and body sensors. The robot can communicate using sound, with a miniature speaker driven by a Polyphonic Sound Chip.

The robot is autonomous with its own rechargeable battery. It uses an internal CPU of a 64-bit 64 MB RISC processor, external memory slot, and wireless LAN card. The software uses the OPEN-R architecture to control the various software modules available for sensing and behavior control.

The AIBO's cognitive architecture was designed to support entertainment interaction and complex behavior using the principles of multiple motivations for actions, a configuration with high degrees of freedom, and the production of nonrepeated behavior (Fujita 2001). The robot's control architecture is based on behavior-based methodology with a hybrid deliberative-reactive control strategy (Arkin 1998; Brooks 1991). A mechanism is implemented to determine the behavioral module to be activated in response to external stimuli and internal states. Each behavior module (for example explore, rest, bark, express one emotion, choose a walking pattern) consists of state-machines with context-sensitive responses. A stochastic method is also used to handle random (thus unpredictable) action generation. During multisession interaction with the user, the AIBO employs a developmental strategy for long-term adaptation to the user based on slow, learned changes of the robot's behavioral tendencies. This is implemented through reinforcement learning.

For cognitive robotics research, the OPEN-R software provides access to the robot's various behavioral and sensory capabilities to produce developmental interaction strategies. For example, the AIBO comes with a variety of walking patterns including slow, steady crawl gaits and fast but unstable trot gait patterns. These patterns can be manually selected, or are autonomously activated by the cognitive architecture. For vision and object perception, the robot has a dedicated large-scale integrated circuit with an embedded color detection engine and a multiresolution image-filtering system (resolutions of 240×120, 120×60, and 60×30 pixels). For example, low-resolution images can be used for fast color filtering to identify the presence of an object, and high-resolution images for object identification and pattern matching. Finally, for auditory interaction with the user, a tonal language is implemented. This tonal system and the internal sound-processing algorithm permit efficient handling of noise and voice interference.

In addition to the OPEN-R software, specific tools for cognitive robotics experiments with the AIBO have been developed, such as the Tekkotsu simulation framework specifically developed for cognitive modeling and for robotics education activities with this robot (Touretzky and Tira-Thompson 2005), and other robot simulators that have a 3D simulation model of the AIBO as standard (see section 2.5).

The AIBO's affordability and ease of access in many labs have resulted in its utilization in many of the early experiments on developmental robotics. In particular, AIBO has been used for studies on intrinsic motivation (Kaplan and Oudeyer 2003; Oudeyer, Kaplan, and Hafner 2007; Bolland and Emami 2007), on joint attention and pointing (see chapter 7), and on word learning experiments (Steels and Kaplan 2002; see also chapter 8). Moreover, some experiments have focused on the teaching of complex behavior through using dog-like training methods (Kaplan et al. 2002) and even for a study on the Turing Test for dogs to understand the reaction of real dogs to this artificial pet (Kubinyi et al. 2004).

Cognitive robotics has also benefited from the use of additional platforms, in particular wheeled robots of different sizes. For example, the *Khepera* and the *e-puck* small, wheeled robots have been extensively used for evolutionary robotics (Nolfi and Floreano 2000). The Khepera II (K-Team Mobile Robotics) is a miniature robot with a round body seven centimeters in diameter and three centimeters high. The e-puck (EPFL Lausanne; Mondada et al. 2009) is an even smaller robot suitable for swarm robot studies.

At the other end of the spectrum of the mobile robot size, the large *PeopleBot* (ActivMedia) is a mobile platform 104 centimeters in height and weighing 19 kilograms, and can carry a payload of 13 kilograms. This is particularly suitable for human-robot interaction experiments, and has been used by Demiris and colleagues for a developmental robotics model of social learning and imitation (chapter 6). Another large mobile robot is the *SAIL* platform used by Huang and Weng (2002) for developmental studies on novelty and habituation.

Finally, other robots uses for developmental robotics are arm manipulators, such as the six-DOFs *Lynx6 robotic arm* (lynxmotion.com) used by Dominey and Warneken (chapter 6) for developmental models of cooperation and altruism.

2.5 Baby Robot Simulators

Robot platforms tend to be expensive research tools, often affordable only to very few laboratories in the world. The cost of a fully equipped iCub in 2012 was about €250,000, and thus only available through participation in large-scale research grants. Other platforms such as the ASIMO and the QRIO are not commercially available, and even if they were available for purchase their cost would be well over a million dollars. More affordable robots, such as the NAO, cost around €6,000, and although they are more affordable for medium-size laboratories and grants, they still require a significant investment in lab facilities. Moreover, the setup and running costs of a robotics lab are again affordable only to a few researchers and laboratories. This is one of the primary reasons why a good, realistic software simulator of existing robotics platforms can be an extremely useful tool to carry out experiments on developmental robotics.

Of course, robot simulators are not intended to be full alternatives to experiments with real robots, given they can only provide a limited realistic simulation of the physical properties of a robot's mechanical structure (not considering the even greater complexity of simulating the external world in a reliable and useful way!). This has important implications for the transferability of results from simulation to physical robots. There are, however, numerous scientifically grounded reasons for why a robot software simulator can be useful in cognitive modeling and cognitive robotics in general (Tikhanoff, Cangelosi, and Metta 2011; Ziemke 2003; Cangelosi 2010). These benefits include (1) robot prototype testing; (2) morphological change experiments; (3) application to multiagent and evolutionary robotics studies, and (4) collaborative research.

The first benefit of using a simulator is for robot prototype testing, especially in the early stage of new platform design. Advanced physics simulator tools exist, such as the freely available Open Dynamics System (ode.org), that permit realistic rendering of the object interaction dynamics for rigid body robot simulation. This allows the designer to test in simulation different configurations of sensors and actuators, before actual mechanical production. The second benefit is that software simulation of virtual robots permits the design of hypothetical morphological configuration without the need to necessarily develop the corresponding structure in hardware. This is the case, for example, of experiments on the evolution of morphology (Kumar and Bentley 2003; Bongard and Pfeifer 2003) as in models on the interaction between the robot's controller and its body and the environment. In developmental robotics, simulations can permit the investigation of maturational phenomena related to body morphology, for example, changing the ratio between limb length and full body size of the growing robot, in order to model known child morphology changes. The third benefit of using robot simulators is in the case of studies on social interaction in multirobot scenarios, where access to multiple robots is not feasible or experimental time costs are high (Vaughan 2008). For instance, in evolutionary robotics the computer simulations can be used to drastically reduce the duration of the experiments for the multiple testing of robotic agents at each generation (Nolfi and Floreano 2000). Finally, there is a more practical benefit of using robot simulator software, for example, to support collaboration between different labs. Researchers in different locations can perform preliminary computational experiments through the same robot simulation software, sharing robot configuration and task setup parameters. This also allows the researchers with no access to the robot to collaborate with a lab owning the physical robot for final, joint adaptation and validation of the simulation work on the physical robot platform.

In the next sections we will look in detail at one example of open source simulator easily available for modeling experiments of the iCub robot (section 2.5.1) and at one example of commercial robot simulation software, the Webots application, which includes body models of various robots (section 2.5.1). One section will also describe the example of computer simulators of the human fetus and newborn infants, developed at the University of Tokyo (section 2.5.3). In addition to these software tools, there are other free and commercially available simulators used in the literature. One is the free software Player (Collett, MacDonald, and Gerkey 2005), which includes the Player module for network interface to a variety of robot sensor hardware, the Stage module for simple 2D environments, and the Gazebo module for 3D robot and environment simulations. Other free robotic simulators are the software for evolutionary robotics EvoRobot* (Nolfi and Gigliotta 2010) and the Simbad robot simulator (Hugues and Bredeche 2006). A variety of simulators for robot soccer also exist, such as Simspark and SimTwo (Shafii, Reis, and Rossetti 2011). Moreover, software like Microsoft Robotics Studio also provides models of existing robots, such as a model of the NAO (www.microsoft.com/robotics).

2.5.1 iCub Simulator

The main open-source iCub simulator (Tikhanoff et al. 2008; Tikhanoff, Cangelosi, and Metta 2011) has been designed to reproduce, as accurately as possible, the physical and dynamic properties of the iCub and further support the aim of using iCub as a benchmark platform for developmental robotics research. The software is openly available in www.icub.org.

The simulator uses Open Dynamics Engine (ODE) for the rendering of the robot's joints as rigid bodies and of the collision detection methods to handle physical interaction between the robot's parts and with objects in the environment. ODE consists of a high-performance library for simulating rigid body dynamics using a simple C/C++ API, and provides a preset variety of joint types, rigid bodies, terrains, and meshes for complex object creation, allowing the manipulation of many object parameters such as mass and friction.

The simulated iCub was created using the specifications of the physical robot and has a total height of around 105 centimeters, weighs approximately 20.3 kilograms, and has the same number of degrees of freedom: fifty-three. The robot body model consists of multiple rigid bodies attached through a number of different joints corresponding to the iCub's actuators. All the sensors were implemented in the simulation on the simulated body, including the hand touch sensors (fingertips and palm) and the force/torque sensors in the shoulders and hips. The simulated robot torque parameters, and their verification in static or motion tasks, have been tested to a degree of acceptable reliability (Nava et al. 2008).

To facilitate the transfer of work from simulation to the physical iCub, the same software infrastructure and interprocess communication based on the YARP middleware was used. The simulator and the actual robot have the same interface either when viewed via the device API or across network and are interchangeable from a user perspective. The simulator, like the real robot, can be controlled directly via sockets and a simple text-mode protocol. All the commands sent to and from the robot are based on YARP scripting instructions. For the vision sensors, two cameras are located at the eyes of the robot. It is possible to let the virtual robot see the real world and interact with human users by connecting a standard camera and projecting its image in a blank screen within the robot's virtual environment. This way the robot's simulated eyes see the external world.

The simulated iCub has full interaction with the virtual world (figure 2.14). The software provides routines to dynamically create, modify, and query objects in the world by simple instructions resembling the YARP syntax. The software also permits the importing of articulated CAD object models using standard 3D file formats. This simulator has been used for a variety of iCub simulation experiments such as language learning (Tikhanoff, Cangelosi, and Metta 2011), mental models for cooperation (Dominey and Warneken 2011, and chapter 6) and models of number cognition and abstract concepts based on this simulator (chapter 8).

An alternative software simulator of the iCub has been developed (e.g., Righetti and Ijspeert 2006a, 2006b), though not to the level of detail of Tikhanoff's simulator. These are however based on proprietary software Webots (see section 2.5.2) that require the purchase of licenses to run the iCub robot. Nolfi and collaborators have developed an additional iCub simulator, FARSA, which is based on the open-source Newton Game Dynamics physics engine and is suitable for evolutionary robotics experiments (Massera et al. 2013; laral.istc.cnr.it/farsa).

2.5.2 Webots

Webots is a commercially available robot simulator developed and distributed by Cyberbotics Ltd (Michel 2004; www.cyberbotics.com) and widely used for research in cognitive robotics. The software permits the simulation of a variety of mobile robots, including wheeled robots, legged humanoid platforms, and flying robots. The Webots standard version by default includes 3D models of the humanoid platforms NAO (figure 2.15) and AIBO used in developmental robotics. Moreover it provides the 3D models of numerous other platforms such as the Katana™ IPR arm robot (Neuronics), the wheeled robots e-puck (EPFL Lausanne) and Khepera III (K-Team Corporation), the Hoap-2™ (Fujitsu Automation) and KHR-2HV™ (Kondo Kagaku Ltd.) humanoid robots, DARwIn-OP™ (Robotis Ltd.), the Pioneer 3-AT™ and Pioneer 3-DX™ (Adept Ltd.) platform, and the KHR-3HV™ and KHR-2HV™ robots.

The software has a robot editor to build novel robot platform configurations, importing VRML file formats, and a world editor to create objects and terrains in the environment (e.g., the properties of objects, such as shape, color, texture, mass, friction), or use default world configurations. For new robot editing, Webots includes an extended library of sensors and actuators. Predefined sensor modules include IR and ultrasound distance sensors, various 1D, 2D, black-and-white, and color cameras for

Figure 2.14
Screenshots from the iCub simulator. Courtesy of Vadim Tikhanoff.

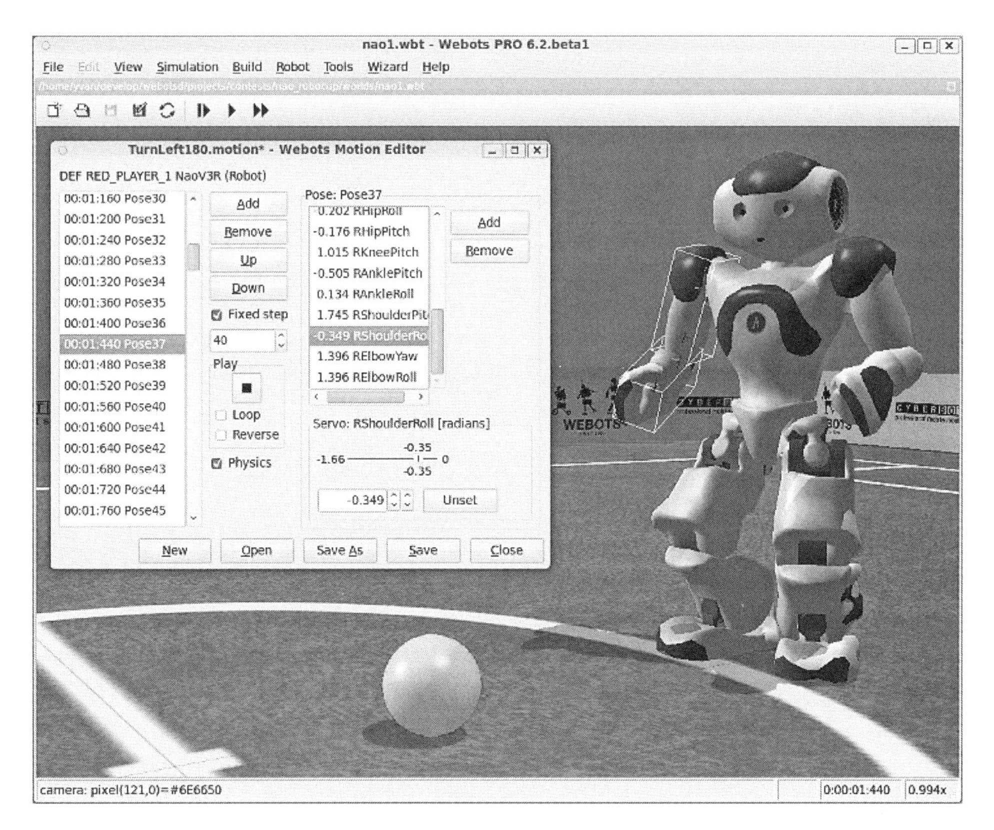

Figure 2.15
3D model of the NAO in the Webots simulator. Courtesy of Cyberbotics.

vision, pressure and bumper sensors, GPS sensors, position and force sensors for servo motors, wheel encoders, and 3D accelerometers and gyroscopes. These sensors can be controlled for range, noise, response, and field of view. The library of actuators includes differential wheel motor units, servo motors (for legs, arms, wheels), grippers, LEDs, and displays.

The robot editor allows the inclusion of receivers and emitters for inter-robot communication, thus simulating multiagent systems and experiments on robot communication and language. The software also has a motion editor, useful to create and reuse motion sequences for articulated robots. Moreover, the software permits the setup of experiment supervision sessions, for example, for the offline execution of a variety of experiments (as in genetic algorithm or training sessions with different task setups). With this supervisor capability the researcher can write a program script including actions such as changing object properties and their locations, sending messages to robots, and recording robot trajectories or videos.

Webots is a multiplatform application that can be used in Windows, Mac OS, and Linux. The simulator is built on the ODE physics simulator engine. It allows a communication interface with programs written in various languages (e.g., C/C++, Java, Python) and in MATLAB™ and other robotic software applications (ROS, URBI™), as well as an interface with third-party software through TCP/IP.

To facilitate the transfer of the robot controller developed in simulation into the physical robot platform, in addition to the availability of 3D models of existing platforms, it provides interfaces for the standard robot's software applications. For example, experiments with the NAO robot can use either URBI for Webots (Gostai SAS) or NaoQi and "NAO_in_webots" (Aldebaran Robotics SA), allowing communication directly with the physical robot or the simulated agent.

Webots is a commercial software application that requires one purchase a license to use all functionalities. A streamlined, free version exists, though with limited editing functionalities for cognitive robotics experiments.

The Webots simulator has been used in cognitive and developmental robotics studies, as in bipedal locomotion experiments with the Hoap-2 robot model (Righetti and Ijspeert 2006b), crawling experiments with the Webots simulated iCub (Righetti and Ijspeert 2006a), and walking with the NAO model (Lee, Lowe, and Ziemke 2011).

2.5.3 Fetus and Newborn Simulators

Two 3D computer simulation models of the fetus and newborn infants were developed by Kuniyoshi and colleagues within the JSP ERATO Asada project. The first model, which we will refer to as the *Fetus Model 1* (Kuniyoshi and Sangawa 2006), is the initial, "minimally simple" body model of fetal and neonatal development. The subsequent "Fetus Model 2" (Mori and Kuniyoshi 2010) provides a more realistic rendering of the fetus's sensorimotor apparatus and a stronger focus on learning experiments.

These models of fetal development were realized through computer simulations given the currently technical impossibility of developing an electromechanical robot fetus floating in liquid. Thus the models offer a useful research tool to investigate prebirth sensorimotor development by providing a realistic representation of the fetus's sensors, and the reaction of the body to gravity and to the womb walls.

The Fetus Model 1 includes a 3D model of an infant that simulates both the fetus in the womb and the newborn at birth (Kuniyoshi and Sangawa 2006; Sangawa and Kuniyoshi 2006) (figure 2.16). The model consists of a musculoskeletal system of nineteen spherical body segments, with 198 muscles attached to the cylinders (finger and facial muscles are not modeled). The simulation is realized through the ODE physics engine, as in the iCub and Webots simulators. The muscle kinetic properties and their size, mass, and inertial parameters were based on the human physiology literature. The joints' settings permit the simulation of the fetus's natural position and movements. A set of 1,448 tactile sensors is distributed along all the body parts to reflect known

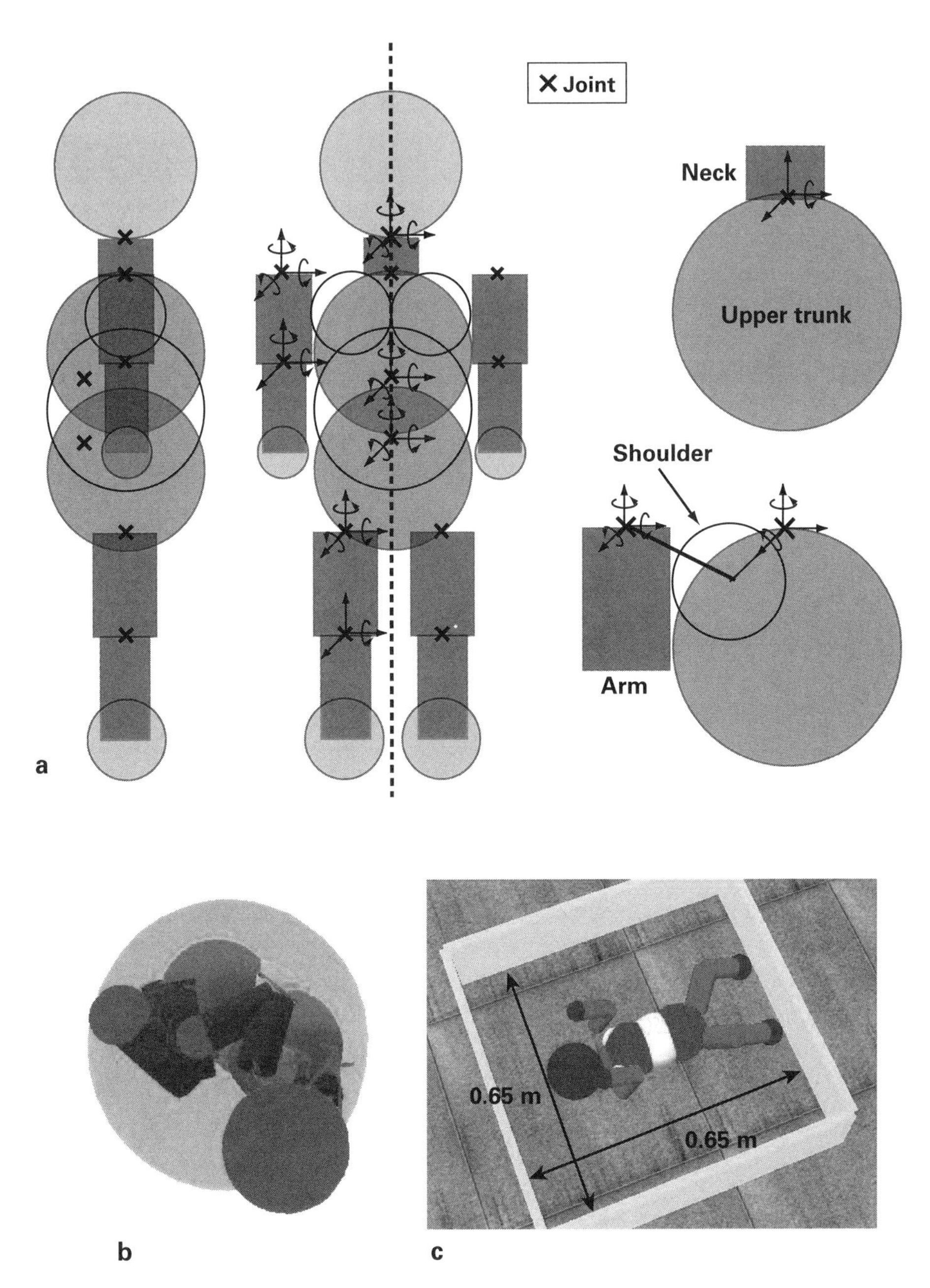

Figure 2.16
Fetus Model 1 with the nineteen body cylinders (a), and 3D visualization of the fetus in the womb (b) and the newborn in its environment (c) in Kuniyoshi and Sangawa 2006. Figures courtesy of Professor Yasuo Kuniyoshi, University of Tokyo. Reprinted with permission from Springer.

tactile sensation sensitivity in the infant's body. The womb environment is constituted by a spherical environment with the simulation of physics parameters such as gravity, buoyancy, and fluid resistance. The navel string attaching the fetus to the wall also affects the fetus's movement. The uterine wall is modeled as a nonlinear spring with damper properties.

This model was designed to study the role of general embodied developmental principles in early sensorimotor learning. In particular it aimed at exploring the hypothesis that partially ordered embodiment dynamical patterns emerge from the chaotic exploration of body-brain-environment interactions during gestation. This will later lead to the emergence of meaningful motor behaviors such as rolling over and crawling-like motion in neonates. This hypothesis is linked to the observation of spontaneous movements reported in the human fetus, such as the *general movements*, that is, smooth movements not reactive to external stimuli, already produced at two months (eight to ten weeks) from gestation (Kisilevsky and Low 1998). The model also implements a detailed neural architecture based on sensory organs, medulla and spine models, and primary somatosensory and motor areas. The model of the muscle sensory organs includes spindles (sensitive to muscle stretch speed) and the Golgi tendon organs (response proportional to muscle tension). The spine model processes paired inputs from spindles and tendons to control stretch reflexes and use inhibition to regulate muscle tension. The medulla model consists of one central pattern generator (CPG) per muscle. Although the CPGs of each of the two muscles are not directly coupled, they are functionally coupled via the body interaction with the physical environment and the muscle's response. The coupled CPGs produce periodic activity patterns, for example, for constant input, and chaotic patterns for nonuniform sensory input. The neural cortical model consists of the interconnected primary somatosensory area (S1) and the motor area (M1), realized though self-organizing maps.

The model has one main parameter related to the developmental timescale, which is the gestational age distinguishing between a fetus and a newborn. The fetus model in the womb represents a thirty-five-week old embryo. The newborn model represents a zero-day old infant. The newborn's environment is modeled as a flat-walled square area, where only gravity is still applied, with no fluid resistance as in the womb.

Simulations on spontaneous movements of the model of the fetus in the womb and of the newborn rolling over in the flat, walled arena show the emergence of a rich variety of meaningful motor patterns and the organization of the sensorimotor areas with a somatotopic structure. The fetus simulation also highlights an important role of spontaneous fetus movements, constrained within the spherical uterine environment. The uterine wall constrains the body mostly to produce rotating movements, which cause the learning of cooperative relationships between legs and neck, with the sensory area S1 driving the organization of the motor area M1. Without such a preliminary fetal environment restrictions and experience, an infant would only experience head-led motor explorations, without coordinated body parts.

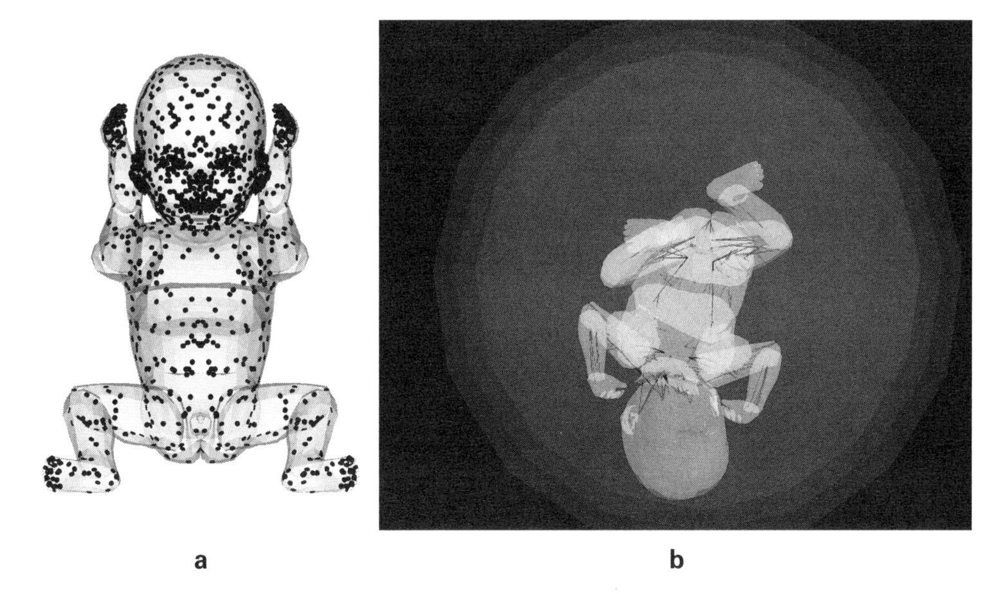

Figure 2.17
Distribution of tactile sensors (a) and 3D rendering of the fetus in the womb (b) in Mori and Kuniyoshi 2010. Figures courtesy of Prof. Yasuo Kuniyoshi, University of Tokyo. Reprinted with permission from IEEE.

A more sophisticated simulation of the embryo, the *Fetus Model 2*, was later developed to investigate the fine mechanisms of intrauterine learning (Mori and Kuniyoshi 2010) (figure 2.17), with the specific contribution of tactile sensation to the learning of behavioral patterns. This more sophisticated fetus has 1,542 tactile sensors (figure 2.17, left), with the greatest concentration in the head (365 tactile points) and hands (173). The rest are distributed in the neck (6), chest and abdomen (54), hip (22), shoulders (15), arms (31), thighs (22), calves (17) and feet (43).

Simulation experiments focus on the learning of more complex behaviors, which in fetal development appear after the general movements, as in the previous fetus model. Specifically the simulations explore the hypothesis that tactile sensation induces motions in the fetus. Results show that the use of human-like tactile distribution in the fetus's body causes the learning of the two reactive movements observed after the embryo's general movements, in other words, (1) isolated arm/leg movements (IALM), a set of jerky movements independent from other body parts, and observed in human fetuses from around ten weeks of gestation; (2) hand/face contacts (HFC), when the hands touch the face slowly, and which have been observed from eleven gestation weeks. These results, in addition to investigating the role of motor learning before birth, also have implications for clinical studies. Specifically they can contribute to the understanding of the learning capabilities underlying developmental care procedures applied to preterm infants in neonatal intensive care units, such as the "nesting" or "swaddling" procedures (Mori and Kuniyoshi 2010; van Sleuwen et al. 2007).

The software of these fetus and newborn simulation models is not yet available for open access, in contrast to the open source iCub simulator or the commercial Webots application. However, they provide an appealing demonstration of the potential of using 3D physics simulation models to study developmental robotics principles even at the embryo level. This permits the testing of hypotheses on the role of spontaneous movements and other cognitive capabilities in the fetus, such as the evaluation of nature vs. nurture origins of behavior, the origins of motor and cognitive primitives, and the contribution of embryonic experience to the bootstrapping of more complex skills after birth.

2.6 Conclusions

This chapter introduced the key concepts and common sensor and actuator technology used in robotics. It also gave an overview of the main humanoid platforms used in developmental robotics, and related simulation software, with particular emphasis on humanoid "baby" robots. The variety and complexity of the developmental humanoid platforms demonstrates the vitality of the field, and the impressive technological achievements reached in less than a decade of developmental robotics research.

The field is still growing quickly, due to continuous improvement in the technologies involved in robot platform design. The impressive developments of new materials, new sensors, more efficient battery systems, and more compact and energy-efficient computer microprocessors all contribute to the design of more advanced, fully autonomous robots.

In the field of material sciences, the development of new soft products usable as robot effectors can contribute to the production of robots made of soft materials (Pfeifer, Lungarella and Iida 2012) and based on anthropomorphic musculoskeletal actuators (Holland and Knight 2006; Marques et al. 2010). Prototypical soft robot applications include pneumatic artificial muscle actuators and electroactive polymers, or in some cases, rigid materials with soft dynamics as with the electromagnetic, piezoactive, or thermal actuators and the variable compliance actuators (Trivedi et al. 2008; Albu-Schaffer et al. 2008). These solutions take direct inspiration from a variety of animals and plants that exhibit complex movements with soft structures, such as hydrostatic muscles in octopus arms and elephant trunks. Great advances are also being made in sensor technology, including the production of robot skin and tactile sensors, also suitable for soft robot covers (Cannata et al. 2008).

Another area of development that will contribute to the design of fully autonomous robots—for example, robots that do not require continuous connection to umbilical cords or continuous recharging of short-lived batteries—is rapid progress on the design of light and efficient electrical-energy batteries. This is mostly driven by R&D on batteries for smart phones, currently leading to the production of even smaller, more efficient, and inexpensive batteries. Innovative solutions are also addressing the design of robots producing their own energy through metabolic digestion of organic material (e.g., microbial fuel cells), as with the slow but fully energy-autonomous fly-eating "EcoBot" at the Bristol Robotics Labs (Ieropoulos et al. 2012).

Finally, another area of technological development contributing to innovation in humanoid robotics is the use of compact, low-energy consumption and affordable many-core microprocessors for the robot's computational processing requirements. This is the case of the growing utilization of GPU (graphic processing units) for parallel processing. GPUs provide an alternative to the common CPUs used in standard computers, which have limited parallel processing capabilities. Instead of being dedicated only to intense computational requirements of graphic tasks, GPUs can be used to perform parallel computation for neural networks, genetic algorithms, and numerical simulations. GPUs also provide an alternative to the expensive, bulky, and high-energy-consuming high-performance computers and clusters, which require the continuous connection of a robot to a central processing unit. For example, GPUs, and their associated programming languages like CUDA, have been used to perform fast computations of large artificial neural networks for robot motor learning in the iCub robot (Peniak et al. 2011).

The chapters that follow will look at the state of the art of the utilization of robotics platforms for developmental studies and the important advancements in the modeling of motivational, sensorimotor, social, and communicative capabilities in baby robots. Continuing advances in robot technology and computational systems will contribute to further bootstrap the design of cognitive capabilities in robots.

Additional Reading

Matarić, M. J. *The Robotics Primer*. Cambridge, MA: MIT Press, 2007.

This is a concise book providing a clear and accessible introduction to robotics concepts and technology for students and readers from nontechnical disciplines. It presents both the basic concepts on the main sensors and actuators for building robots, and an overview of current progress in robot behavior modeling for locomotion, manipulation, group robotics, and learning. The book is complemented with a free robot programming exercise workbook. This book is recommended to students and researchers from human science disciplines who want to have a more technical introduction to the main concepts and technologies in robotics.

Siciliano, B., and O. Khatib, eds. *Springer Handbook of Robotics*. Berlin and Heidelberg: Springer, 2008.

This is the most comprehensive handbook on robotics. It includes sixty-four chapters written by the international leaders in the field. The various chapters are grouped into seven parts covering (1) robotics foundation, (2) structures, (3) sensors and perception, (4) manipulation and interfaces, (5) mobile and distributed robotics, (6) applications in field and service robotics, and (7) human-centered and life-like robotics. Part 7 is the most relevant to the readers of this book because it covers the various areas of cognitive and bio-inspired robotics such as humanoid robots, evolutionary robotics, neurorobotics, human-robot interaction, and roboethics.

3 Novelty, Curiosity, and Surprise

A key design principle in developmental robotics is *autonomy*, which means that the developing robot, machine, or agent is free to interact with and explore its environment. In contrast to robots whose decisions and actions are rigidly programmed in advance, or who are driven by a remote controller, autonomous robots *self-select* their actions adaptively, in response to internal states and external environment sensing (e.g., Nolfi and Parisi 1999; Schlesinger and Parisi 2001). In this chapter, we focus in particular on the autonomy of *learning*, that is, on the agent's freedom to choose *what*, *when*, and *how* it will learn.

A fundamental question that arises from this freedom is: what is the best strategy for exploring the environment? When confronted with a novel experience, and a variety of options for probing or examining this experience, how should the robot decide which actions or options to try first, and when should it move from one option to the next? Conventional AI approaches to this question typically treat the issue as an optimality problem (e.g., energy minimization, exploration rewards, etc.), and consequently, they often focus on analytical and learning methods that are designed to identify an optimal exploration strategy. While research in developmental robotics relies on many of the same computational tools, the way in which the problem is framed, and the theoretical perspectives that inspire and guide these models differ from conventional approaches. In particular, in this chapter we highlight the emerging area of *intrinsic motivation*, which provides robots with a form of "artificial curiosity." Thus, intrinsically motivated robots are not focused on solving a particular problem or task, but rather on the process of learning itself (e.g., Oudeyer and Kaplan 2007).

The application of intrinsic motivation (IM) as a mechanism to drive autonomous learning, not only in developmental robotics, but also more broadly within the field of machine learning, offers three important advantages over conventional learning methods (e.g., Mirolli and Baldassarre 2013; Oudeyer and Kaplan 2007). First, IM is *task-independent*. This means that the robot or artificial agent can be placed in a completely new environment—with which the model-builder may have no prior knowledge or experience—and through self-directed exploration, the robot will potentially learn not only the important features of the environment, but also the behavioral skills necessary for dealing with that environment. Second, IM promotes the hierarchical *learning and reuse* of skills. Learning is directed toward acquiring knowledge or skill or both, rather than solving a specific, predefined task. Thus the intrinsically motivated robot may acquire an ability in one context (or developmental stage) that has no immediate benefit, but which then becomes a critical building block for later, more complex skills. Finally, IM is *open ended*. Thus, learning in a particular environment is determined by the robot's level of skill or knowledge, rather than by its progress toward a predetermined, externally imposed goal. Indeed, as we will highlight, there are several IM models that illustrate this principle: as the robot achieves mastery in one area, it can efficiently shift its focus toward new features of the environment or new skills that it has not yet learned.

As a research topic within developmental robotics, the study of IM is inspired by two closely related areas of work. First, there is a wide array of theories and empirical studies, primarily within psychology, that explore how IM develops in both humans and nonhumans (e.g., Berlyne 1960; Harlow 1950; Hull 1943; Hunt 1965; Kagan 1972; Ryan and Deci 2000; White 1959). Second, there is also a considerable amount of work within neuroscience, which seeks not only to identify the neural substrates for IM, but also to explain how these biological mechanisms operate (e.g., Bromberg-Martin and Hikosaka, 2009; Horvitz, 2000; Isoda and Hikosaka 2008; Kumaran and Maguire 2007; Matsumoto et al. 2007; Redgrave and Gurney 2006).

In contrast to other research areas in developmental robotics (e.g., motor-skill development or language acquisition), the study of IM in robots and artificial agents is still at a comparatively early stage. As a result, there are a few key differences between how this chapter is organized and the themes that appear throughout the rest of the volume. First, there is not yet a clear correspondence between the major studies and experimental paradigms used to study infants and children, and robotics models of IM. In particular, much of the research to date within developmental robotics on IM has predominantly focused on designing effective algorithms and architectures, while there is comparatively less work that points directly toward human development (e.g., the self-other distinction; Kaplan and Oudeyer 2007). In section 3.3.1, we therefore provide a detailed description of the class of architectures that are available for simulating IM. Second, much of the modeling work that has been conducted thus far focuses on simulation studies. Consequently, there are comparatively less data available from real-world robot platforms. Nevertheless, there is a growing trend toward using robots to study IM, and in the second half of the chapter, we highlight several examples of both simulation and real-world studies.

3.1 Intrinsic Motivation: A Conceptual Overview

As we will highlight, the concept of *IM* is heavily influenced by both data and theory from psychology. Nevertheless, and perhaps surprisingly, the use of the terms "extrinsic" and "intrinsic motivation" in developmental robotics (and more generally, machine learning) differ from their use in psychology. Thus, psychologists typically define intrinsically motivated behaviors as actions that are chosen by the organism "freely," that is, without any external incentives or consequences, while extrinsically motivated behaviors are produced in response to external prompts or cues. By this view, a child can draw simply for fun (intrinsic) or, instead, for a reward such as money or candy (extrinsic). In contrast, we adopt the view proposed by Baldassarre (2011) that extrinsically motivated behaviors are those that directly serve the needs of basic biological functions (e.g., thirst or hunger), while intrinsically motivated behaviors have no clear goal, purpose, or biological function, and are therefore presumably performed *for their own sake*.

3.1.1 Early Influences

Early approaches to understanding intrinsic motivation were influenced by existing theories of behavior, and in particular, by drive-based, homeostatic theories. A well-known example of the homeostatic approach is Hull's theory (1943), which proposes that all behaviors can be understood as the result of either (a) "primary" physiological drives such as hunger or thirst, or (b) "secondary" or psychological drives, which are acquired in the process of satisfying the primary drives. There are two critical components of the Hullian view. First, primary drives are innate and therefore biologically specified: they are evolved for the purpose of protecting or promoting the organism's survival. Second, they are homeostatic: this means that, for a given physiological system, there is an ideal "set point," and the primary drive then serves to keep the organism as close to this point as possible. For example, when an animal becomes cold, it may shiver or move toward sunlight in order to increase its temperature. In other words, homeostatic drives function to bring an animal "back into balance" or equilibrium when the environment alters or disrupts the organism's state.

Several researchers asked whether Hull's drive theory could be applied to behaviors like play and object exploration, especially in nonhumans. For example, Harlow (1950; Harlow, Harlow, and Meyer 1950) observed the behavior of rhesus monkeys that were presented with a mechanical puzzle illustrated in figure 3.1 (e.g., including a lever, hinge, and chain). The monkeys typically became engrossed by the puzzle and played with it extensively. Notably, their exploratory behavior was not dependent on the presentation of an external reward (e.g., food), and over repeated experiences, the monkeys became progressively more effective at solving the puzzle. Thus, the monkeys

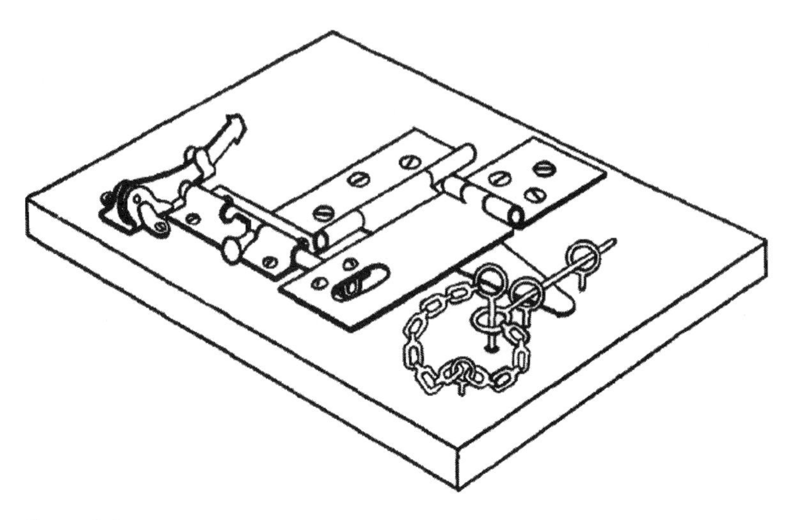

Figure 3.1
Mechanical "puzzle" studied by Harlow (1950). Public domain figure, American Psychological Association.

in the Harlow study not only learned to solve the puzzle, but also, more important, their attempts to manipulate and explore the puzzle also seemed to be directed toward the goal of understanding or *figuring out how it worked.*

Other researchers observed similar instances of exploratory behavior (e.g., Butler 1953; Kish and Anonitis 1956). One way to account for these behaviors is to incorporate them within the Hullian framework, that is, as drives or instincts for "manipulation," "exploration," and so on (e.g., Montgomery 1954). However, as White (1959) notes, exploratory or play behaviors like those observed by Marlow are not homeostatic, and therefore differ from classical drives in two fundamental ways. First, they are not a response to an environmental perturbation or disturbance, such as deprivation of food or water. Second, performing these behaviors does not bring the organism "back" to a desired physiological state. Instead, they seem to be open ended, with no obvious goal or immediate benefit to the organism.

3.1.2 Knowledge and Competence

Subsequent approaches to IM focused on addressing the limitations of drive-based, homeostatic theories. These approaches can be divided into two broad theoretical views, the knowledge-based and the competence-based IM views (e.g., Baldassarre 2011; Mirolli and Baldassarre 2013; Oudeyer and Kaplan 2007). First, the *knowledge-based view* proposes that IM is a cognitive mechanism that enables the organism to detect novel or unexpected features, objects, or events in the environment. According to this view, IM is a product of the organism's current state of knowledge. In particular, the organism is motivated to expand its knowledge base (i.e., learn) by systematically exploring the environment, and searching for experiences that are unfamiliar or poorly understood.

Knowledge-based IM includes two variants or subclasses: novelty-based and prediction-based IMs. *Novelty-based IM* reflects the principle, proposed by several developmental theorists, that experience is organized into cognitive structures, which are used to interpret new information (e.g., Fischer 1980; Kagan 1972; Piaget 1952). Novel situations produce a mismatch or incongruity between an ongoing experience and stored knowledge, which results in an effort to resolve the discrepancy (e.g., by increasing attention to the object or situation; see "comparator theory," to follow). The novelty-based approach also suggests that there is a critical difference between situations with low, moderate, and high levels of novelty: a low level of novelty maps to a familiar experience, while highly novel experiences may not be interpretable within the organism's current knowledge base. In contrast, moderate novelty may be optimal for learning, as it is both comprehensible and unfamiliar (e.g., Berlyne 1960; Ginsburg and Opper 1988; Hunt 1965). The other variant of knowledge-based IM is *prediction-based IM*. This approach emphasizes the role of organism-environment interaction, and characterizes the organism as actively exploring the unfamiliar "edges" of its knowledge base. Prediction-based IM is consistent with the concepts of *curiosity* and *surprise*: in probing its environment, the organism implicitly predicts how objects or events will respond to its actions, and when unexpected outcomes occur, additional energy or attention is devoted to further probe the situation (e.g., Piaget 1952).

While novelty-based and prediction-based IM both generate new knowledge about the environment, there is a subtle but important difference between the two learning mechanisms. In the case of novelty-based IM, the agent is viewed as generally *passive*, as its primary tool for seeking out novel experiences is movement through space (e.g., head and eye movements). In contrast, prediction-based IM is comparatively *active*, as the agent can systematically operate on the environment and observe the outcomes of its actions (e.g., grasp, lift, or drop an object). However, this distinction is somewhat arbitrary, and it should be stressed that novelty seeking and action prediction are not mutually exclusive, and in fact can frequently co-occur.

Knowledge-based IM focuses on the properties of the environment, and how the organism gradually comes to know and understand these properties (i.e., objects and events)—and in the case of prediction-based IM, how these properties might change as a consequence of the organism's actions. An alternative approach, the *competence-based view*, focuses instead on the organism and the particular abilities or skills it possesses. There are several theoretical motivations for competence-based IM. For example, White (1959) highlights the notion of "effectance," which is the subjective experience that one's actions will influence the outcome of a situation (see Bandura 1986 for a related term, "self-efficacy"). Similarly, de Charms (1968) proposes the concept he terms "personal causation." More recently, *self-determination theory* (Deci and Ryan

1985) has elaborated on these ideas, not only by linking the concept of IM to the subjective experiences of autonomy and competence, but also by arguing that competence manifests itself as a tendency toward improvement and increasing mastery. A closely related phenomenon, described by Piaget (1952; see also Ginsburg and Opper 1988), is *functional assimilation*, which is the tendency for infants and young children to systematically practice or repeat a newly emerging skill (e.g., learning to grasp or walk). Therefore, a fundamental implication of competence-based IM is that it promotes skill development by leading the organism to seek out challenging experiences.

3.1.3 Neural Bases of IM

The ideas and approaches described thus far represent one of two fundamental influences on IM in developmental robotics, that is, the observation and analysis of behavior, coupled with psychological theory. As we noted at the start of the chapter, however, there is another research area that has provided a second fundamental influence: neuroscience. In particular, we briefly highlight here how activity in specific brain regions has been linked to each of the types of IM outlined in the previous section.

First, an important region for novelty detection is the hippocampus, which not only plays a fundamental role in supporting long-term memory, but also in the process of responding to new objects and events (e.g., Kumaran and Maguire 2007; Vinogradova 1975). In functional terms, when a novel experience is encountered, the hippocampus activates a recurrent pathway between itself and the ventral tegmental area (VTA), which uses the release of dopamine to establish new memory traces in the hippocampus. This mechanism continues to operate over repeated presentations, eventually resulting in a diminished response (i.e., habituation; see Sirois and Mareschal 2004). More generally, dopamine release in the mesolimbic pathway (including the VTA, hippocampus, amygdala, and prefrontal cortex) is associated with the detection of salient or "alerting" events (e.g., Bromberg-Martin and Hikosaka 2009; Horvitz 2000). A comprehensive theoretical account that integrates several of these areas, in addition to the hypothalamus, nucleus accumbens, and other nearby structures has been proposed by Panksepp (e.g., Wright and Panksepp 2012), who argues that activation of this network motivates "SEEKING" behaviors (i.e., curiosity, exploration, etc.).

Next, there are a number of brain regions involved in sensorimotor processing that may provide a substrate for prediction learning and the detection of unexpected objects or events. For example, the frontal eye field (FEF), which is associated with the production of voluntary eye movements, plays a critical role during visuomotor scanning are (e.g., Barborica and Ferrera 2004). Single-cell recordings from FEF cells in monkeys suggest that FEF activity is anticipatory (e.g., while tracking a target that briefly disappears). In addition, FEF activity also increases if there is a mismatch between the expected and observed location of the target when it reappears (e.g., Ferrera and Barborica 2010). Thus, neural activity in this region is not only predictive, but it also provides a "learning signal" that may modulate future sensorimotor predictions.

Finally, a brain region that has been implicated as a potential substrate for competence-based IM is the superior colliculus (SC). A recent model, proposed by Redgrave and Gurney (2006), suggests that unexpected events, such as a flashing light, activate the SC and result in a short-term (i.e., phasic) increase in dopamine release. It is important to note, however, that this pathway is not simply a "novelty detector." In particular, Redgrave and Gurney propose that a phasic dopamine increase—produced as a result of SC activation—strengthens the association between ongoing motor and sensory signals, which converge in the striatum. Thus, the SC functions as a "contingency" or "causal" detector, which not only signals when the organism's ongoing actions produce salient or unexpected consequences, but more important, also increases the likelihood that the given action will be repeated in the corresponding context.

3.2 The Development of Intrinsic Motivation

We next turn to the question of how IM develops during infancy and early childhood. It is important to note that IM is not a distinct research topic or knowledge domain, but rather, is part of a general set of issues that cut across multiple research areas (e.g., perceptual and cognitive development). We begin by describing work that overlaps with the topic of knowledge-based IM (i.e., novelty and prediction) and then focus on competence-based IM.

3.2.1 Knowledge-Based IM in Infants: Novelty

From a behavioral perspective, the process of identifying novel objects or events in the environment can be deconstructed into two key abilities or skills. The first is *exploratory behavior*, that is, searching or scanning the environment for potential areas of interest. The second is *novelty detection*, which is identifying or recognizing that a situation is novel, and focusing attentional resources on that object or event. It is important to note that both of these phenomena can be manifested in a variety of ways. For example, exploration and novelty can be generated as a product of the child's own behavior, such as through babbling or visual regard of the hand as it moves. Alternatively, they can also result from searching the environment. As a particular example, we focus here specifically on the phenomenon of visual exploration in young infants.

How does visual exploration develop? One way to address this question is to measure infants' scanning patterns during free viewing of simple geometric shapes. For example, figure 3.2 presents two samples of gaze behavior from a study of visual encoding in two- and twelve-week-old infants (Bronson 1991). During the study, infants viewed an inverted-V figure while their eye movements were recorded with a camera located beneath the figure. For each infant, the initial fixation is indicated by a small dot, and subsequent gaze shifts are indicated by the links between the dots. The size of each dot indicates the relative dwell time of the fixation, with larger dots representing longer dwell times. The sample on the left (figure 3.2a, produced by a two-week-old)

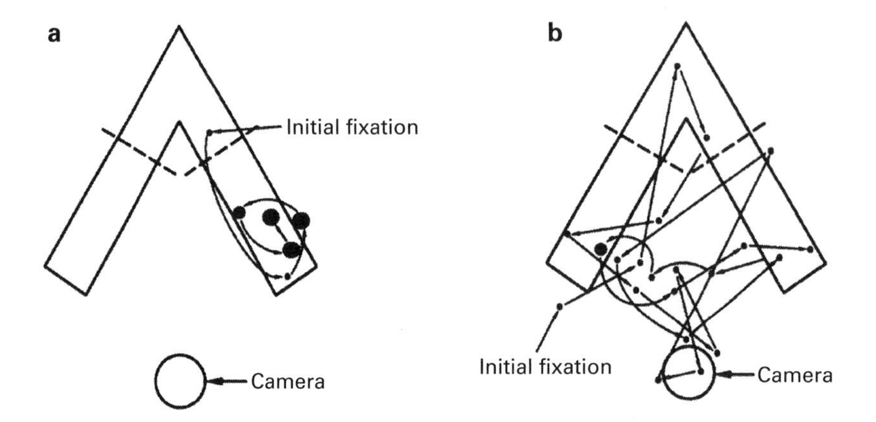

Figure 3.2
The scan pattern on the left (a) was produced by a two-week-old and has a few, long fixations clustered in one corner of the stimulus, while (b) was produced by a twelve-week-old and has many brief fixations, evenly distributed over the stimulus. From Bronson 1991. Reprinted with permission from Wiley.

highlights two characteristic aspects of scanning behavior that are observed in very young infants: (1) individual fixations (the large dots) are not evenly distributed over the shape, but instead are clustered over a small area; and (2) dwell or fixation times are fairly long (e.g., several seconds). In contrast, in the scanning pattern produced by a twelve-week-old (figure 3.2b), fixations are much more brief and more evenly distributed. Other studies, looking at the same age range, report comparable findings. For example Maurer and Salapatek (1976) analyzed the gaze patterns produced one- and two-month-olds as they viewed pictures of faces. Younger infants tended to focus on a portion of the outside edge of a face, while the older infants more systematically scanned the entire face, including the eyes.

One way to account for this developmental pattern is to describe it as a shift from *endogenous* to *exogenous orienting* (e.g., Colombo and Cheatham 2006; Dannemiller 2000; Johnson 1990). Thus, during the first few weeks of postnatal life, infants' orienting behaviors are dominated by salient sensory events. Between one and two months, infants begin to develop more control over their visual exploratory behavior, and gradually acquire the ability to deploy their attention from one location to another in a more deliberate or strategic manner. We return to this issue in the next chapter, where we discuss the development of *visual selective attention*.

The second, core component of identifying novel objects or events is novelty detection. Research on this ability has focused on two questions: (1) at what age do infants begin to respond to novel objects and events, and (2) how is novelty detection manifested in their behavior? While several methods are available to investigate these questions, the dominant approach has been to narrow the scope of the problem in two specific ways. First, most studies present infants with visual images or objects, and measure the amount of time that infants spend looking at the visual stimuli (i.e., *looking time*). Second, changes in looking time—specifically, systematic increases or decreases across presentations of an image or object—are used as an index or proxy for novelty detection (e.g., Colombo and Mitchell 2009; Gilmore and Thomas 2002). The most common paradigm used to study novelty perception in infants is *habituation-dishabituation*. In chapter 4, we describe this paradigm in detail and highlight its use as a research tool for studying perceptual development (e.g., face perception). Here, in contrast, we focus on the more general question of how habituation-dishabituation provides a window into infants' cognitive processing, and specifically, how it is used to measure infants' preference for novel objects and events.

In a habituation-dishabituation study, infants are shown an object or event over a series of trials (i.e., discrete presentations of the visual stimulus), and their looking time is recorded during each trial. Infants gradually lose interest in the object or event—that is, they *habituate*—which is reflected by a decrease in their looking time across trials. At this point, infants are then presented with a visual stimulus that is similar to the one seen during habituation, but that differs along one or more critical dimensions. For example, female faces may be presented during the habituation phase, followed by male faces in the post-habituation phase. A statistically significant increase in looking time to the novel, post-habituation stimuli is then interpreted as reflecting a *novelty preference*: infants not only detect that a new object or event has been presented, but they also increase their attention toward it.

Using this paradigm, developmental researchers have asked: how does novelty preference develop during the first twelve months of infancy? Initial work on this question suggested a surprising answer. In particular, rather than showing a novelty preference, very young infants (i.e., between birth and two months) *tended to prefer familiar objects* (e.g., Hunt 1970; Wetherford and Cohen 1973). Between three and six months, this preference appears to shift toward novel objects and events, and from six to twelve months, a robust and consistent preference for novel stimuli is observed (e.g., Colombo and Cheatham 2006; Roder, Bushnell, and Sasseville 2000).

The observed shift in early infancy from familiarity preference to novelty preference can be understood by applying Sokolov's *comparator theory* (Sokolov 1963). In particular, Sokolov proposed that as infants view an object, they gradually create an internal representation (or internal "template"). The process of habituation is then interpreted as the time spent by the infant constructing an internal representation: as the internal copy comes to match the external object, looking time (i.e., visual attention) declines. When a new object is presented—which creates a mismatch between itself and the internal representation—the infant dishabituates, that is, looking time is increased as the internal representation is updated with new information.

Subsequent work on novelty perception has used comparator theory to help explain the (apparent) shift from familiarity preference to novelty preference in young infants. Specifically, comparator theory proposes that because younger infants have limited visual experience, as well as limited visual processing ability, they are less skilled at encoding objects and events, and their internal representations are less stable or complete. As a result, they tend to focus on familiar visual stimuli. According to this view, infants at all ages should show a novelty preference when the stimuli are presented in a manner that accounts for infants' processing speed and visual experience (e.g., Colombo and Cheatham 2006). Thus, novelty detection and novelty preference appear to be present in human infants as early as birth, and familiarity preference is now understood as the product of partial or incomplete visual encoding (e.g., Roder, Bushnell, and Sasseville 2000).

3.2.2 Knowledge-Based IM in Infants: Prediction

The development of infants' ability to predict has been primarily studied by investigating their anticipatory reactions to the outcomes of simple dynamic events (e.g., a ball that rolls down a track). In this context, "prediction" is specifically defined as a sensorimotor skill, where an action is performed by the infant, such as a gaze shift or a reach, in anticipation of the event outcome.

A well-established technique for measuring predictive behavior in young infants is the *visual expectation paradigm* (VExP; see Haith, Hazan, and Goodman 1988; Haith, Wentworth, and Canfield 1993), in which infants are presented with a sequence of images at two or more locations, in a consistent pattern, while the location and timing of infants' gaze patterns to the images are recorded. In box 3.1, we provide a detailed description of the VExP, including the major developmental findings that have been revealed with this technique. It is important to note that VExP is typically used to study

Box 3.1

The Visual Expectation Paradigm (VExP)

One of the earliest and most basic forms of predictive behavior observed in young infants is the ability to view a sequence of events—such as images that appear on a video screen, one after another—and to anticipate the location of the events before they appear. In a groundbreaking study, Haith, Hazan, and Goodman (1988) designed the visual expectation paradigm (VExP) to examine this ability. In the VExP, infants view a series of images that appear at two or more locations, following either a regular spatial pattern (e.g., A-B-A-B-A-B) or an irregular sequence (e.g., A-B-B-A-B-A). As we highlight here, the core finding from this work is that infants as young as age two months quickly learn to predict the upcoming locations of images in the regular pattern. Subsequent work by Haith and colleagues demonstrates that young infants are not only sensitive to the spatial locations of the images, but also to the timing and content of the images as well.

Procedure

The figure below illustrates the apparatus used by Haith, Hazan, and Goodman (1988) to study infants' gaze patterns during the VExP. In the experiment, three-and-a-half-month-old infants rested on their back, while viewing a series of video images reflected through a mirror ("Mirror Y"). At the same time, a video camera ("eye camera") recorded a close-up view of one of the infant's eyes, which was illuminated with an infrared light source ("light collimator"). Test trials were divided into two sequences: a *regular-alternating sequence*, in which images alternated on the left and right sides of the screen, and an irregular sequence, in which images appeared in the two locations randomly (see figures below). All infants viewed both sequences, with the order counterbalanced across infants.

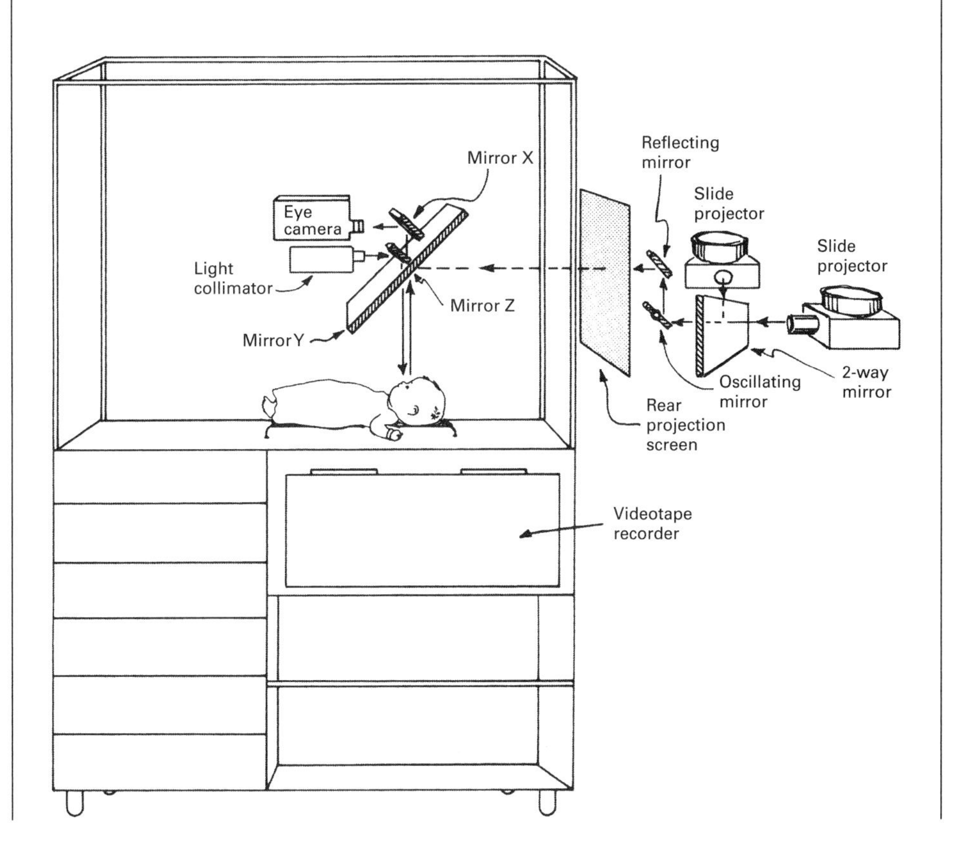

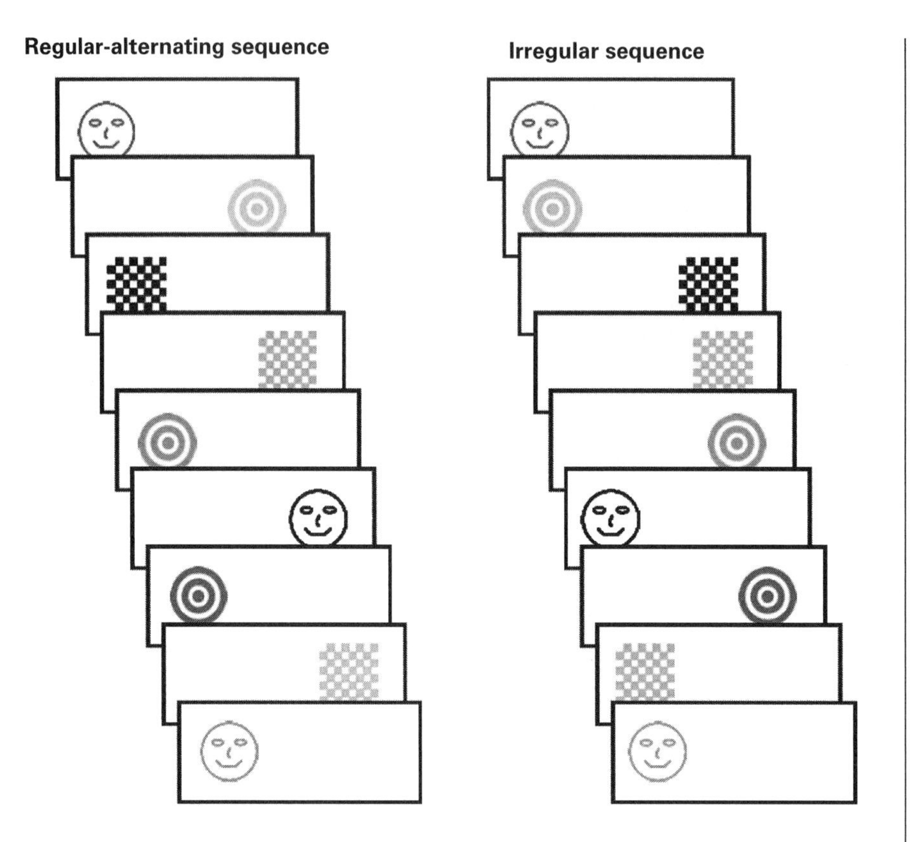

Note: Apparatus used in the VExP of Haith, Hazan, and Goodman (1988) (a), and diagrams of regular-alternating (b) and irregular image sequences (c), respectively. Figure reprinted with permission from Wiley.

Results

Haith, Hazan, and Goodman (1988) systematically analyzed infants' reaction times by computing the average difference in time between the appearance of each image, and the corresponding fixation to that location. There were two major findings. First, infants' overall reaction times were faster to the regular-alternating sequence than to the irregular sequence (i.e., 391 vs. 462 milliseconds). Second, *anticipations* were defined as eye movements to the image location within 200 milliseconds of its appearance. Using this criterion, infants were twice as likely to anticipate an image during the regular-alternating sequence than during the irregular sequence (i.e., 22 vs. 11 percent).

Subsequent Findings

In a follow-up study, Canfield and Haith (1991) extended the original findings with three-and-a-half-month-olds to infants as young as two months, by demonstrating that the younger infants also learn to anticipate two-location alternating image sequences. In contrast, a limitation at this age is that two-month-olds are unable to learn asymmetric sequences, such as A-A-B-A-A-B; however, this limitation is lifted by age three months. Several other impressive abilities appear to be in place by age three months. For example, Wentworth, Haith, and Hood (2002) found that three-month-olds not only learn three-location sequences, but they can also use the specific content of an image at one location (e.g., the center) to correctly predict the location of the next image (e.g., right vs. left). Finally, Adler et al. (2008) systematically varied the timing (instead of the locations) between images, and found that three-month-olds successfully learned to use the temporal interval as a predictive cue for the appearance of the upcoming image.

infants within a narrow age range—that is, between ages two and four months—which suggests that predictive or anticipatory visual activity not only is rapidly developing at this age period, but also is consistent with independent estimates of FEF maturation in human infants (e.g., Johnson 1990).

A more challenging example of a task that measures prediction involves tracking a ball that moves in and out of sight, as it travels behind an occluding screen. Successfully anticipating the reappearance of the ball requires two skills: first, the infant must hold the occluded object "in mind" while it is behind the screen (i.e., what Piaget calls *object permanence*), and second, they must prospectively control their eye movements by directing their gaze to the point of the ball's reappearance, before it arrives there. In chapter 4, we highlight the first component of the task, as a fundamental step in the development of object perception. Here, meanwhile, we focus on the latter skill, that is, how anticipatory or prospective gaze develops during occluded object tracking.

At age four months, infants successfully track the moving ball while it is visible (Johnson, Amso, and Slemmer 2003a). In contrast, infants at the same age are unable to track the ball once it is occluded: in particular, they do not anticipate its appearance, and only direct their gaze toward the ball after it reappears. However, when additional support is provided—for example, the width of the occluding screen is narrowed—four-month-olds produce anticipatory eye movements. Thus it appears that the basic perceptual-motor mechanism for predictive object tracking is present by age four months, but it may require perceptual support before it can be reliably produced. By age six months, infants reliably anticipate the reappearance of the ball, even when it is occluded by the wider screen.

Finally, the predictive tracking task can be made even more challenging by placing the infant next to a real track, along which the ball rolls. Again, a portion of track is occluded, so that the ball briefly passes out of sight and then reappears. In this case, predictive action can be measured in two ways at the same time: that is, by predictive eye movements in addition to predictive reaching movements. Berthier et al. (2001) studied how nine-month-olds performed on this task, under two tracking conditions. As figure 3.3 illustrates, in the "no wall" condition, infants watched as the ball rolled down the track, behind the screen, and out the other side. In the "wall" condition, an obstacle was placed on the track, behind the screen, which prevented the ball from reappearing after occlusion. (Note that the "wall" extended several inches higher than the screen, so that it was clearly visible to infants.)

Three major findings were reported. First, when analyzing infants' gaze behavior, Berthier et al. (2001) noted that nine-month-olds consistently anticipated the reappearance of the ball, only in the no-wall condition. In contrast, infants quickly learned that the wall prevented the ball from reappearing, and thus, they did not direct their gaze at the usual point of reappearance in the wall condition. Second, when analyzing infants' reaching behavior, Berthier et al. (ibid.) also reported that infants reliably produced more reaches during the no-wall condition than the wall condition. Infants therefore also appeared to use the presence of the wall as a cue for whether or not to reach for the ball. Third, however, as figure 3.3 illustrates, on some trials infants reached for the ball regardless of whether or not the wall was present. Interestingly, Berthier et al. (ibid.) found that the kinematic properties of these reaches did not vary as a function of the presence of the wall. In other words, the reaching behavior appeared to be somewhat ballistic, and once initiated, though anticipatory it was not influenced by additional visual information. In particular, they propose that these occasional reaching errors reflect a partial integration of visual and visual-motor skill, which continues to develop in the second year.

3.2.3 Competence-Based IM in Infants

As we noted earlier, competence-based IM differs from knowledge-based IM by focusing on the developing organism and the acquisition of skills, rather than on acquiring knowledge or information about the environment. An early-emerging component of competence-based IM in human infants is the discovery of *self-efficacy* (or *effectance*), that is, the recognition that one's own behavior has an effect on the objects and people around them.

One way that self-efficacy has been investigated in infants is through *behavior-based contingency perception* (or simply, contingency perception). The general design of a contingency-perception experiment is to place an infant in a situation where perceptually salient events occur (e.g., images appear on a screen, or sounds are played on a speaker), and to "link" the occurrence of the events to the infant's ongoing behaviors. A well-known example, developed and studied by Rovee-Collier (e.g., Rovee-Collier and Sullivan 1980), is the "mobile" paradigm, in which an infant is placed in a crib, and a

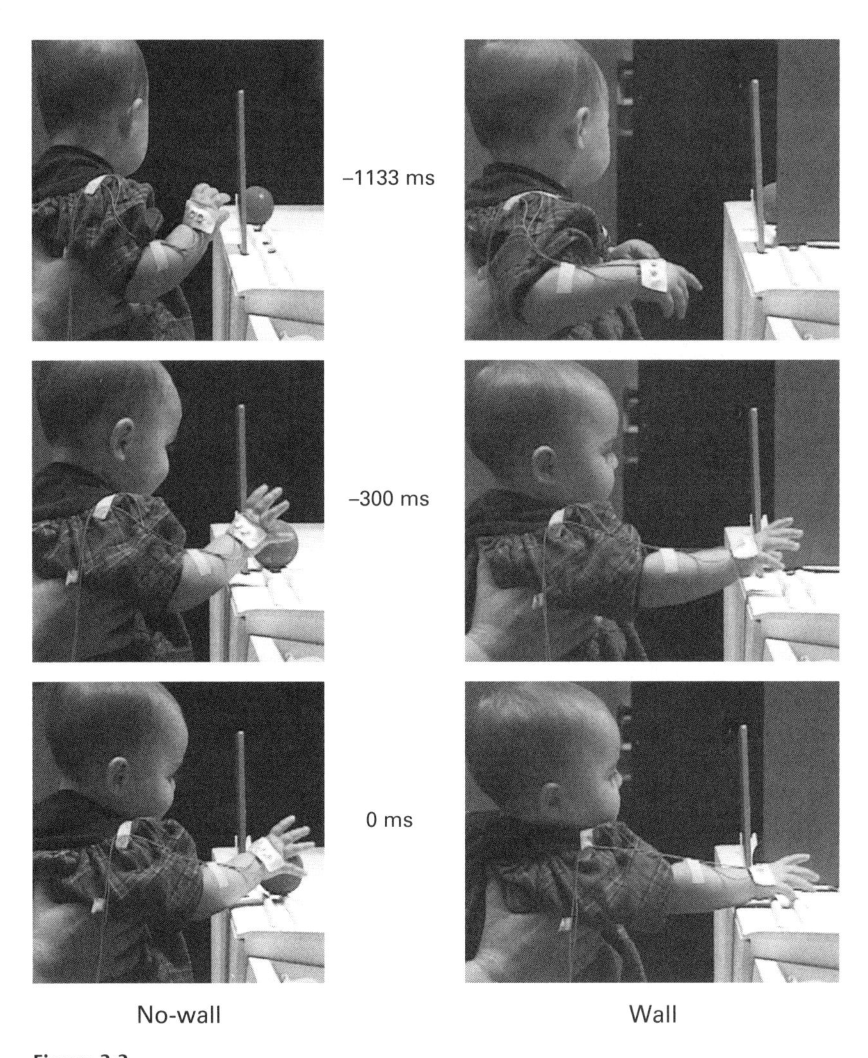

Figure 3.3
Experiment on predictive tracking task with eye and reaching predictive movements. From Berthier et al. 2001. Reprinted with permission from Wiley.

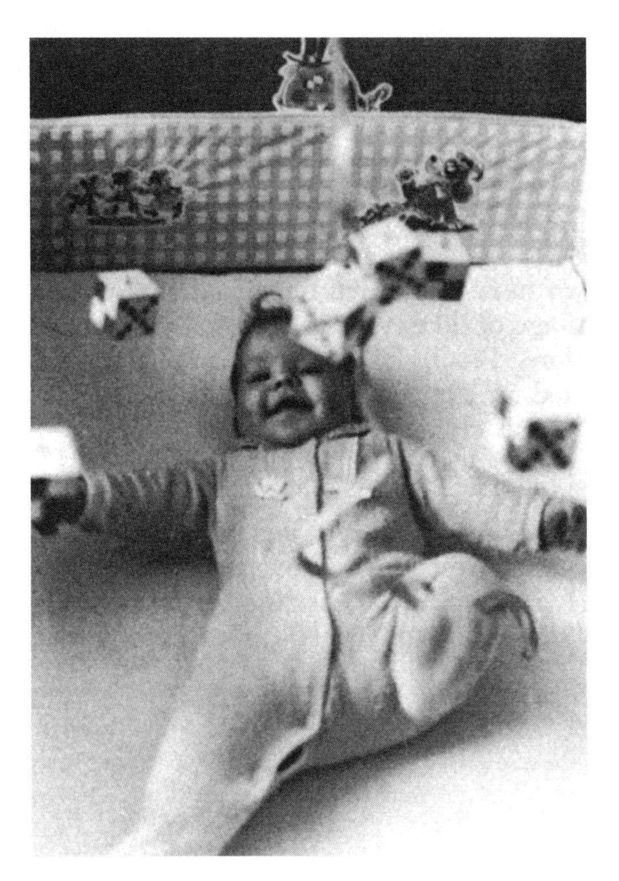

Figure 3.4
Rovee-Collier and Sullivan's (1980) "mobile" paradigm, in which an infant is placed in a crib, and a ribbon is used to connect the infant's leg to a mobile hanging overhead. Reprinted with permission from Wiley.

ribbon is used to connect the infant's leg to a mobile hanging overhead (see figure 3.4). In behavioral terms, the mobile provides a form of *conjugate reinforcement*: it is reinforcing, because the infant quickly learns to kick the leg attached to the mobile (while the other leg is typically still), and it is a form of conjugate reinforcement because the reinforcer (presumably, sight of the mobile moving) occurs continuously and is proportional to the amount of leg movement.

Over numerous studies, Rovee-Collier and colleagues have systematically investigated learning in the mobile paradigm. Interestingly, the age group that has received the most attention with this paradigm is three-month-olds. A key finding from this work is that after learning to control the mobile, three-month-olds retain a memory of the experience for several days (e.g., Rovee-Collier and Sullivan 1980). While infants typically forget after a week, their memory can be extended by as much as four weeks if they are briefly exposed to a moving mobile without actually controlling it (Rovee-Collier et al. 1980). An important feature of the paradigm—though one that has not been systematically studied—is that infants appear to enjoy controlling the mobile, often cooing and smiling while they kick their legs (see figure 3.4).

A related paradigm for investigating contingency perception involves presenting the infant with a video display of her legs on one video screen, while a second screen displays the legs of another infant (e.g., Bahrick and Watson 1985). In order to discriminate between the two displays (i.e., self vs. other), the infant must be able to match the proprioceptive feedback generated by moving her legs to the corresponding visual display. Two interesting questions suggested by this paradigm are: at what age do infants discriminate between the two displays, and which do they prefer? Bahrick and Watson (ibid.) first studied five-month-olds, and found that at this age infants look significantly longer at the non-self video display (i.e., the display of another infant's legs). When the experiment was repeated with three-month-olds, a bimodal distribution emerged: roughly half of the infants preferred viewing their own legs, while the other half looked longer at the non-self display. Gergely and Watson (1999) later proposed that during the first three months of life, infants focus on how their body movements create sensory consequences (what Piaget called *primary circular reactions*). This focus leads to a preference for perfect contingencies. Gergely and Watson hypothesize that after age three months, events that are perfectly contingent become aversive, causing infants to shift their preference toward events that are highly (but not perfectly) contingent, such as the social responses of caretakers (see Kaplan and Oudeyer 2007, for an IM model that captures an analogous transition). An intriguing implication of this account is that infants at risk for autism spectrum disorders (ASDs) may not make the same shift. According to Gergely and Watson, this preference for perfect contingencies in autistic children may then manifest in repetitive behaviors, self-stimulation, and an aversion to changes in the everyday environment or routine.

Finally, a somewhat different perspective on competence-based IM in infants and children is the study of *spontaneous play behavior*. Table 3.1 outlines a series of increasingly complex levels of play, which emerge over infancy and early childhood (Bjorklund and Pellegrini 2002; Smilanksy 1968). These levels do not represent discrete stages, but rather, overlapping developmental periods during which various styles of play behavior begin to appear. The first level is *functional play* (also called sensorimotor or locomotor play), which is typically expressed as gross motor activity, such as running, climbing, digging, and other similar whole-body actions. Functional play may also involve object manipulation, such as swinging on a swing, kicking a ball, or dropping blocks. This level of play begins to appear during the first year, and continues through early childhood and becomes increasingly social. It is noteworthy that this type of play

Table 3.1
Levels of play behavior (adapted from Smilansky 1968)

Age	Type of play	Example
0–2 years	Functional or sensorimotor	Running, climbing a ladder, digging in the sand
1–4 years	Constructive	Stacking blocks, connecting pieces of a train track, drawing with a crayon
2–6 years	Pretend or symbolic	Flying in an airplane, cooking breakfast, commanding a pirate ship
6+ years	Games with rules	Kickball, four-square, checkers, hopscotch

behavior (in particular, so-called rough-and-tumble play) is seen in both humans and nonhumans, and likely has an evolutionary basis (e.g., Siviy and Panksepp 2011). The second level is *constructive* (or object) *play*, which emerges during the second year and also continues to develop into early childhood. Constructive play typically involves the use of fine motor skills and the manipulation of one or more objects, with the implicit goal of building or creating. Examples include stacking blocks, arranging the pieces of a puzzle, or painting with a paintbrush.

Constructive play overlaps extensively with *pretend play*, which emerges around age two years and becomes one of the dominant forms of play during early childhood. During pretend play (also called fantasy, symbolic, or dramatic play), children use their imagination to transform their actual environment into something make-believe, such as pretending to be on a rocket or pirate ship, or imagining that they are making a meal in a real kitchen. While early forms of pretend play tend to be solitary, by age four it often becomes socio-dramatic, and not only includes elements of fantasy and pretend, but also cooperation among multiple "players" in order to coordinate and maintain the play theme. The trend toward social interaction and cooperation culminates in formalized *games with rules*, like baseball and checkers, which appear near age six, as children begin to transition into a conventional school setting.

3.3 Intrinsically Motivated Agents and Robots

We now shift to the topic of intrinsically motivated machines, artificial agents, and robots. We begin this section by first providing a conceptual overview of how IM is approached as a computational problem, and in particular, we describe a set of basic architectures for simulating knowledge-based and competence-based IM. We then survey how these architectures have been implemented in a variety of simulations, models, and real-world robotic platforms.

3.3.1 A Computational Framework for IM

Most research to date on IM within the developmental robotics approach has tended to focus on the use of reinforcement learning (RL). As Barto, Singh, and Chentanez (2004) note, RL provides a versatile framework for studying not only how the environment (i.e., "external" motivations) influences behavior, but also how internal or intrinsic factors influence behavior as well. We refer the interested reader to Sutton and Barto's (1998) comprehensive introduction to RL, and describe here the basic elements that are most relevant to modeling IM.

The core components of an RL model are an autonomous agent, an environment, a set of possible sensory states that can be experienced within the environment, and a function (i.e., policy) that maps each state to a set of possible actions. Figure 3.5a illustrates the relations between these elements: the agent senses its environment, uses the policy to select an action, and performs the selected action. The environment then provides two forms of feedback: first, it delivers a reward signal, which the agent uses to modify its policy so as to maximize future expected reward (i.e., the particular action chosen is incremented or decremented in value), and second, the sensory signal is updated. In the mobile-robot domain, the agent would be the mobile robot, the environment might be a cluttered office space, possible actions would include rotation of the robot's wheels (e.g., movement forward, backward, and turning left or right), and the sensory signal could include readings over an array of infrared sensors and the state of a "gripper" or touch sensor. Imagine in this example that the engineer would like the robot to return to its charger so that it can recharge its power supply. A simple reward function might be to give the robot a reward of 0 after each action (e.g., "step" in one of eight directions), and otherwise a 1 when it reaches the charger.

As figure 3.5a highlights, because the reward signal originates in the environment, the agent is externally motivated. In contrast, imagine instead that the reward signal occurs within the agent itself. Figure 3.5b illustrates such a scenario, which is almost identical to the previous configuration, with the exception that now the environment is limited to providing only the sensory signal, while an internal motivation system (within the agent) provides the reward signal. In this case, the agent is now internally motivated.

However, it is important not to confuse *internally* with *intrinsically* motivated behavior. In particular, when the mobile robot just described computes its own reward, it is internally motivated. Nevertheless, because its goal is to maintain a charged battery, it is also extrinsically motivated, specifically because successful behavior (i.e., reaching the charger) directly benefits the robot. On the other hand, when the robot wanders the workspace and generates an internal reward each time it succeeds, for example, at predicting which areas of the workspace have not previously been visited, it is intrinsically motivated. This is because the reward signal is not based on satisfying a homeostatic

a

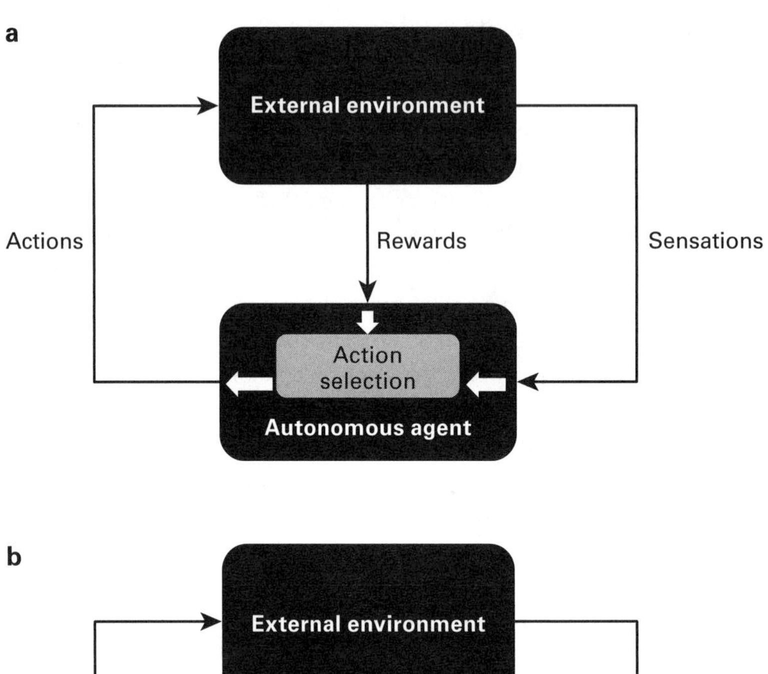

b

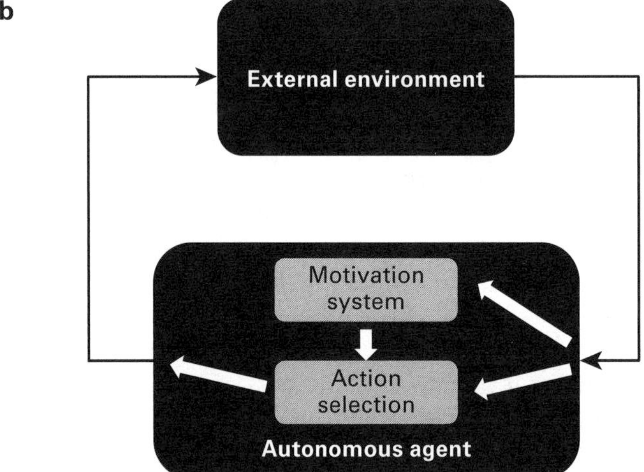

Figure 3.5

The two alternative external/internal motivation systems proposed by Oudeyer and Kaplan (2007): (a) the environment provides the reward signal and the agent is externally motivated, and (b) the agent has internal motivation system and provides its own reward signal, and the environment only provides the sensory signal. Figure courtesy of Pierre-Yves Oudeyer.

"need," but rather is based on the flow of information over time, and in particular, on the robot's current level of knowledge and experience.

Working within this computational framework, Oudeyer and Kaplan (2007) describe a systematic taxonomy of possible architectures for simulating IM. The basic scenario captured by the taxonomy is based on a single autonomous agent that interacts with its environment over time, and which receives an intrinsically motivated reward signal after each action in addition to having its sensors updated. For simplicity and brevity, we focus here on the case where the sensory or state space is treated as discrete, and the transitions between the states after each action are deterministic. Nevertheless, it should be stressed that each of the reward functions that we describe can also be generalized and applied to continuous spaces as well as partially observable or stochastic environments, or both.

To begin, we define the event e^k as the k th sample from the set of all possible events E. Typically, e^k is assumed to be a vector of sensor readings (as described above) at a particular moment, but is otherwise general enough to accommodate the specific cognitive architecture being employed. Next, we define $r(e^k, t)$ as the discrete scalar reward obtained by the agent at time t as it observes or experiences event e^k. Given these basic constructs, we can then explore each of the IM learning mechanisms described thus far.

Knowledge-Based IM: Novelty

A simple and straightforward method for motivating the agent to seek out novel events is to first provide it with a function $P(e^k, t)$ that returns the estimated probability of an event e^k being observed at time t. One strategy is to assume that this function is initially unknown (e.g., the distribution of events is uniform; that is, all events are equally probable), and that the agent uses its experience to tune P as a model of the environment is gradually acquired. Next, given P and the constant C,

$$r(e^k, t) = C \cdot (1 - P(e^k, t)) \tag{3.1}$$

results in a reward that increases proportionally as the probability of the given event decreases. When this reward function is embedded within a conventional RL problem (i.e., in which the goal is to maximize the cumulative sum of rewards), the agent should then preferentially choose actions that lead to infrequent or low-probability events. One problem with this formulation, however, is that events that are maximally novel or improbable are also maximally rewarding. As we saw earlier, this may be at odds with both the theoretical view and ample empirical data, both suggesting that *moderately novel events* are maximally interesting or rewarding. We return to this issue later.

Oudeyer and Kaplan (2007) describe reward function (3.1) as *uncertainty motivation*: the agent is intrinsically motivated to seek novel or unfamiliar events. An alternative formulation is *information-gain motivation*, in which the agent is instead rewarded for observing events that increase its knowledge. In this case, we first define $H(E,t)$ as the total entropy over all events E at time t. Thus:

$$H(E,t) = -\sum_{e^k \in E} P(e^k,t)\ln(P(e^k,t)) \tag{3.2}$$

where $H(E)$ characterizes the shape of the probability distribution $P(e^k)$. Extending this measure of information, the reward function is then defined as

$$r(e^k,t) = C\cdot(H(E,t) - H(E,t+1)) \tag{3.3}$$

that is, the agent is rewarded when consecutive events result in a decrease in entropy. In contrast to uncertainty motivation, which links IM to the absolute probabilities of events in the environment, a potential strength of information-gain motivation is that it allows IM to vary as a function of the agent's knowledge state.

Knowledge-Based IM: Prediction

Rather than learning a static world-model P, the agent can instead actively learn to predict future states. Differences between predicted or expected events and those that actually occur can then provide a basis for prediction-based IM. A key element of this formulation is the function $SM(t)$, which represents the current sensorimotor context at time t and encodes a generalized notion of *event* that includes contextual information, e.g., the robot's current camera image and IR sensor readings, the state of its motors, and so on. We use here the notation $SM(\rightarrow t)$, which incorporates not only the state information at time t, but also information from past contexts, as needed. Next, Π is a prediction function that uses $SM(\rightarrow t)$ to generate a prediction of the event $\tilde{e}^k$ estimated or expected to occur on the next timestep:

$$\Pi(SM(\rightarrow t)) = \tilde{e}^k(t+1) \tag{3.4}$$

Given the prediction function Π, the prediction error $E_r(t)$ is then defined as

$$E_r(t) = \|\tilde{e}^k(t+1) - e^k(t+1)\| \tag{3.5}$$

that is, as the difference between the expected and observed events at time $t+1$. Finally, a particularly compact reward function defines the reward r as the product of a constant C and the prediction error E_r at time t:

$$r(SM(\rightarrow t)) = C\cdot E_r(t) \tag{3.6}$$

Interestingly, Oudeyer and Kaplan (2007) refer to reward function (3.6) as *predictive-novelty motivation*. In this case, the agent is rewarded for seeking out events that it predicts "badly," that is, where prediction errors are greatest. Note, however, that like uncertainty motivation, this formulation also suffers from the fact that reward increases monotonically with novelty. One way to address this limitation is by assuming a "threshold" of moderate novelty E_r^{σ} that maps to maximal reward, and around which all other prediction errors are less rewarding:

$$r(SM(\rightarrow t)) = C_1 \cdot e^{-C_2 \cdot \left\| E_r(t) - E_r^{\sigma} \right\|^2} \tag{3.7}$$

An alternative to (3.6) proposed by Schmidhuber (1991) is to reward improvements between consecutive predictions $E_r(t)$ and $E_r'(t)$:

$$r(SM(\rightarrow t)) = E_r(t) - E_r'(t) \tag{3.8}$$

where

$$E_r'(t) = \left\| \Pi'(SM(\rightarrow t)) - e^k(t+1) \right\| \tag{3.9}$$

Thus (8) compares two predictions made with respect to time t. The first prediction $E_r(t)$ is made before $e^k(t+1)$ is observed, while the second $E_r'(t)$ is made after the observation and the prediction function has been updated to Π' and becomes the new prediction model.

Competence-Based IM

As we noted earlier, a unique feature of competence-based IM is its focus on skills, rather than on environmental states or knowledge of the environment. As a consequence, computational approaches to modeling competence-based IM also differ from those used to simulate knowledge-based IM. An important component of this approach is the notion of a goal g_k, which is one of several k goals or options. A related concept is that goal-directed behaviors occur in discrete episodes, which have a finite duration t_g (i.e., the budget of time allotted to achieve goal g), and may include as a low-level building block various methods for learning forward and inverse models and planning strategies to use them (e.g., see Baranes and Oudeyer 2013, for discussion of such hierarchical exploration and learning architectures). Finally, the function l_a computes the difference between the expected goal and the observed result:

$$l_a(g_k, t_g) = \left\| \widetilde{g_k(t_g)} - g_k(t_g) \right\| \tag{3.10}$$

where l_a denotes the level of (mis)achievement of the intended goal. Given these components, a reward function can be designed that favors large differences between attempted and achieved goals:

$$r(SM(\rightarrow t), g_k, t_g) = C \cdot l_a(g_k, t_g) \quad (3.11)$$

Oudeyer and Kaplan (2007) give reward function (3.11) the amusing description of *maximizing-incompetence motivation*. Indeed, such a function rewards the agent for selecting goals that are well beyond its skill level (and then failing miserably to accomplish them!).

To address this problem, a strategy similar to the one employed in reward functions (3.3) or (3.8) can be used, in which consecutive attempts are compared. Oudeyer and Kaplan (2007) refer to this as *maximizing-competence progress*, as it rewards the agent for subsequent goal-directed behaviors that improve over previous attempts:

$$r(SM(\rightarrow t), g_k, t_g) = C \cdot (l_a(g_k, t_g - \Theta) - l_a(g_k, t_g)) \quad (3.12)$$

where $t_g - \Theta$ indexes the previous episode in which g_k was attempted.

3.3.2 Knowledge-Based IM: Novelty

The set of architectures in section 3.3.1 represents an idealized abstraction of IM at the computational level. In particular, it is important to note that they have not yet been systematically evaluated or compared. Nevertheless, in recent years several researchers have begun to implement these and other related architectures, in both simulation and on real-world robot platforms. We briefly highlight here recent work from each of the classes of IM, beginning with novelty-based IM.

As we noted in section 3.2.1, novelty-seeking behavior can be divided into exploration and novelty detection. A model that integrates both of these components is proposed by Vieira-Neto and Nehmzow (2007), who investigate visual exploration and habituation in a mobile robot. First, the robot explores its environment by implementing a basic obstacle-avoidance strategy. As it wanders the environment, the robot acquires a salience map of the visual input. Next, highly salient locations on the map are selected for additional visual analysis. Specifically, each of these regions is processed through a visual novelty filter that takes as input the color values in the corresponding region and projects them onto a self-organizing map (SOM) of features. A habituation mechanism is implemented in the model by gradually lowering the connection weights of nodes in the SOM that are continuously active.

Figure 3.6a presents a snapshot of the robot's input as it begins to explore the environment: the numbered patches correspond to the salient locations (with 0 marking the most salient location) and the circles around each location represent the output of

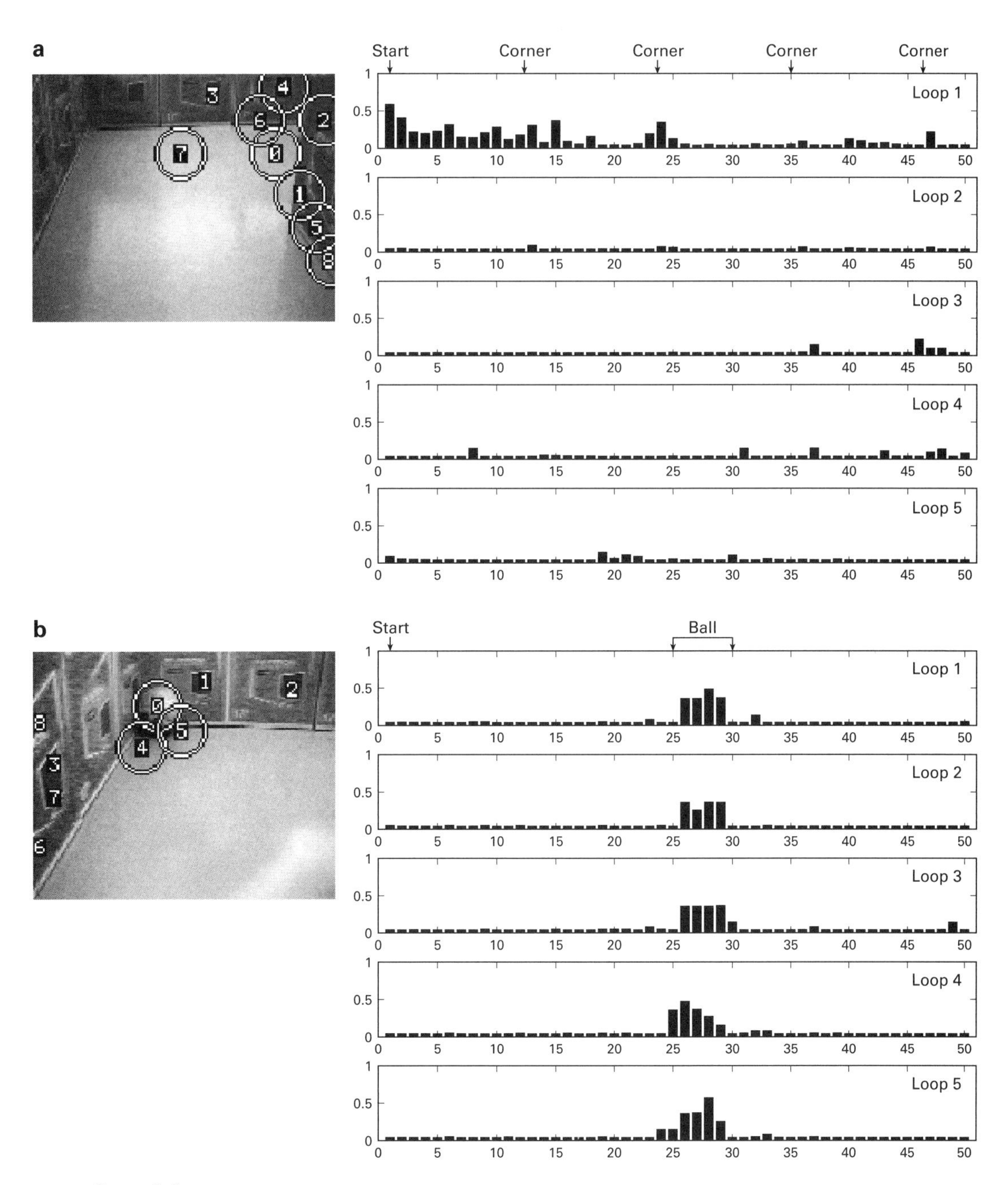

Figure 3.6
Results from Vieira-Neto and Nehmzow (2007) experiment on visual exploration and habituation in a mobile robot. Snapshot of the robot's input as it begins to explore the environment (a). Robot's response to a new object (b). Reprinted with permission from Springer.

the novelty filter (a circle identifies locations with novelty levels that exceed a threshold). As the plot on the right illustrates, the output of the novelty filter is initially large, but slowly falls over time. By the fifth pass through the environment, the robot has become familiar with the sides and floor of the arena, and the output of the novelty filter is consistently low.

Figure 3.6b illustrates how the robot responds when a new object—a red ball in the second corner—is introduced into the environment. When the robot reaches the red ball, three maxima in the salience map occur at the ball's location (i.e., labels 0, 4, and 5), which results in further processing of this location by the novelty filter. Consequently, the output of the novelty filter abruptly increases. Note that in this implementation, the habituation mechanism is deactivated during the novelty phase, so that the novelty of the ball can be estimated over several encounters (i.e., Loops 2–5). Although the model is not explicitly designed to capture the process of visual exploration and learning in a human infant or child, it not only provides a parsimonious explanation for how a basic behavioral mechanism such as obstacle avoidance can drive visual exploration, but also demonstrates how novel objects or features in the environment can be detected and selected for further visual processing.

A related approach is proposed by Huang and Weng (2002), who investigate novelty and habituation in the SAIL (Self-organizing, Autonomous, Incremental Learner) mobile-robot platform (see figure 3.7a). In contrast to the Vieira-Neto and Nehmzow (2007) model, which uses visual salience and color histograms to determine novelty, the Huang and Weng model defines novelty as the difference between expected and observed sensory states. In addition, the model combines sensory signals across several modalities (i.e., visual, auditory, and tactile). Another important feature of the Huang and Weng model is that it implements novelty detection and habituation within an RL framework. In particular, the model includes both intrinsic and extrinsic reinforcement signals: the intrinsic training signal is provided by sensory novelty, while the external signal is provided by a teacher who occasionally rewards or punishes the robot by pushing either the "good" or "bad" buttons on the robot.

Figure 3.7b illustrates the model's cognitive architecture. Sensory input is propagated into the IHDR (incremental hierarchical decision regression) tree, which categorizes the current sensory state, in addition to updating the model's estimate of the current sensory context (which is stored as a context prototype). The model then uses the combined context + state representation to select an action, which is evaluated by the value system. Next, the Q-learning algorithm is used to update the context prototypes, which combine sensory data, actions, and Q-values.

$$r(t) = \alpha p(t) + \beta r(t) + (1 - \alpha - \beta) n(t) \tag{3.13}$$

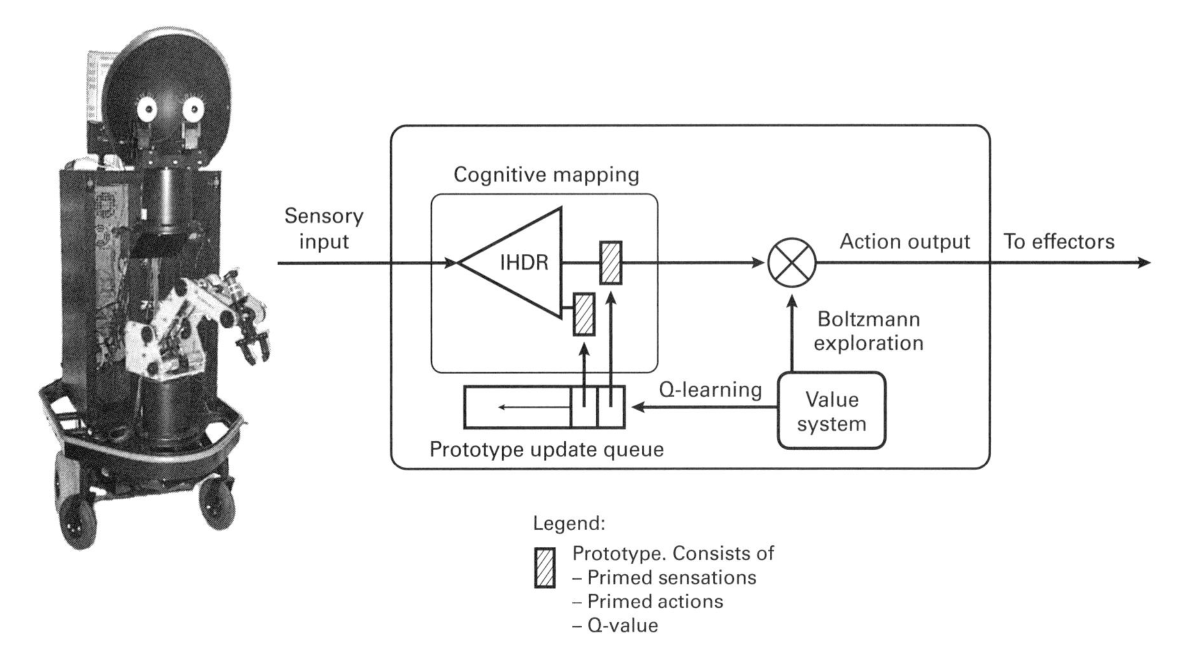

Figure 3.7
The SAIL robot (a) and the model's cognitive architecture (b) (Huang and Weng 2002). Courtesy of John Weng.

Equation (3.13) presents the reward function implemented in the Huang and Weng model. Reward on each timestep is the sum of three components: punishment p and positive reinforcement r, which are generated by the teacher, and novelty n, which is generated by the robot intrinsically. Note that the values α and β are adjustable parameters that weight the relative contributions of each of the three reward signals. In addition, Huang and Weng assume that punishment has a stronger effect on behavior than reinforcement, and that each of the two external reward signals is greater than the intrinsic reward (i.e., $\alpha > \beta > 1 - \alpha - \beta$).

Habituation in the model is an emergent feature: because novelty is determined relative to the robot's experience, initially all sensory inputs are novel, and as a result, the robot indiscriminately explores its environment. After several minutes, however, the discrepancy between expected and observed sensory inputs begins to decrease, and as a result, the robot orients selectively toward visual stimuli that have been infrequently experienced or are visually complex (e.g., a Mickey Mouse toy). Thus, the robot gradually habituates to well-learned objects, and strategically selects actions that will lead to novel sensory input. In addition, Huang and Weng also show that this orienting response toward novel stimuli can be selectively shaped or overridden by the feedback provided by the teacher.

Hiolle and Cañamero (2008) investigate a related model, in which a Sony AIBO robot visually explores its environment in the presence of its caretaker. An interesting twist in the Hiolle and Cañamero model is that exploration is explicitly driven by an arousal mechanism, which can be modulated not only by looking at the caretaker, but also by when the caretaker touches the robot. Like the Huang and Weng model, the robot in the Hiolle and Cañamero study learns an internal model of its environment (represented by a Kohonen self-organizing map). During visual exploration, the robot compares visual input with its internal representation, and its looking behavior is guided by a three-choice decision rule: (1) for low levels of novelty, the robot turns its head away from the current direction, (2) for moderate levels of novelty the robot remains still (and continues to encode its current experience), and (3) for high levels of novelty, the robot makes a barking sound and searches for the caretaker. In the latter case, sight of the caretaker lowers arousal a moderate amount, while touch by the caretaker lowers the arousal level significantly.

During training, Hiolle and Cañamero compare the performance of the robot in two learning contexts. In the high-care context, the caretaker is continually present and actively soothes the robot when it expresses distress. In the low-care context, meanwhile, the caretaker is only sporadically present and does not actively respond to the robot. Two key findings emerge from this comparison. First, in both contexts, the robot learns to adapt its visual exploratory behavior as a function of the availability of the caretaker. Thus, in the high-care context, when under high arousal the robot frequently barks and searches for the caretaker. In contrast, in the low-care context, the robot instead learns to shift its focus away from visual experiences that are highly novel and therefore produce distress. Second, the robot also develops a more robust internal model of its environment in the high-care context, resulting in lower average long-term levels of novelty. Taken together, these findings not only highlight the role of the caretaker during visual exploration, but more important, they also illustrate the relation between novelty seeking and arousal modulation and social interaction.

Finally, a fourth approach is proposed by Marshall, Blank, and Meeden (2004), who also define novelty as a function of expected and observed sensory inputs. Indeed, the Marshall, Blank, and Meeden model explicitly incorporates both novelty-based and prediction-based computations. In particular, they simulate a stationary robot that is situated in the center of circular arena and watches a second robot that travels from one side of the arena to the other. As the stationary robot observes its environment and views the moving robot, it generates a prediction of the upcoming visual input. This anticipated input is then compared with the observed input, and used to derive a prediction error. Similar to equation (3.8), novelty is then defined as the change in prediction error over two consecutive timesteps. Two key findings emerge from the

Marshall, Blank, and Meeden model. First, in the process of orienting toward visual stimuli that are highly novel, the stationary robot learns to track the motion of the moving robot "for free" (i.e., without any explicit extrinsic reward). Second, and perhaps more important, after achieving high accuracy in learning to track the moving robot, the stationary robot gradually spends less time orienting toward it. Thus, as in the Huang and Weng model, a habituation mechanism emerges as a consequence of learning to detect novel events.

3.3.3 Knowledge-Based IM: Prediction

As the previous example illustrates, there is some overlap between novelty-based and prediction-based models of IM. Thus, while we focus in this section on prediction-based models, it is important to note that these models not only share a number of features with the novelty-based approach, but in some cases, they also combine elements of prediction and novelty within the same computational framework.

An ambitious and intriguing model of prediction-based IM is proposed by Schmidhuber (1991, 2013), who argues that the cognitive mechanisms that underlie IM are responsible for a diverse range of behaviors, including not only novelty seeking, exploration, and curiosity, but also problem solving, artistic and musical expression (or more generally, aesthetic experience), humor, and scientific discovery. Schmidhuber's "Formal Theory of Creativity" is comprised of four major components:

1. *A world model.* The purpose of the world model is to encompass and represent the totality of the agent's experience. In effect, the model encodes the full history of the agent, including the actions taken and the sensory states observed. At the functional level, the world model can be viewed as a *predictor* that compresses raw data into a compact form, by detecting patterns or regularities in experience.
2. *A learning algorithm.* Over time, the world model improves in its ability to compress the agent's history of experience. This improvement is the result of a learning algorithm that identifies novel sensory data or events, which when encoded, increase the compression rate of the data stored in the world model.
3. *Intrinsic reward.* Improvements in the world model (i.e., increased compression) provide a valuable learning signal. In particular, Schmidhuber proposes that novelty or surprise can be defined as the magnitude of improvement in the world model. Thus, each time the learning algorithm detects an experience that increases compression of the model, the action that produced the experience is rewarded.
4. *Controller.* The final component is the controller, which learns on the basis of feedback from the intrinsic reward signal to select actions that produce novel experiences. The controller therefore functions as an exploratory mechanism that generates new data for the world model, and in particular, is intrinsically motivated to produce experiences that increase the predictive power of the world model.

An important feature of Schmidhuber's theory, which distinguishes it from other prediction-based models, is that rather than generating an IM reward signal on the basis of the prediction error itself, the reward is based instead on changes in the prediction error over time, that is, on *learning progress* (see section 3.1.1). For discrete-time models, learning progress can be measured as the difference in consecutive prediction errors (see equation 3.8). First, given the current state and planned action, the world model generates a prediction of the next state, which is then observed and used to compute the initial prediction error. The world model is updated on the basis of this error, and then, using the original state and action, a subsequent prediction is generated by the updated model. A prediction error is again calculated, and this error is subtracted from the initial one. A positive change in the prediction error over consecutive predictions corresponds to learning improvement, and generates an IM signal that increases the value of the action selected by the controller. In psychological terms, this action is rewarded because it leads to an improvement in the predictor. Alternatively, negative or null changes in prediction error lead to a decrease in the value associated with the corresponding action.

Oudeyer and his collaborators have employed a related approach in a series of computational architectures, beginning with Intelligent Adaptive Curiosity (IAC; Oudeyer, Kaplan, and Hafner 2007; Oudeyer et al. 2005; Gottlieb et al. 2013). These architectures were specifically designed to study how the IM system could be scaled to allow efficient life-long autonomous learning and development in real robots with continuous high-dimensional robotic sensorimotor spaces. In particular, the IAC architecture and its implications were studied in a series of experiments, called the *Playground Experiments* (Oudeyer and Kaplan 2006; Oudeyer, Kaplan, and Hafner 2007). Figure 3.8a illustrates the cognitive architecture employed by the IAC. Similar to the Schmidhuber model, prediction learning plays a central role in the IAC architecture. In particular, there are two specific modules in the model that predict future states. First, the "Prediction learner" M is a machine that learns a forward model. The forward model receives as input the current sensory state, context, and action, and generates a prediction of the sensory consequences of the planned action. An error feedback signal is provided on the difference between predicted and observed consequences, and allows M to update the forward model. Second, the "Metacognitive module" metaM receives the same input as M, but instead of generating a prediction of the sensory consequences, metaM learns a metamodel that allows it to predict how much the errors of the lower-level forward model will decrease in local regions of the sensorimotor space, in other words, modeling learning progress locally. A related challenge for this approach is the task of partitioning this sensorimotor space into a set of well-defined regions (Lee et al. 2009; Oudeyer, Kaplan, and Hafner 2007).

In order to evaluate the IAC architecture in a physical implementation, the Playground Experiments were developed (Oudeyer and Kaplan 2006; Oudeyer, Kaplan, and

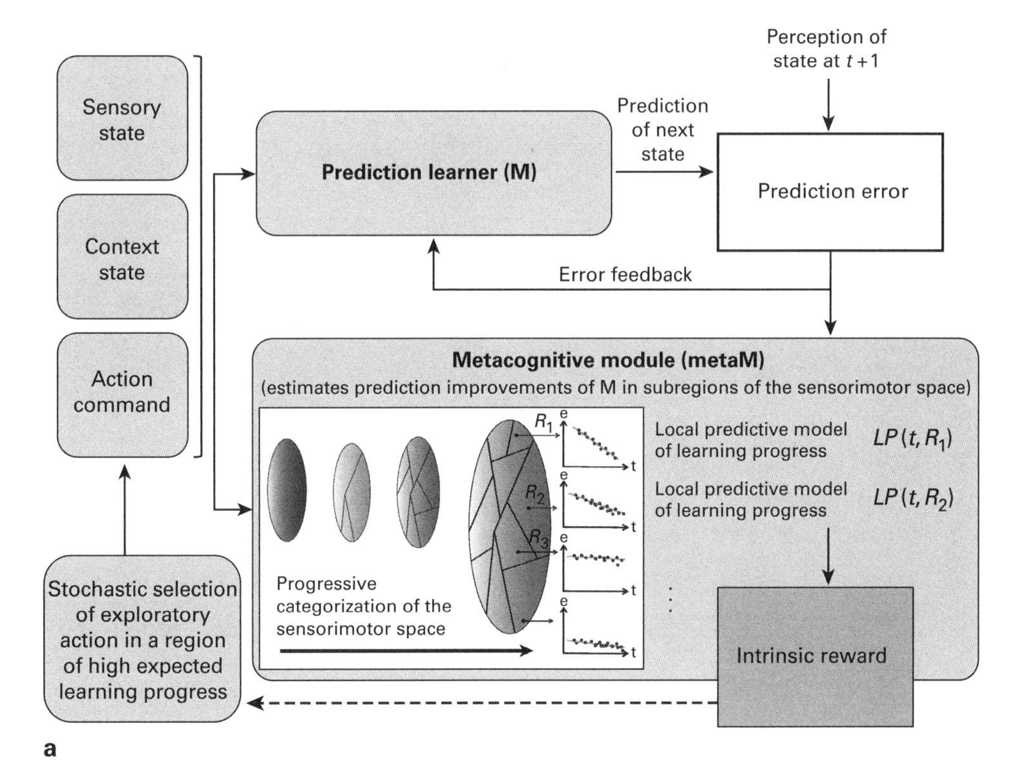

Figure 3.8
The modeling architecture (a) and the robot platform used in the Playground Experiments (b) (Gottlieb et al. 2013). Figure courtesy of Pierre-Yves Oudeyer; reprinted with permission from Elsevier.

Hafner 2007). During the experiment, a Sony AIBO robot is placed on an infant play mat and presented with a set of nearby objects, as well as an "adult" robot caretaker (see figure 3.8b). The robot is equipped with four kinds of motor primitives with parameters denoted by several continuous numbers and which can be combined, thus forming an infinite set of possible actions: (1) turning the head in various directions; (2) opening and closing the mouth while crouching with varying strength and timing; (3) rocking the leg with various angles and speed; (4) vocalizing with various pitches and lengths. Similarly, several kinds of sensory primitives allow the robot to detect visual movement, salient visual properties, proprioceptive touch in the mouth, and pitch and length of perceived sounds. For the robot, these motor and sensory primitives initially are black boxes and it has no knowledge about their semantics, effects or relations. The IAC architecture is then used to drive the robot's exploration and learning purely by curiosity, that is, by the exploration of its own learning progress. The nearby objects include an elephant (which can be bitten or "grasped" by the mouth), a hanging toy (which can be "bashed" or pushed with the leg) and an adult robot "caretaker" preprogrammed to imitate the learning robot when the latter looks at the adult while vocalizing at the same time.

A key finding from the Playground Experiments is the self-organization of structured developmental trajectories, where the robot explores objects and actions in a progressively more complex stage-like manner, while acquiring autonomously diverse affordances and skills that can be reused later on. As a result of a series of runs of such experiments, the following developmental sequence is typically observed:

1. The robot achieves unorganized body babbling.
2. After learning a first rough model and metamodel, the robot stops combining motor primitives, exploring them one by one, but rather explores each primitive in a random manner.
3. The robot begins to experiment with actions toward areas of its environment where the external observer knows there are objects (the robot is not provided with a representation of the concept of *object*), but in a nonaffordant manner (e.g., it vocalizes at the nonresponding elephant or bashes the adult robot, which is too far away to be touched).
4. The robot explores affordance experiments: it focuses on grasping movements with the elephant, then shifts to bashing movements with the hanging toy, and finally shifts to exploring vocalizations to imitate the adult robot.

Two important aspects of this sequence should be noted. First, it shows how an IM system can drive a robot to learn autonomously a variety of affordances and skills (see Baranes and Oudeyer 2009 for the reusability for control) for which no engineer provided in advance any specific reward functions. Second, the observed process spontaneously generates three properties of infant development so far mostly unexplained:

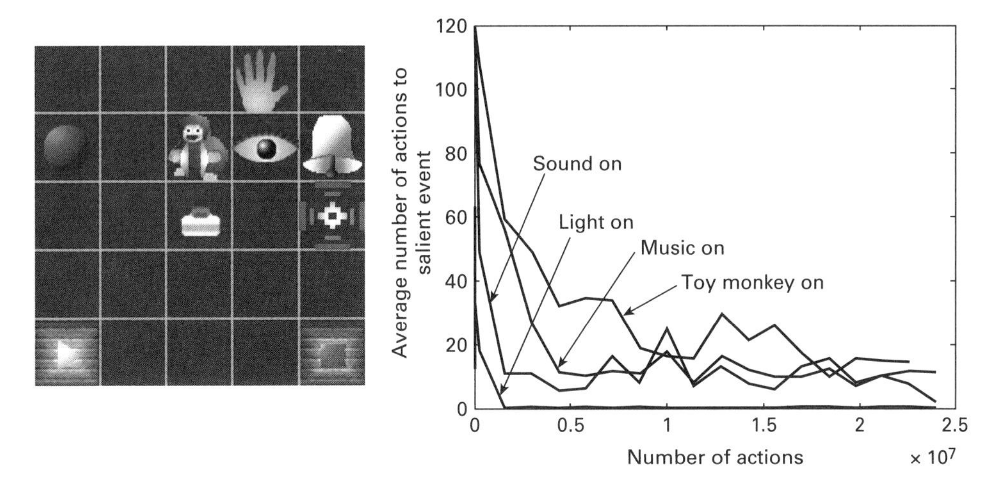

Figure 3.9
The toy-world environment (a) and pattern of skill learning (b) in Barto, Singh, and Chentanez 2004. Figure courtesy of Andy Barto.

(1) qualitatively different and more complex behaviors and capabilities appear along with time (i.e., stage-like development); (2) a wide range of developmental trajectories emerge, with both shared and unique temporal patterns; and (3) communication and social interaction emerge autonomously (i.e., without explicit direction from the model builder).

Barto, Singh, and Chentanez (2004) propose a third model that investigates prediction-based IM. An important feature of the Barto, Singh, and Chentanez model is that it employs the *options framework*, which includes both *primitive actions* that occur once per timestep, as well as *options*, which are composed of multiple primitive actions and occur on a longer (variable) timescale. Figure 3.9 illustrates the task domain for the model: a 5 × 5 grid, in which the agent can move its hand or eye, and also mark locations with the t-shaped (crosshair) icon. Each object in the grid produces a salient response when manipulated by hand. For example, one of the blocks plays music when touched, while touching the other block turns the music off. Some objects, however, only respond when an appropriate sequence of actions is performed. The bell, for instance, rings when the ball is rolled toward it. Thus, some objects can be explored by producing a primitive action, while others only respond when the correct option is performed.

The reward signal for intrinsic motivation in the model is linked to the outcomes of the options. In particular, when an option is selected, the model predicts the outcome of the final action in the option. Intrinsic reward is then proportional to the magnitude of the prediction error. During early learning, the agent occasionally “unintentionally” triggers a salient event, which produces an unexpected outcome and motivates the agent to then focus on reproducing the event. As prediction errors decline reward also diminishes, and thus the agent shifts its attention to other objects in the task space. Like Oudeyer's IAC model, the Barto, Singh, and Chentanez (2004) model also acquires a set of stable actions in a regular order: it first learns to produce simple events, such as turning on the light, which then become component skills that are integrated into options that produce more complex events (e.g., turning the monkey on, which requires a sequence of fourteen primitive actions!).

3.3.4 Competence-Based IM

The modeling work described thus far focuses on what the robot or autonomous agent learns about its environment, either by seeking out novel or unfamiliar experiences, or by predicting how its actions will transform its ongoing stream of sensory data. In this section we highlight models that emphasize the competence-based approach, in which the exploration and learning process is focused on *discovering what the robot can do.*

While robotics researchers have not yet attempted to simulate most of the developmental phenomena that we described in section 3.2, an important topic that has the potential to link the two disciplines is *contingency perception.* Recall that contingency perception begins to develop during early infancy, and is manifested in the infant's ability to detect the influence that their actions have on events in the environment. As we noted in section 3.1.3, a neural mechanism that may help account for this capacity is proposed by Redgrave and Gurney (2006), who suggest that behaviors which produce novel or unexpected sensory events are followed by a burst of dopamine from superior colliculus cells. Baldassarre (2011; Mirolli and Baldassarre 2013) proposes that these phasic bursts may serve as a learning signal that supports two related functions. First, the bursts provide a "contingency signal" that binds or links the organism's actions with the concurrent sensory consequences. Second, they provide an intrinsic reinforcement signal that rewards the corresponding action.

In order to evaluate the proposal, Fiore et al. (2008) implemented the model in a simulated robot rat that is placed in a box with two levers and a light (see figure 3.10a). The rat has a number of built-in behaviors that enable it to explore its environment, including pushing the levers and avoiding contact with walls. In this environment, pressing lever 1 causes the light to turn on for two seconds, while pressing lever 2 has no effect. Figure 3.10b illustrates a diagram of the modeling architecture that determines the robot's responses to the two levers: visual input is projected through an associative layer to the basal ganglia, which then project to the motor cortex. Transient light produces activity in the superior colliculus, which results in a dopamine (see DA in figure 3.10b) signal that is combined with an efferent copy of the motor signal and modulates the strength of the connections between the associative cortex and basal ganglia. Fiore et al. (2008) show that within twenty-five minutes of simulated time

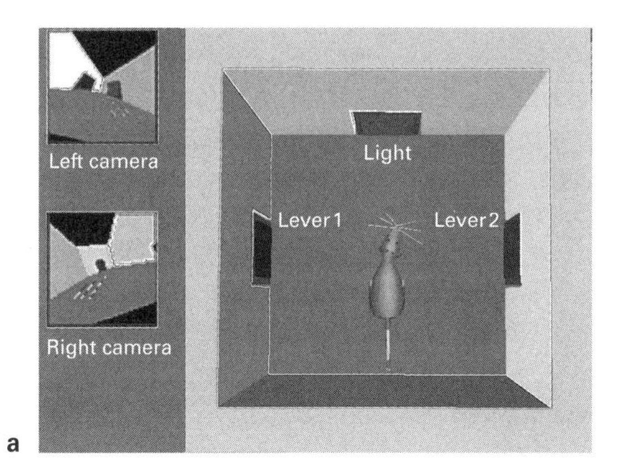

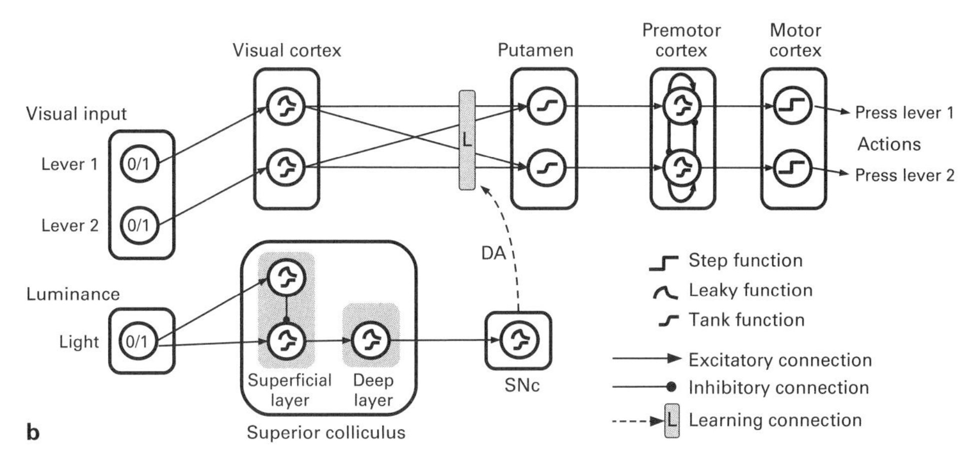

Figure 3.10
The simulated robot rat (a) and modeling architecture (b) used by Fiore et al. (2008) to investigate contingency detection and intrinsic motivation. Figure courtesy of Gianluca Baldassarre.

in the environment, the robot rat acquires a bias toward pressing lever 1 roughly four times as often as lever 2. Thus, their model not only demonstrates the feasibility of the Redgrave and Gurney learning mechanism, but it also provides a behavior-based implementation on a simulated-robot platform.

While the Fiore et al. model highlights the role of perceptually salient events from a neurophysiological perspective, a more general question concerns how a given neural signal gains the ability to function as an intrinsic reward. Schembri, Mirolli, and Baldassarre (2007) investigate this question by simulating a population of mobile robots that learn to solve a spatial navigation task. There are two important features of the model. First, the lifespan of each robot is divided into a childhood phase and an adult phase;

during childhood, the robot wanders through and explores its environment, while during adulthood it is evaluated on its ability to reach a designated target. The reward signal is intrinsically generated during childhood and extrinsically generated during adulthood. Second, the population of robots evolves over multiple generations. In particular, the robots that are most successful at reaching the target are selected to produce the next generation (variation is introduced with a mutation operator during reproduction).

Using the actor-critic architecture, Schembri, Mirolli, and Baldassarre (2007) ask whether evolution can produce an internal critic that effectively evaluates and motivates the robot's exploratory behavior during childhood. In other words, can an IM system evolve, and if so, will it result in improved performance on the navigation task during adulthood? The simulation results provide strong support for this idea. Not only do adults easily learn to solve the navigation task, but also the IM system rapidly emerges during evolution and leads to a pattern of exploratory behavior in young robots that facilitates learning in adults. The model suggests two important implications for competence-based IM. First, it illustrates how intrinsically motivated behaviors or skills—which may have no immediate benefit or value to the organism—can be exploited at a later stage of development. Second, it also shows how evolution can help establish the capacity for intrinsically motivated learning that has a measurable, though indirect, impact on long-term fitness.

Another important question raised by competence-based IM models is how they perform compared to other approaches, and in particular, relative to knowledge-based IM models. Merrick (2010) addresses this question by proposing a general neural network architecture that can accommodate both approaches. Figure 3.11a illustrates the version of the network used to model competence-based IM. First, sensory input from a Lego "crab" robot (see figure 3.11b) projects to the *observation layer*, which classifies the input. Next, the *observation layer* projects to the *error layer*, in which each of the observation values are weighted by a corresponding error weight. The activation values from this layer then project to the *action* or reinforcement learning layer, which produces a set of potential actions. Alternatively, the *error layer* can be replaced with a set of novelty units, which are used to estimate the novelty for each observation, or interest units, which are modified novelty units that respond maximally to moderate novelty. In each case, the corresponding network controls the legs of the robot; the reinforcement learning rule used to train the network varies as a function of the model's respective IM motivation system. In particular, the competence-based IM is rewarded for selecting actions that are associated with high learning errors (i.e., TD or temporal-difference errors).

Merrick (2010) compared the performance of four versions of the model: novelty, interest, and competence, and as a baseline, a model that selects actions at random. One important measure of learning is the frequency with which the model repeats a behavior cycle, such as lifting a leg, lowering the leg, and then lifting it again. Figure 3.11c presents the mean frequency of consecutive repetitions, across all behavior

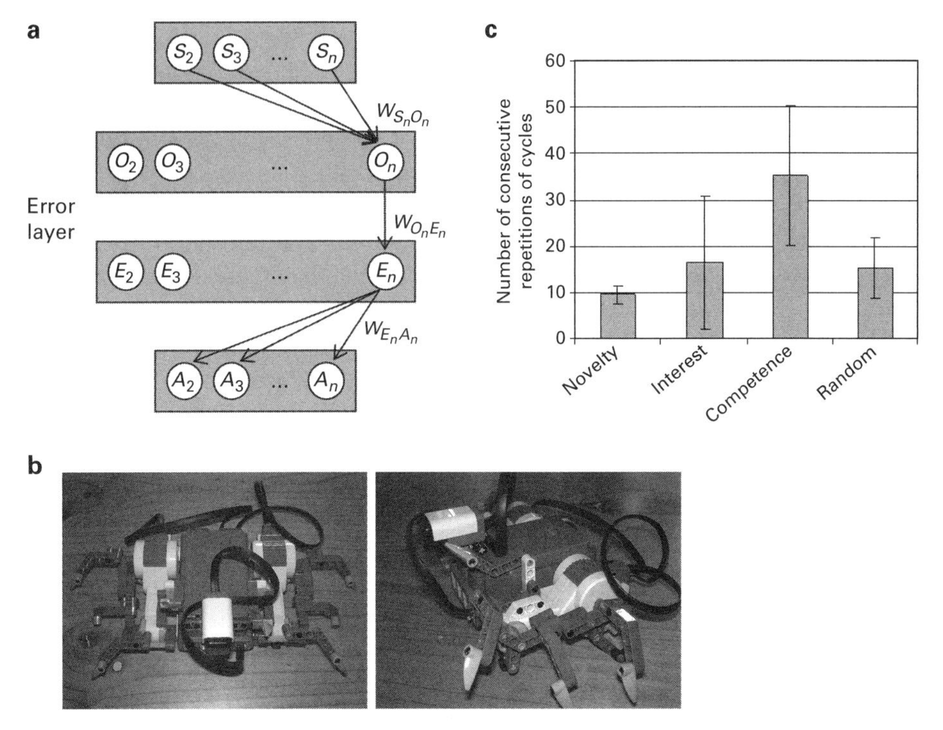

Figure 3.11
The neural architecture (a) and robot "crab" platform (b) used by Merrick (2010) to compare competence-based and knowledge-based IM. The number of consecutive behavior cycles produced by the different models is presented in (c). Figure courtesy of Kathryn Merrick. Reprinted with permission from IEEE.

cycles. As figure 3.11c illustrates, repetitions were significantly more frequent in the competence model. In addition, Merrick (ibid.) found that the duration of these behavior cycles was significantly longer in the competence model as well. On the one hand, this pattern of findings is somewhat expected, given that the competence model is specifically designed to focus on the process of skill-development and improvement. On the other hand, however, it is important to note that the pattern also demonstrates a key developmental principle highlighted by Piaget, that is, *functional assimilation*, which we noted earlier in the chapter is the tendency for infants and young children to practice or repeat an emerging skill.

3.4 Conclusions

Intrinsic motivation is a relatively new area of research in developmental robotics. In contrast, the idea of an intrinsically motivated infant or child has deep roots in the history of psychology, including connections with the concepts of curiosity, exploration, surprise, and the drive to understand. Harlow's work with rhesus monkeys suggests an important question: while traditional drives satisfy a biological need like hunger, what need is satisfied by solving a puzzle or exploring a novel object? In response to drive-based theories of intrinsic motivation, a more elaborate view has been proposed in recent years, including (1) knowledge-based IM, which can be subdivided into novelty- and prediction-based IM; and (2) competence-based IM. Each of these variants can also be linked to functional properties of the mammalian brain, including regions specialized for detecting novelty, for predicting upcoming events, and for detecting contingencies between self-produced actions and environmental consequences.

In addition, there is also extensive behavioral evidence for novelty-based, prediction-based, and competence-based IM in infants and young children. First, infants' visual activity represents an early form of exploration, in which visual attention is biased toward objects and events that are relatively unfamiliar. Second, by age two months infants generate actions that are directed toward future events, such as shifting their gaze to a location where an object is about to appear. Over the next several months, more sophisticated forms of predictive or future-oriented behaviors emerge, such as anticipatory reaching. Third, also as early as age two months, infants quickly detect objects and events in their environment that respond contingently to their actions. Indeed, infants show a strong attentional bias toward contingent events, which suggests that perceived control of the environment is a highly salient stimulus.

A major focus in developmental robotics has been to establish a taxonomy of architectures for modeling IM at the computational level. Significant progress toward this goal is offered by Oudeyer and Kaplan (2007), who present a systematic framework for organizing a wide array of architectures within the two classes of knowledge- and competence-based IM.

While there are not yet any IM models that are designed to capture a particular infant behavior or stage of development, there are a number that fit well within the novelty-, prediction-, and competence-based approaches. First, several researchers have proposed novelty-based strategies for motivating exploration in mobile robots. A common theme across these models is the use of a habituation mechanism, which computes novelty as the difference between the current state and a memory trace of recent experiences. Within this family of models, there a number of unique features, such as combining novelty with: (1) visual salience, (2) an external reinforcement signal, (3) a social cue from a caretaker, or (4) sensory prediction.

Second, there are also several models that focus on prediction as a core learning mechanism. A significant contribution to the prediction-based approach is offered by Schmidhuber (1991), who proposes an ambitious and encompassing theoretical framework in which prediction (and compression of knowledge) plays a central role. In addition, work by Oudeyer, Kaplan, and Hafner (2007) on the AIBO platform illustrates a key consequence of prediction-based IM: by linking IM to learning progress, the robot shifts its attention and actions from one region of the environment to another in a progressive and systematic manner. A related approach is proposed by Barto, Singh, and Chentanez (2004), who describe a toy-world model of prediction-based IM in which the agent's actions become hierarchically organized as a result of learning to predict the consequences of its actions.

Finally, competence-based models of IM have also received some support. A noteworthy model is proposed by Fiore et al. (2008), who investigate Redgrave and Gurney's (2006) theory of contingency detection and dopamine release in the superior colliculus. In particular, Fiore et al. demonstrate that the theory can be successfully implemented in a simulated rat, which learns that one of two levers in its environment controls a light. Merrick (2010) offers another valuable contribution by systematically comparing a variety of IM architectures that control the movement of a robot crab. A key finding from this work is that competence-based IM produces a developmentally relevant pattern, that is, repetition of a newly learned behavior.

We conclude by noting that because modeling of IM in robots and artificial agents is a relatively new area of research, there are a number of important and interesting behaviors that have not yet been simulated, and therefore deserve further study. These include not only experimental paradigms like VExP and habituation-dishabituation experiments, but also phenomena such as pretend play and intrinsically motivated problem solving. Indeed, a potential long-term goal may be to design a robot that is capable of engaging in a wide range of intrinsically-motivated behaviors, such as painting and drawing, musical composition, daydreaming, and so on.

Additional Reading

Baldassarre, G., and M. Mirolli, eds. *Intrinsically Motivated Learning in Natural and Artificial Systems*. Berlin: Springer-Verlag, 2013.

Baldassarre and Mirolli's edited volume presents a wide variety of approaches to understanding intrinsic motivation, including not only neurophysiology and behavior in real organisms, but also computational models and robotics experiments. An important feature of the book is a focus on both theory and empirical data, as well as a comprehensive discussion of the open challenges for researchers who investigate intrinsic motivation. It represents the state of the art in this area and will continue to be an influential work as new ideas and approaches emerge.

Berlyne, D. E. *Conflict, Arousal, and Curiosity*. New York: McGraw-Hill, 1960.

Berlyne's text—now more than fifty years old—is an essential read for students of intrinsic motivation. A key feature of his book is a chapter that analyzes the strengths and limitations of drive-reduction theory, which ultimately provides a foundation for an alternative approach that highlights the adaptive roles of curiosity and learning. In other chapters he focuses on additional issues that are central to developmental robotics, including novelty, uncertainty, and exploration. Another valuable feature is Berlyne's use of data from animal-behavior experiments, which helps bridge the gap between artificial systems and humans.

Ryan, R. M., and E. L. Deci. "Self-Determination Theory and the Role of Basic Psychological Needs in Personality and the Organization of Behavior." In *Handbook of Personality: Theory and Research*, 3rd ed., ed. O. P. John, R. W. Robins, and L. A. Pervin, 654–678. New York: Guilford Press, 2008.

Though written primarily for the psychology community, Ryan and Deci's chapter provides a detailed introduction to self-determination theory (SDT), which emphasizes the role of autonomy and competence in human experience. Their approach also addresses a fundamental dimension of experience that is often overlooked or neglected by behavioral researchers, that is, the qualitative, personal, or subjective aspect. Understanding the nature of subjective experience, and in particular, the developmental significance of self-efficacy, may provide valuable insights for the study of intrinsic motivation in robots and other artificial systems.

4 Seeing the World

Students in the field of child psychology are often introduced to the subject with a well-known quote from William James's *Principles of Psychology*, in which he describes the young infant's first experiences of the world "as one great blooming, buzzing confusion" (James 1890, 462). Well over a hundred years later, James's observation continues to leave parents, philosophers, and scientists wondering how infants make sense of their world. Do infants begin life overwhelmed and confused by their first perceptual experiences? If James is correct, then how do infants learn to perceive objects, space, sounds, tastes, and all the many other forms of sense data?

The study of perception conventionally includes the five major senses—vision, hearing, touch, taste, and smell—and also extends beyond sensation by defining perception as the cognitive process that helps organize sensory experience. Under this framework, then, perception is a process of transforming sensory information into higher-level patterns that can be identified, recognized, categorized, and so on. While a comprehensive survey of the perceptual systems is beyond the scope of this chapter, we refer the interested reader to Chaudhuri 2011 for a superb introduction. In addition, we also limit our focus in this chapter to the discussion of visual development, which not only is the predominant area of research in early perceptual development, but also is comparatively well studied in developmental robotics, relative to the other four sense modalities.

As we highlight elsewhere in this volume, an important issue to first consider is the theoretical landscape. In particular, what are the primary theoretical perspectives that have emerged in the study of perceptual development? One view emphasizes the role of endogenous or biological influences on the developmental process. This approach, in contrast to the perspective advocated by James, argues that infants *are not* born with completely disorganized sensory systems that must somehow become structured over time. Instead, several theorists have proposed theories of perceptual development that suggest human infants are in fact well prepared for the sensory world, and that they are able to begin interpreting sense data shortly after birth. Some theorists have focused on innate or preprogrammed forms of perceptual activity. For example, Haith

(1980) describes a set of innate principles that guide visual activity in young infants. This set includes the "high firing-rate principle," a simple exploratory heuristic that works by driving young infants' attention toward areas of the visual field with high contrast. Other theorists have focused on innate structures or processing systems. An area of research that highlights this second approach is the study of face perception in early infancy. A robust finding from this work is that newborn infants show a strong tendency to prefer face-like stimuli (e.g., de Haan, Pascalis, and Johnson 2002; Maurer and Salapatek 1976). This finding has been interpreted to suggest that infants have an innate or prewired capacity for perceiving faces, which emerges prenatally and is functional by birth (e.g., Valenza et al. 1996). In a similar vein, other studies have investigated the ability of neonates to integrate or "match" sensory experiences across multiple modalities (e.g., vision and touch), and have drawn the conclusion that intermodal (or cross-sensory) perception is an innate capacity (e.g., Meltzoff and Borton 1979; Meltzoff and Moore 1977).

A second, alternative theoretical perspective emphasizes the interaction between low-level, innate processing biases (e.g., genetically specified neural connections or pathways), which are subsequently shaped by sensory experience. A fundamental concept from this perspective, articulated by Greenough and Black (1999), is the notion of *experience-expectant* development. According to this view, specific neural structures reliably emerge across individuals and environments through a two-step process: first, a coarse, initial neural pattern is established during prenatal development, and then, second, this pattern becomes fine-tuned as a function of sensory input after birth. Deviations in the typical pattern of development occur, however, when the "expected" environmental experiences are not present at the normal time. Another critical issue raised by interactionist theories of perceptual development is the role of the child as an active participant in the developmental process. Piaget (1952), for example, emphasizes the *moderate novelty principle* during perceptual and cognitive development. According to this principle, infants' exploratory activity is determined by their current level of perceptual knowledge, and more specifically, by the tendency to seek out or orient toward sensory experiences that are moderately novel or unfamiliar. Piaget suggests that this tendency serves to "stretch" or "exercise" infants' perceptual capacities, by placing them in circumstances that require modest changes or modifications to their existing sensorimotor schemes.

While both of these theoretical perspectives can be found in robotics research, the interactionist (or constructivist) approach complements several important themes in developmental robotics (e.g., Guerin and McKenzie 2008; Schlesinger and Parisi 2007; Weng et al. 2001). For example, rather than programming in specific knowledge of the environment, learning and development are understood to occur as an interaction between primitive or basic action tendencies (e.g., modifiable reflexes) with the physical environment. An additional influence on development, which corresponds to early biases in neural processing, is the specific form of computational structure or cognitive architecture that is implemented in the robotic model (e.g., artificial neural network, production system, swarm model, etc.). While these structures may begin "life" with no prior sensory data—there is initially no content in their perceptual systems—the mere fact that they are organized in a particular way provides an important constraint on what is experienced and how it becomes organized. Elman et al. (1996) refer to this as a weak form of innate constraint on development that they call *architectural innateness*. Another important theme from the interactionist perspective that complements developmental robotics and is apparent in the study of perceptual development is the notion of "uncertainty." In particular, this view emphasizes that sensory data are noisy and stochastic, and that instead of programming in advance specific perceptual categories, it may be more effective to allow the robot or autonomous agent to build up or discover internal representations through interaction with the environment.

We now take an in-depth look at the development of perception, as it is studied in human infants as well as in artificial systems and developing robots. As we noted earlier, rather than providing exhaustive coverage across each of the major sensory systems, we focus our discussion on the development of visual perception. In addition, it is important to note that the study of perceptual development in humans—in contrast to, for example, motor development or language acquisition—is a comparatively broad and diverse topic. As a result, in section 4.1 we highlight a selection of five fundamental perceptual capacities and trace their developmental trajectories in human infants: objects, faces, space, self-perception, and affordances. The remainder of the chapter (sections 4.2–4.6) then presents a systematic overview of how each of these five capacities has been studied within the field of developmental robotics.

4.1 Visual Development in Human Infants

In this section, we first familiarize the reader with two of the primary experimental methods that are used to study the development of visual perception in young infants, and then survey the development of five essential perceptual abilities.

4.1.1 Measuring Infant Perception

Studying perceptual development in infancy is a unique challenge for developmental researchers, in particular because it takes several months for young infants to begin understanding spoken language, and up to a year before infants start to speak. How can an infant's perceptual experiences be assessed, if the infant cannot be explicitly instructed to perform in a particular way on a task (e.g., look at a picture, search for a

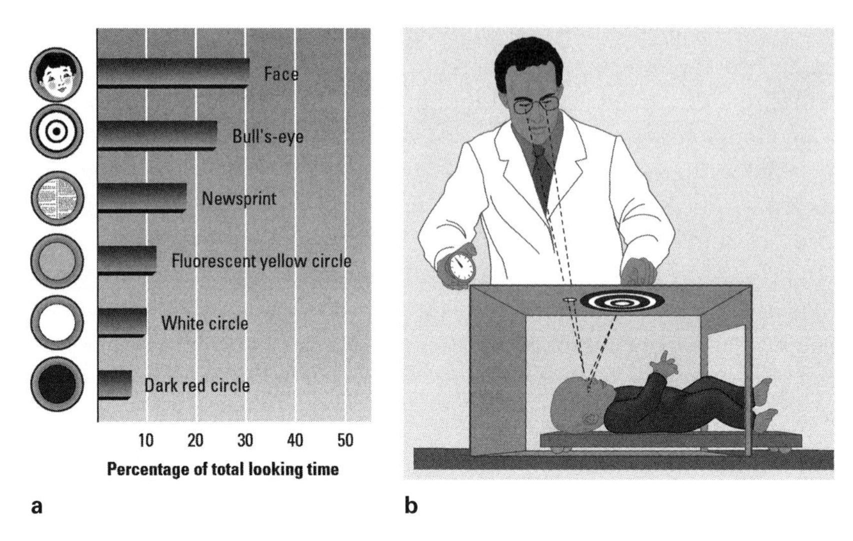

Figure 4.1
Examples of the visual stimuli used by Fantz (1956) (left), and an infant being observed within the looking chamber (right). From Cook and Cook 2014. Reprinted by permission of Pearson Education, Inc., Upper Saddle River, NJ.

target object, etc.)? Similarly, if an infant cannot provide a verbal response, how can the researcher presume to know what the infant perceives?

These questions challenged researchers for decades, until a major methodological advance occurred. In particular, Robert Fantz (1956) designed an apparatus called the *infant looking chamber*, in which infants were placed and observed while they looked at a variety of images. As figure 4.1 illustrates, the infant rests comfortably on her back, looking toward the top of the chamber, while a trained observer views the infant's face through a small peephole. Using this apparatus, Fantz presented young infants with a wide variety of visual images, including colored shapes, geometric patterns, and simple faces. When Fantz measured the amount of time that infants spent looking at these images, he found that infants showed a consistent preference for certain types of visual stimuli over others. That is, young infants tended to look longer at stimuli with patterns, contrast, and varying details, compared to stimuli that were relatively plain or homogeneous.

The infant looking chamber, and the data provided by Fantz, helped generate a crucial insight: if infants prefer looking at certain types of stimuli over others, this tendency could be used as a nonverbal method for testing infants' visual perception. As a result, Fantz's method for studying perception in young infants, including newborns, quickly became the predominant research paradigm, and the rationale that guided

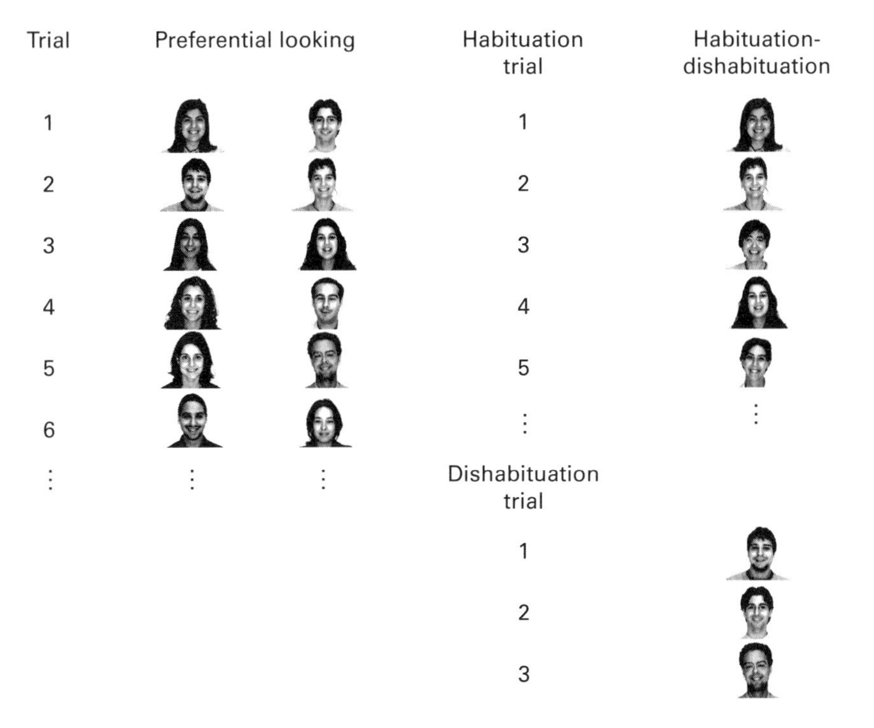

Figure 4.2
Illustration of the preferential looking (left) and habituation-dishabituation (right) paradigms. The sample stimuli test for the perception of female versus male faces.

the design and use of the looking chamber became known as the *preferential-looking technique*.

As a concrete example of this technique, imagine that a researcher wants to determine whether young infants can discriminate between male and female faces. The left half of figure 4.2 illustrates how the preferential-looking paradigm can be used to investigate this question. Over a series of trials, infants are shown a pair of faces. Note that male and female faces are randomly presented, on both the left and right sides of the infant's midline. During each trial, the amount of time that the infant spends looking at each face is recorded (e.g., by an observer viewing the infant via a remote camera, by an automated eye-tracking system, etc.). As a control condition, on some trials two male or two female faces are presented at the same time. After the data are collected, the hypothesis that infants can discriminate between male and female faces is then tested by computing the difference in looking time to each set of faces. Interestingly, when preferential looking is used at the testing method, it is not necessary for all infants to prefer the same set of stimuli. In this case, for example, some infants may prefer female faces, while others may prefer male faces. As a result, the appropriate sample statistic for preferential looking is the absolute value of the difference in looking time between the two categories of stimuli, averaged across the infants in the sample.

Preferential looking is one of two behavioral paradigms that have helped developmental researchers systematically study and map the development of visual perception in young infants. The second technique is *habituation-dishabituation*, which also uses infants' looking times across a set of stimuli as an index of perceptual processing. However, there is an important distinction between the two methods. While preferential looking depends on infants having a preference for one set of visual stimuli over another, habituation-dishabituation does not require a prior preference. Instead, it assumes—if infants can distinguish between two sets of visual stimuli—that after they are familiarized or habituated to one set, they should then show a preference for the new set (i.e., dishabituate) by increasing their looking time. In other words, habituation-dishabituation assumes that infants will be more interested in (i.e., look longer at) novel stimuli. However, as we noted in chapter 2, it is important to remember that the preference for novelty may not be constant over infancy, but instead may vary across infants at different ages (e.g., Gilmore and Thomas 2002).

The right half of figure 4.2 illustrates an example of the habituation-dishabituation paradigm, applied to the question of gender perception in infants. Note that there are two important differences between this paradigm and preferential looking. First, only one stimulus or image is typically presented on each trial during habituation-dishabituation, in contrast to the pair of stimuli that are presented on each trial during preferential looking. Second, during the habituation phase, all of the stimuli are sampled from the same class (e.g., female faces). In this case, infants first view a series of female faces; although a new face is presented on each trial, infants should gradually lose interest and look less during successive trials. This loss of interest is assumed to reflect a perceptual encoding process, in which infants detect a common feature or element across the faces (e.g., Charlesworth 1969; Sokolov 1963). Once infants have adequately encoded the class of faces, infants become "bored" or more formally, *habituated* to female faces (note that in a second condition, another group of infants would be habituated first to male faces). Infants are then presented with the male faces (i.e., during the posthabituation phase), and a statistically significant increase in looking time at this point is interpreted as support for the hypothesis that infants can discriminate between male and female faces.

While preferential looking and habituation-dishabituation are the predominant methods used to study perceptual development (especially in young infants), there are a number of related methods and paradigms that are also employed. For example, infants' head and eye movements can be recorded. In addition, physiological measures such as heart rate, respiration, and pupil dilation can also be measured as an index of attention. In the next section, as we survey the development of the perception of faces, objects, space, affordances, and the self, we briefly describe some of these alternative methods.

4.1.2 Faces

At what age do infants begin to perceive faces? There are two lines of research that provide support for the claim that the capacity is present at birth. First, Meltzoff and Moore (1977) demonstrated that newborn infants imitate simple facial gestures produced by an adult model, such as sticking out their tongue. This behavior, they argue, is due to an innate internal representation of faces, and a matching mechanism that allows an infant to recognize the correspondence between another person's face and their own. However, this is a somewhat controversial claim, and subsequent studies have provided mixed support. The second line of work measures face perception using a more direct method. In particular, Bushnell (2001; Bushnell, Sai, and Mullin 1989) used the preferential-looking method to study face perception in two- and three-day-old infants. After a few days of interaction with their mothers, infants in Bushnell's study viewed their mother and a similar-looking stranger, while an observer recorded the amount of time that infants looked at each face. A comparison of infants' looking times not only supported the conclusion that infants could discriminate between the two faces, but also showed that infants significantly preferred their mother's face over the stranger's.

A potential limitation of Bushnell's study, however, is that each infant viewed a different pair of faces. In order to more systematically control the infant's visual experience, other researchers have presented infants with simplified, schematic faces. This approach has two advantages. First, it ensures consistency across infants in terms of the visual stimuli (i.e., all infants view the same set of faces). Second, it also enables the researcher to manipulate specific visual aspects of the faces, in order to determine which features are detected and exploited by infants during the perceptual process. For example, figure 4.3 illustrates the face stimuli used by Maurer and Barrera (1981), who used the preferential-looking method to study face perception in one- and two-month-olds: (a) a normal schematic face, (b) a scrambled face that preserves left-right

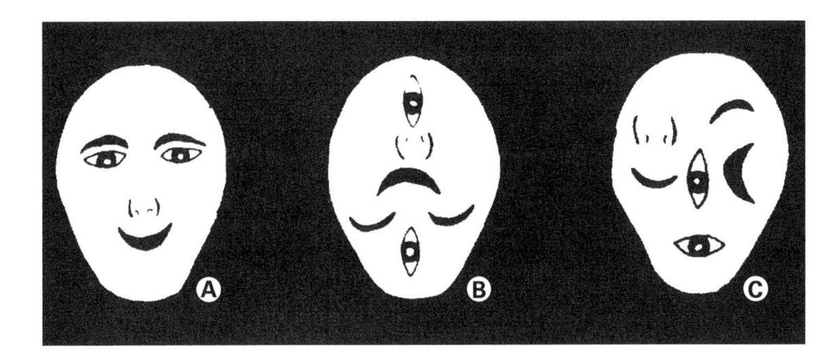

Figure 4.3
Examples of the stimuli used by Maurer and Barrera (1981) to study face perception in young infants. Figure reprinted with permission from Wiley.

symmetry, and (c) a scrambled face without symmetry. Maurer and Barrera (ibid.) found that one-month-olds looked equally long at all three of the schematic faces, while two-month-olds significantly preferred the normal face. This finding was replicated in a follow-up study by Morton and Johnson (1991).

Taken together, the two sets of findings appear difficult to reconcile. In order to address the apparent conflict between studies, Morton and Johnson (1991) proposed a two-process model of face perception development in human infants, in which a subcortical visual pathway is present at birth, followed by a second, cortical pathway that develops roughly a month after birth. Morton and Johnson (ibid.) reasoned that if their two-process model is correct, then behavioral measures that tap the subcortical and cortical structures, respectively, should provide differing estimates of when face perception first emerges. In particular, they hypothesized that measuring infants' tracking of moving face stimuli—which relies on subcortical processing—should be present in newborns, while a preference for looking at static face stimuli (over non-faces)—which relies on cortical processing—should not be present until age two months. Morton and Johnson (ibid.) tested their prediction by systematically comparing infants' tracking of moving face and non-face stimuli with their looking time to the same stimuli while stationary, and found the expected pattern.

Thus, in a sense, *both* the nativist and constructivist views are supported by these findings. Infants are born with a basic visual mechanism that allows them to detect face-like features. This initial ability is complemented by a second, more advanced mechanism that enables infants to learn about specific features (e.g., the relative location of the eyes, nose, etc.), and that begins to function within the first and second month after birth.

In addition, there are a number of important and interesting ways that face perception continues to develop in the months following birth that highlight the role of social interaction with the caretaker and others. For example, the ability to distinguish between male and female faces is present by age three months (e.g., Leinbach and Fagot 1993; Quinn et al. 2002). However, Quinn et al. (ibid.) demonstrate that this ability is shaped in part by extensive interaction with the primary caretaker: infants raised by a female caretaker tend to prefer looking at female faces, while infants raised by a male caretaker tend to prefer looking at male faces. A second, fascinating developmental milestone is the emergence of the *other-race effect* in infants. The other-race effect is a well-documented visual-processing bias, in which observers are more accurate at recognizing faces from their own racial group than faces from other groups (e.g., Meissner and Brigham 2001). Interestingly, while three-month-old Caucasian infants can recognize and differentiate between similar pairs of faces within four racial groups (i.e., African, Middle Eastern, Chinese, and Caucasian), by nine months this ability is lost, and Caucasian infants can only differentiate between faces within their own racial group (Quinn et al. 2002).

Figure 4.4
The visual cliff, an apparatus used to study depth perception and the fear of heights. From Santrock 2011.

4.1.3 Space

While the development of face perception begins shortly after birth, space perception is comparatively slower. For example, it is not until approximately age two months that infants begin to use binocular visual information to perceive depth (e.g., Campos, Langer, and Krowitz 1970). One way to assess this skill is with an apparatus called the *visual cliff* (Gibson and Walk 1960). As figure 4.4 illustrates, the visual cliff is a large platform that is covered with a glass surface. One side of the platform has a patterned surface directly beneath it, while the other side appears to have a large drop. Campos, Langer, and Krowitz (1970) investigated whether two-month-olds could distinguish between the shallow and deep sides of the cliff by placing infants in a harness, and then slowly lowering them down onto one side of the cliff or the other. As a measure of infants' perception, their heart rate was monitored while they were being lowered. Campos, Langer, and Krowitz (ibid.) found that infants who were lowered onto the shallow side did not change their heart rate. Infants lowered on the deep side, meanwhile, experienced a deceleration in heart rate, which is consistent with the experience of mild curiosity or interest.

Interestingly, it is not until age nine months that infants begin to show fear and distress when placed near the deep side of the cliff. What is the developmental "trigger" that leads to the onset of the fear of heights? Across a series of studies, Campos and colleagues (e.g., Campos, Bertenthal, and Kermoian 1992; Kermoian and Campos 1988) have demonstrated that the emergence of self-produced movement (i.e., crawling) has a profound influence on the development of spatial perception. Thus, even when matched by age, infants who have just begun to crawl will cross the visual cliff, while those who have a few weeks of crawling experience will show expressions of fear and refuse to cross. A similar pattern occurs during the development of the perception of *optic flow*, which is the ability to detect the direction of self-movement through space. In particular, infants with locomotor experience adjust their posture in response to optic flow cues that imply self-movement, while infants who have not begun to crawl do not respond to the same cues (e.g., Higgins, Campos, and Kermoian 1996).

A related area of spatial perception development is the transition from an *egocentric* to an *allocentric* (or *world-centered*) spatial frame of reference. In particular, the *egocentric bias* is a tendency to code spatial locations in terms of their position relative to the observer, rather than their absolute location. A series of studies by Acredolo (1978; Acredolo, Adams, and Goodwyn 1984; Acredolo and Evans 1980) highlight two important findings. First, young infants tend to code spatial locations egocentrically, but this bias decreases with age. For example, after seeing a person appear in a window on their left, six-month-olds will continue looking toward their left when they are rotated 180 degrees; nine-month-olds, in contrast, compensate for the rotation and look toward the right after being moved (Acredolo and Evans 1980). Second, Acredolo, Adams, and Goodwyn (1984) also show that the emergence of self-produced movement is associated with an increase in allocentric spatial coding. In their study, infants view an object hidden at one of two locations (see figure 4.5). The infant is then moved to the opposite side of the apparatus. Acredolo, Adams, and Goodwyn (ibid.) found that twelve-month-olds continued to look at the hiding location as they were being

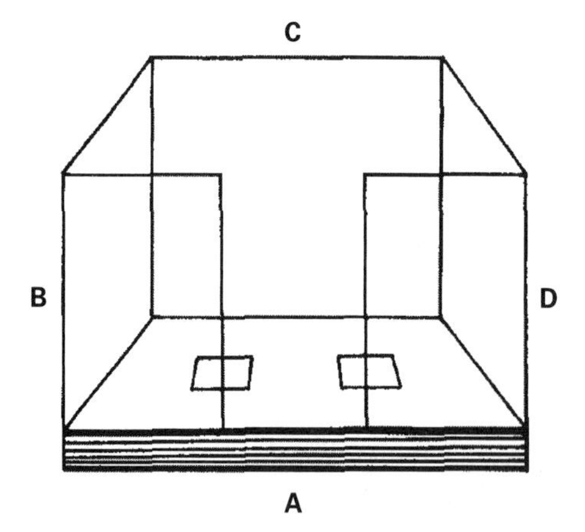

Figure 4.5
The two-location apparatus used by Acredolo, Adams, and Goodwyn (1984) to study the development of spatial-coding strategies in infants. Figure reprinted with permission.

moved, and searched correctly. However, if the apparatus was covered while the infant was moved—which prevented tracking of the target location—they searched in the egocentric location. At age eighteen months, meanwhile, infants searched in the correct location even when the apparatus was covered. Acredolo et al. (ibid.) suggest that the older infants "mentally tracked" the target location, a skill that develops as infants learn to walk.

4.1.4 Self-Perception

An important skill that crosses the boundary between perceptual and social development is the ability to perceive or recognize the self. The primary method for studying self-perception is to place the child in front of a mirror, picture, or video display, and to record and analyze his behavior. In one version of this paradigm, the infant views a single display (which may vary across trials or conditions). Alternatively, the infant may be presented with two video displays side by side (e.g., a live video of the infant on one display, and a delayed video of the infant—or another child—on the other display), while his looking time to each is recorded. Although there is considerable debate concerning the underlying mechanisms that make self-perception possible during infancy, there is relatively strong agreement on the developmental milestones (e.g., Bahrick, Moss, and Fadil 1996; Courage and Howe 2002; Lewis and Brooks-Gunn 1979; Rochat and Striano 2002). Table 4.1 presents a developmental timescale that highlights these milestones during the first two years.

The earliest evidence of self-perception is at age three months (e.g., Butterworth 1992). When placed in front of a mirror, three-month-olds increase their frequency of self-exploratory behaviors, such as looking at their hands (e.g., Courage and Howe 2002). In addition, infants at this age are capable of visually differentiating between themselves and another person. In particular, when presented with two video images, three-month-olds prefer to view a non-self display that is paired with a live display of themselves. The capacity to distinguish between "self" and "other" continues to develop between five and six months, when infants begin to use intermodal contingencies as

Table 4.1
Timescale and major milestones of self-perception (adapted from Courage and Howe 2002 and Butterworth 1992)

Age (months)	Features
3 months	Mirror self-exploration; self/other differentiation (preference for "other")
5–6 months	Detection of intermodal contingency cues
9 months	Social behaviors directed to "other" (e.g., smiling, vocalizing, etc.)
15–18 months	Mirror self-recognition (i.e., "mark test")
22–24 months	Correct labeling of self in an image

a basis for self-perception (e.g., Bahrick and Watson 1985; Rochat 1998). For example, Bahrick and Watson (ibid.) presented five-month-old infants with a live video display of their legs, next to a second display that showed either (a) a tape-delayed view of their legs, or (b) the legs of another infant. In both cases, infants looked significantly longer at the noncontingent display. While infants show a preference for viewing displays of others as early as three months, it is not until age nine months that infants begin to specifically direct social-communicative behaviors toward the "other" during tests of self-perception (e.g., smiling, vocalizing; Rochat and Striano 2002).

Mirror self-recognition, a particularly salient and important capacity, emerges between fifteen and eighteen months. In contrast to self-perception, which is a more basic ability to distinguish between the self and others, self-recognition involves a more advanced awareness that the person in the mirror or video display is "me" (i.e., an enduring entity that persists through time). The predominant method for measuring this capacity is the *mark test*. The mark test was first studied by Gallup (1970), who placed a red mark on chimpanzees' faces, and then observed the chimps' behavior in front of a mirror (see figure 4.6). Gallup (ibid.) found that while chimps noticed and touched the mark, two monkey species (i.e., macaques and rhesus) did not show self-directed behavior in front of the mirror. Human infants between fifteen and eighteen months old not only orient toward the mark like chimpanzees, but they also show signs of embarrassment (e.g., Lewis and Brooks-Gunn 1979). This behavior suggests that infants have begun to develop a self-concept, and are aware of how they appear to others. A subsequent development in self-recognition is the ability to label oneself in an image, which emerges between twenty-two and twenty-four months (e.g., Courage and Howe 2002).

Figure 4.6
The mark test. A chimpanzee detects a red mark on its brow, and uses its reflection in the mirror to explore the mark (Gallup 1970). Figure reprinted with permission.

4.1.5 Objects

Like the development of space perception, how infants learn to perceive objects is influenced by their ability to manipulate and interact with their environment (e.g., Bushnell and Boudreau 1993). In general, the developmental pattern over the first year of life is a constructive one: infants first learn to detect an important feature or dimension of three-dimensional objects, and then they integrate this feature into their existing perceptual skill set.

While the overall pattern of perceptual development is consistent with the constructivist view, an early exception is the possibility that very young infants are able not only to detect a basic property of objects (i.e., texture), but more important, to also match the perception of that property across sensory modalities. Support for this claim comes from a study by Meltzoff and Borton (1979), who presented one-month-old infants with either a smooth or "nubbed" pacifier. After exploring the pacifier with their mouths, infants were then shown the two pacifiers side by side. During the preferential-looking test, infants looked significantly longer at the familiar pacifier, that is, the one that matched the pacifier that they had previously mouthed. Subsequent work with newborn infants (e.g., Sann and Streri 2007) has replicated this finding, providing additional support for Meltzoff and Borton's claim that intermodal object perception is present at birth. However, it is interesting to note that while Sann and Streri (ibid.) found a bidirectional transfer between sight and touch for *texture*, there was only a one-way transfer (from touch to sight) for *shape*. Thus, while these findings are suggestive, it is not clear how much of the ability is present at birth.

Fortunately, there is wider consensus among researchers regarding subsequent developments in object perception (e.g., Fitzpatrick et al. 2008; Spelke 1990). For example, one of the key skills that have been investigated is how infants learn to parse or segregate objects in the visual scene. An important visual feature is the presence of joint or common surfaces, which serves as a cue for a single, bounded object (e.g., Marr 1982). By age three months, infants not only are able to detect common surfaces, but they also exploit such a surface as a cue for perceiving objects (e.g., Kestenbaum, Termine, and Spelke 1987). However, a fundamental limitation at this age is that infants lack *perceptual completion*, that is, the ability to perceive a partially occluded object as an integrated whole. In box 4.1, we provide a detailed description of the *unity-perception task*, which is used to study the development of perceptual completion in young infants. As we highlight in box 4.1, results from the unity perception task demonstrate that by age four months, infants have learned to use a variety of visual cues to aid in the perception of partially occluded objects, including common motion and alignment of the visible object surfaces.

Object perception continues to develop between four and six months. In particular, during this period infants acquire the ability to perceive objects that briefly undergo

Box 4.1
The Unity-Perception Task

A fundamental step in the development of object perception is *perceptual completion*, which is the ability to perceive a partially occluded object as an integrated, coherent whole. At a functional level, perceptual completion involves "reconstructing" an object that is partially occluded, by bridging or linking surfaces that are separated by an occluder. Is perceptual completion an innate skill, or is it acquired through a constructive, learning process? Kellman and Spelke (1983) designed the *unity-perception task* to investigate this question. Infants in this task view a rod that moves laterally, while the center portion of the rod is occluded by a large, stationary screen. Infants then view two new events—a solid rod in one display, and a broken or two-segment rod in the other—and it is their reaction to these two events that provides a nonverbal index into how the occluded-rod display is experienced. We briefly highlight here the task, as well as the developmental pattern that has emerged by studying infants at different ages.

Procedure

The unity-perception task employs the habituation-dishabituation paradigm. Infants are first habituated to the occluded-rod display (see figure below). In particular, the display is repeated until the infant's looking time falls below a predetermined threshold. Once habituated, infants view the solid-rod and broken-rod displays, on alternating trials (see figures below). Longer looking to one of the two posthabituation displays is assumed to reflect a novelty preference. Therefore, infants who respond to the broken rod as the novel object are called "perceivers," as this preference suggests that they perceive the occluded rod as a single, integrated object. Alternatively, infants who respond to the solid rod as the novel object are called "nonperceivers," as this preference suggests instead that they perceive the occluded rod as two separate objects.

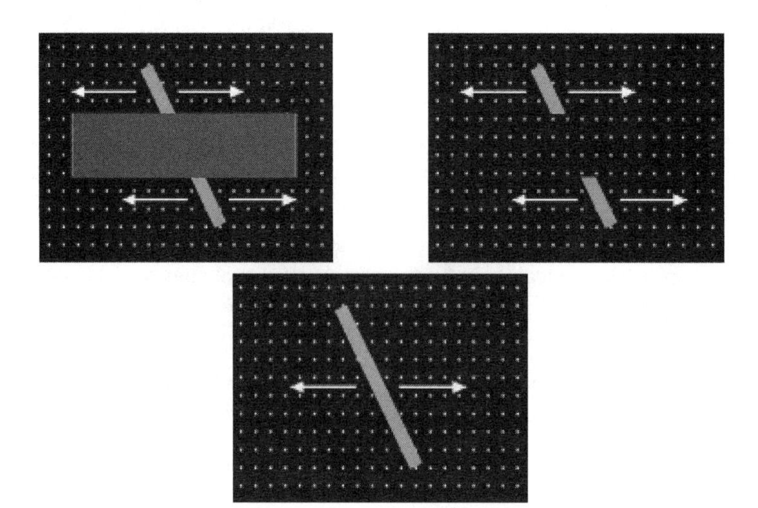

Stimuli used in the unity-perception task: (a) an occluded rod, (b) a solid rod, and (c) a broken rod.

Development of Perceptual Completion

Perceptual completion is not present at birth (e.g., Slater et al. 1996). During the posthabituation phase, newborn infants look longer at the solid rod, indicating that the broken-rod is relatively familiar (i.e., more similar to the occluded-rod display than the solid rod). This pattern persists until age four months, at which point infants begin to reliably prefer the broken-rod display during the posthabituation phase (e.g., Johnson 2004). These data support the conclusion that while young infants can perceive disjoint object surfaces that are separated by occlusion, they are not able to integrate these surfaces into a perception of a solid object until approximately age four months.

Selective Attention as a Mechanism for Perceptual Completion

What cognitive-perceptual mechanism drives the development of perceptual completion? Consistent with the notion of *active vision* (e.g., Ballard 1991), Johnson and colleagues have proposed that progressive improvements in visual-motor skill, and in particular, *visual selective attention*, support the discovery of relevant visual cues for perceptual completion. Support for this proposal is provided by Amso and Johnson (2006), who recorded three-month-olds' eye movements during the occluded-rod display. Infants were subsequently presented with the solid- and broken-rod test displays, and categorized as either perceivers or nonperceivers as a function of their posthabituation preference. The figure below (left) illustrates the cumulative scanplot of a three-month-old perceiver, while the right figure presents the scanplot of a three-month-old nonperceiver. Note that most of the perceiver's fixations are focused on the rod, while the nonperceiver's fixations tend to fall on less informative areas, such as the occluding screen. This pattern was confirmed by a quantitative analysis of all twenty-two infants in the sample (ibid.).

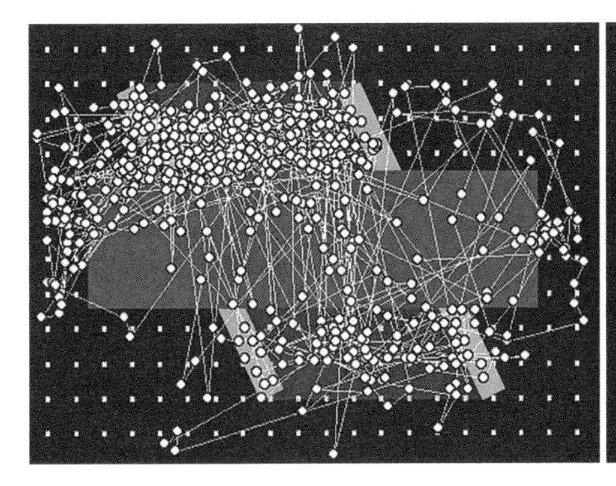

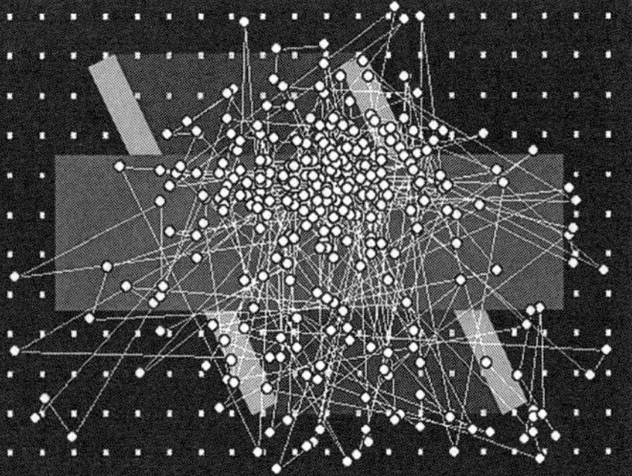

total occlusion. For example, six-month-olds will track an object that moves behind an occluder, and anticipate its reappearance (e.g., Johnson, Amso, and Slemmer 2003a). In contrast, four-month-olds do not spontaneously anticipate an occluded object's reappearance. Similarly, six-month-olds perceive the trajectories of occluded objects as moving along a continuous path (e.g., Johnson et al. 2003b). However, at age four months, infants appear to perceive the same trajectories as discontinuous (i.e., as if there are gaps in the motion of the object). Interestingly, Johnson et al. (ibid.) demonstrate that while four-month-olds appear to lack the ability to perceive fully occluded objects, their performance on these tasks can be improved to the level of six-month-olds by either briefly presenting infants with the motion of a visible object (which facilitates tracking behavior), or by reducing the width of the occluding screen (which decreases demand on spatial-working memory).

4.1.6 Affordances

While self-perception provides a link between perceptual and social development, the perception of affordances bridges perceptual and motor-skill development. The term "affordance" was proposed by Gibson (1979), who theorized that objects are perceived as a function of the goals and abilities of the observer. In particular, the affordances of an object are the perceptual qualities that specify potential action. As a concrete example, consider a chair. In the conventional sense, a chair is "something to sit on." However, depending on the circumstances, it may also be "something to stand on while changing a light bulb," or alternatively, "something to hide behind," in the presence of a wild animal. In this sense, Gibson argued that how one perceives the affordances of an object such as a chair varies with the intentions and plans in mind at that moment.

Unlike the forms of perceptual skill described thus far, which typically involve measuring infants' visual preference or attention pattern, investigating the perception of affordances requires measuring not only infants' perceptual reactions but also their motor responses. One area in which the perception of affordances has been investigated is the development of infants' tool use. For example, eight-month-old infants quickly learn to use a variety of tools, such as a supporting cloth, a string, or even a large hook-shaped stick, to retrieve an out-of-reach toy (e.g., Schlesinger and Langer 1999; van Leeuwen, Smitsman, and van Leeuwen 1994). Interestingly, infants' sensitivity to the spatial relationship between the tool and toy varies with their tool-use skill. Thus, eight-month-olds will only pull a cloth—which is a relatively easy tool—when a toy is placed on it, but not when the toy is placed beside it. When presented with a hook—which is a more difficult, complex tool—infants at the same age will pull the hook regardless of the position of the toy. It is not until age twelve months that infants use the relative position of the hook and the toy to guide their retrieval behavior (Schlesinger and Langer 1999). Similarly, van Leeuwen, Smitsman, and van Leeuwen (1994) studied infants' sensitivity to the spatial relationship between the tool and goal

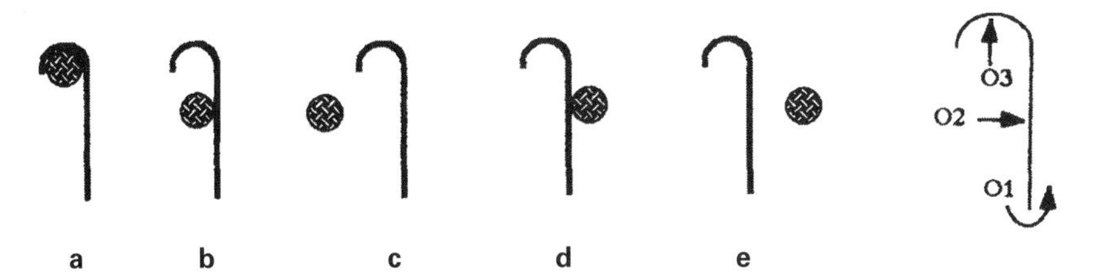

Figure 4.7
Van Leeuwen, Smitsman, and van Leeuwen (1994) studied the perception of tool-use affordances by varying the position of the goal object with respect to the tool. Copyright © 1994 by the American Psychological Association. Reproduced with permission.

object, by systematically varying the location of the goal. For example, figure 4.7 illustrates how the position of the goal object can be varied in relation to a hook. In particular, van Leeuwen, Smitsman, and van Leeuwen (ibid.) found that younger infants tended to use the tool indiscriminately, while older infants and children adapted their tool-use strategies as a function of the particular affordance between the tool and goal object.

A comparable developmental pattern also occurs with other tools. For instance, McCarty, Clifton, and Collard (2001b) presented infants between nine and twenty-four months with a series of objects such as spoons and hairbrushes, either oriented to the left or right. While younger infants tended to reach automatically with their dominant hand, regardless of the object's orientation, older infants adapted their reaching and grasping movements, prior to contacting the tool, as a function of its orientation. In addition, McCarty, Clifton, and Collard (ibid.) also found that infants were more successful at adapting their movements toward tools when the intended action was self-directed (e.g., spoon-to-self) than when it was other-directed (e.g., hairbrush-to-doll).

A second area of research on the perception of affordances is the development of crawling and walking. For example, as figure 4.8 illustrates, an infant may encounter a sloped surface and need to adjust her posture, depending on the degree of slope. A series of studies by Adolph (1997, 2008; Joh and Adolph 2006) illustrates two important and interesting developmental patterns. First, new crawlers are relatively poor at perceiving sloped surfaces. Thus, regardless of whether a slope is slight or dangerously steep, infants who have recently begun to crawl do not adjust their posture or crawling strategy as a function of the sloped surface. With subsequent days or weeks of crawling experience, however, infants will effectively approach and navigate sloped surfaces. Second, and perhaps more important, the same developmental pattern occurs a second time as infants begin to walk. In other words, at the onset of walking infants again behave indifferently to the angle of sloped surfaces, and learn anew to perceive the slope. These findings provide strong support for Gibson's notion of affordances, and in

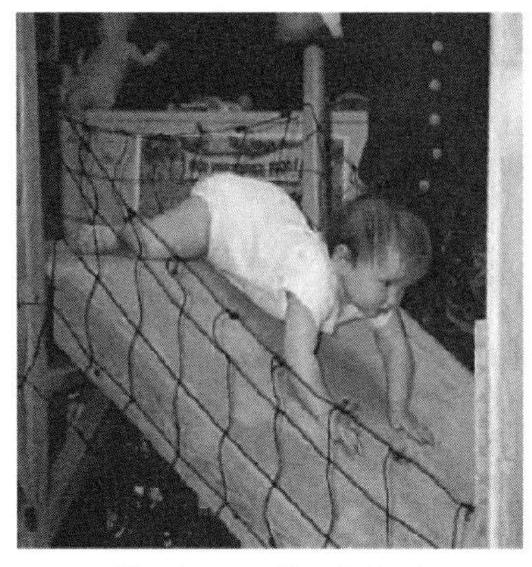

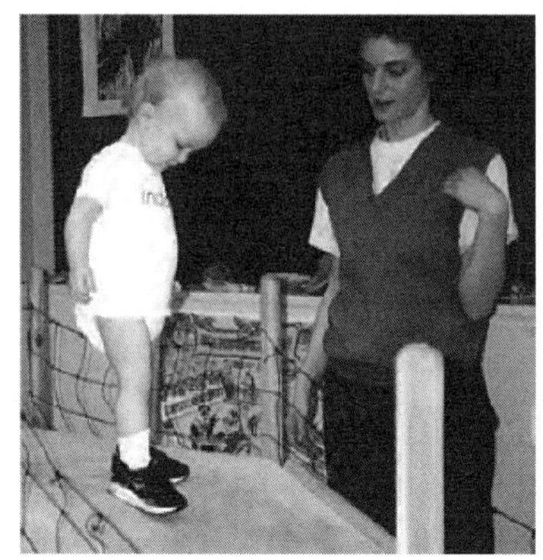

Newly crawling infant Experienced walker

Figure 4.8
Development of locomotion down sloped surfaces. From Santrock 2011.

particular, highlight the dynamic relationship between the development of perception and action and goal-directed behavior.

4.2 Face Perception in Robots

Developmental models of face perception have largely relied on simulation methods, rather than the use of embodied or humanoid robot platforms. An important exception to this trend is work by Fuke, Ogino, and Asada (2007), who investigate the question of how infants acquire a representation for their face when it is effectively "invisible." While the Fuke, Ogino, and Asada model does not directly focus on face perception or neonatal imitation, it is still relevant to the Meltzoff and Moore (1977) study, which found that newborn infants imitated facial gestures such as tongue protrusions; in particular, as we noted earlier, the development of a multimodal face representation is a prerequisite ability for facial imitation. The core idea of the Fuke, Ogino, and Asada model is that while infants cannot directly view their own faces, they have three important sources of sensory data that can provide information about the state of the face. First, as figure 4.9 illustrates, infants can see their hand when it is front of and near their face. Second, infants also receive proprioceptive information indicating the position of their hand. Third, although neither the hand nor face is visible when infants touch their own face, they then receive tactile information during this self-exploration. Fuke, Ogino, and Asada (ibid.) show that by integrating these three sources of information, the simulated robot can learn to estimate its hand position, and via bootstrapping of the estimated hand position, develop an internal representation of the face.

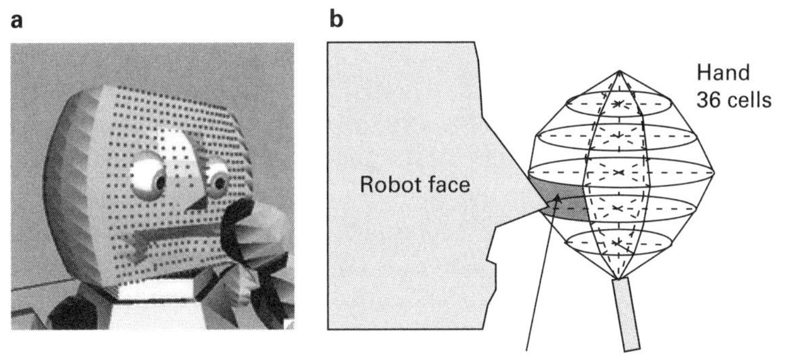

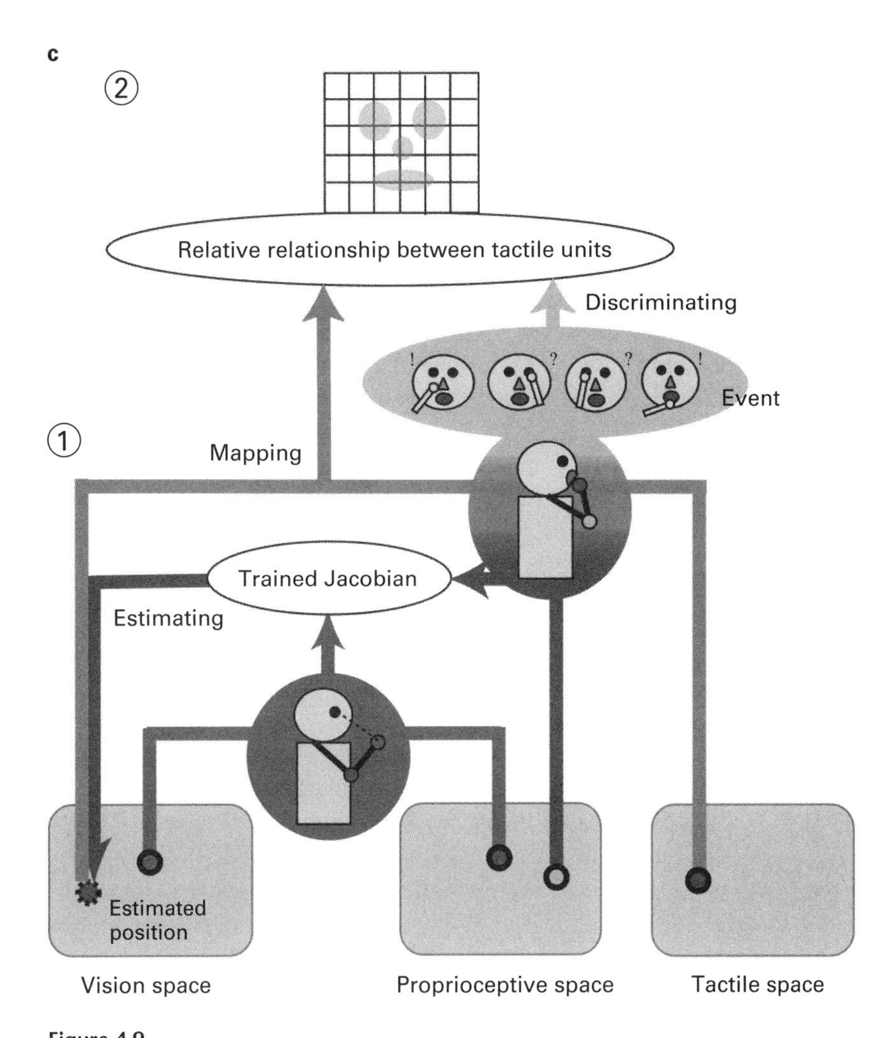

Figure 4.9
Development of face representation in a simulated infant: (a) front view; (b) the robot's hand model; (c) architecture of the model. From Fuke, Ogino, and Asada (2007). Figure courtesy of Sawa Fuke.

Related work by Bednar and Miikkulainen (2002, 2003) focuses on a more fundamental problem: why do newborn infants show a preference for orienting toward face-like stimuli? Bednar and Miikkulainen address this question by proposing that during prenatal development, internally generated spontaneous neural activity shapes ascending connections from the lateral geniculate nucleus (LGN) forward to face-selective areas (FSAs). Figure 4.10a highlights two stages of development. During prenatal development, the pathway is first stimulated by the PGO (ponto geniculo occipital) pattern generator; after birth, input from the PGO decreases and is replaced by visual stimulation from the retina. Figure 4.10b illustrates receptive field activity in the FSA under two training conditions. Specifically, in the prenatally trained network, face-like receptive fields have spontaneously formed, while in the naïve network, the receptive fields have a classical Gaussian response pattern. Figure 4.10c illustrates activity in the model—in response to both schematic and real faces—in the retina, LGN, and FSA. There are two key findings, which closely replicate the developmental pattern described by Morton and Johnson (1991). First, note that while the FSA responds to both schematic and real faces at the end of the prenatal training period (FSA-0; i.e., "birth"), it does not respond to schematic non-faces. Second, over 1,000 iterations of postnatal learning, the innate response to schematic faces declines, while the response to real faces is maintained (FSA-1000). By manipulating the frequency of particular faces that are experienced during the postnatal period, the Bednar and Miikkulainen model is also able to capture the preference for "mother's" face reported by Bushnell, Sai, and Mullen (1989).

An additional question concerns the development of the other-race effect. Furl, Phillips, and O'Toole (2002) explore this issue by comparing the performance of several face-perception models. In particular, they propose the *developmental contact hypothesis*: while early face perception reflects an unbiased weighting of face features, increased experience with one's own race "warps" this feature space, and results in a gradual decline in the ability to differentiate faces of other races. Furl, Phillips, and O'Toole (ibid.) tested their hypothesis by systematically biasing the set of faces experienced by each model during early learning, and successfully reproduced the development of the other-race effect. In contrast, the performance of an alternative model that is designed to represent the computational processes of early visual processing, but is not tuned by experience with specific faces, failed to capture the phenomenon. Taken together, these findings provide support for the idea that the other-race effect emerges "for free" as a consequence of perceptual specialization within a particular population of faces.

4.3 Perception of Space: Landmarks and Spatial Relations

In contrast to other areas of perceptual learning and development, there is comparatively less robotics research from the developmental perspective on spatial perception. One possible reason for this discrepancy is that some tasks (e.g., object recognition or

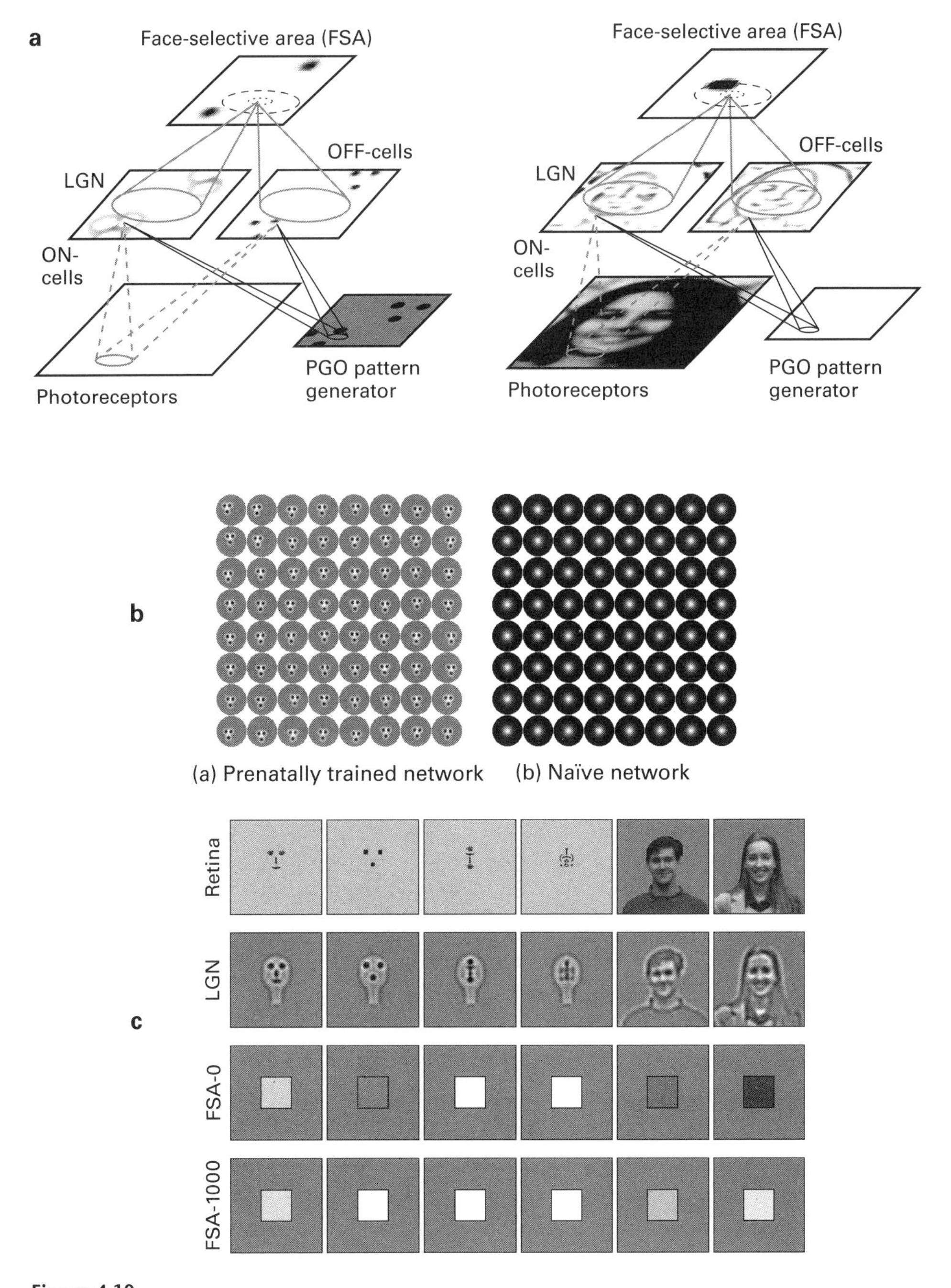

Figure 4.10

Neonatal model of face processing (Bednar and Miikkulainen 2002) highlighting the two stages of development (a), the receptive field activity in the FSA under two training conditions (b), and activity in the model—in response to both schematic and real faces—in the retina, LGN, and FSA (c). Figure courtesy of James Bednar. Reprinted with permission.

natural language processing) are inherently difficult, due to computational intractability, information-processing bottlenecks, sensor noise or uncertainty, and so on. As a result, researchers may look for new approaches such as developmental robotics to help study these tasks. Other tasks may receive less attention from the developmental perspective because they can largely be solved with the use of conventional AI methods. Navigation and spatial perception may fall in this latter category, due to the fact that it is relatively straightforward for the robot designer to provide the mobile robot with near-perfect location sensing. If the robotics researcher decides to adopt a developmental perspective, they must then rely on less accurate local-sensing methods, and approximate localization strategies such as dead reckoning.

Nevertheless, there is a good example of a robotics study of spatial perception that explicitly adopts a developmental approach. In particular, Hiraki, Sashima, and Phillips (1998) investigate the performance of a mobile robot on the search task described by Acredolo, Adams, and Goodwyn (1984), in which an observer is asked to find a hidden object, after being moved to a new location. What is especially appealing about the Hiraki, Sashima, and Phillips study is that it not only uses an experimental paradigm comparable to the task studied by Acredolo, Adams, and Goodwyn (ibid.), but it also proposes a model that is specifically designed to test Acredolo's hypothesis: that is, that self-produced locomotion facilitates the transition from egocentric to allocentric spatial perception.

The Hiraki, Sashima, and Phillips robot uses a combination of forward and backward models, in order to keep a target location in the center of its visual field. The *forward model*, which is acquired first, maps each planned movement to the predicted change in the location of the target within the visual field (see the right half of figure 4.11a). Next, the *inverse model* is trained to map the current sensory state to the optimal movement (the left half of figure 4.11b). After the forward and inverse models are acquired, the robot is then able to perform *mental tracking* by replacing the visual input from the robot's camera with the expected visual input that is generated by the forward model (see figure 4.11b). In other words, when visual information about the target location is not available, the Hiraki, Sashima, and Phillips model uses an internal representation—while compensating for its own movements—to estimate the location of the target.

In order to test the hypothesis that self-produced locomotion leads to allocentric spatial perception, the robot is trained in three stages. In stage 1, only head movements are produced. In stage 2, head and body movements are produced. Finally, in stage 3, the robot produces head, body, and wheel (i.e., locomotor) movements. Note that training of the forward and inverse models occurs within each of these stages. At the end of each stage, the robot is tested on the hiding task, in which the target object is hidden, and the robot is moved to a new location while being prevented from visually tracking the hiding location.

During stages 1 and 2 the robot performed at chance level on the search task (i.e., it searched for the hidden object in the egocentric location). In stage 3, however,

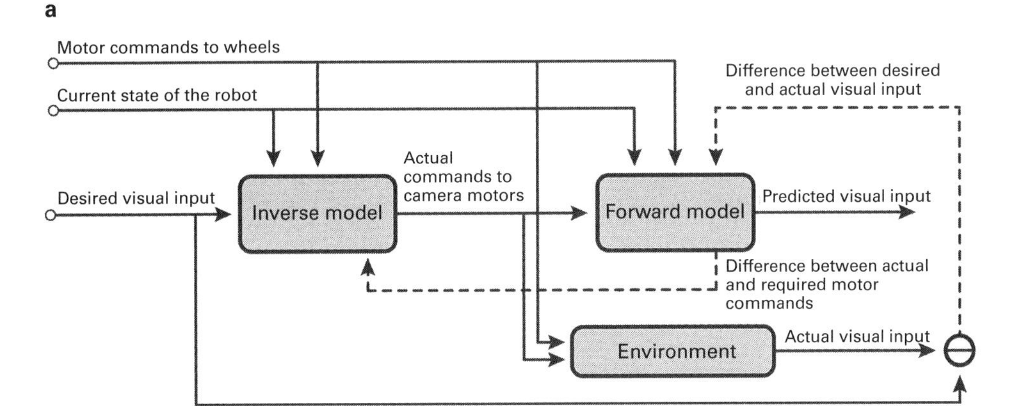

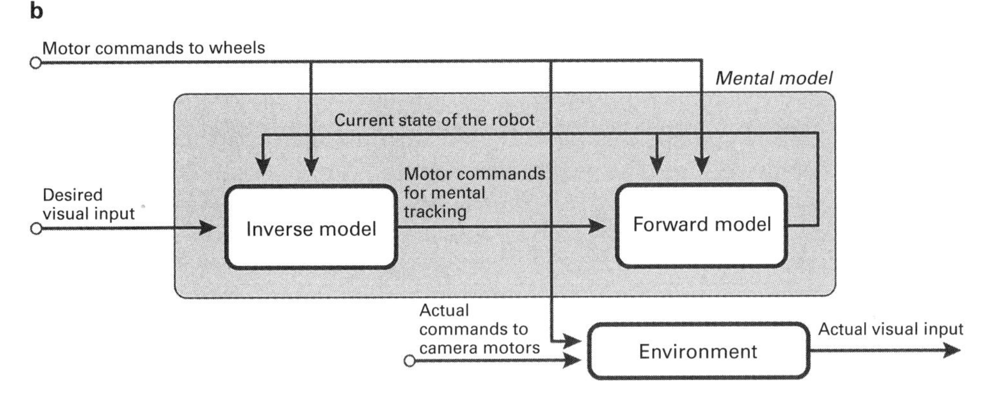

Figure 4.11
Diagrams of the models of visual tracking (a) and mental tracking (b) investigated by Hiraki, Sashima, and Phillips (1998). Figure adapted from ibid.

where self-produced movement became possible, an allocentric spatial coding strategy emerged, and the robot searched in the correct location on over 80 percent of the test trials. These findings not only support Acredolo's hypothesis regarding the relation between locomotion and allocentric perception of space, but they also suggest a specific cognitive-perceptual skill (i.e., mental tracking), which can serve as the developmental mechanism that makes this relation possible.

4.4 Robot Self-perception

As we noted in our review of self-perception in human infants, a key step in the development of self-perception is detecting the correlation or temporal contingency between self-produced movements and the sensory consequences of those movements. As a result, most robotic models of self-perception have applied the same general detection strategy: generate movement, observe the sensory consequences that follow the movement, and then label or categorize as "self" those sensory events or consequences that occur above some threshold level of chance. An example of this modeling approach is provided by Stoytchev (2011), who employed a two-stage process for simulating the development of self-perception. Colored markers were placed on the robot's 3-DOF arm, and during the first stage (i.e., "motor babbling"), the robot moved its arm randomly while it viewed its arm movements via a remote video camera (see figure 4.12b). The discovery strategy exploited by the model is illustrated in figure 4.12a. In particular, the model assumes that "self" is any visually detected motion that is observed within a precomputed delay after a self-produced movement (i.e., efferent and afferent signals, respectively). Following this logic, then, only the blue feature would be classified as "self" because change in this feature occurs within the efferent-afferent delay threshold. In contrast, the red feature moves too late, while the green feature moves too soon.

The goal of the first stage is to observe the distribution of efferent-afferent delays during the process of motor babbling. Figures 4.12b illustrates the movement of a robot that is alone in the environment, and the histogram of efferent-afferent delays that is observed by the robot in this movement condition. This histogram is used to compute an average efferent-afferent delay, which defines a temporal threshold or window and provides a basis for self-perception. During the second stage (not illustrated here), Stoytchev tested the robot's ability to accurately detect itself across a variety of new contexts, and found that it was highly accurate. An additional key finding concerns the context in which the efferent-afferent delay data are acquired. In particular, figure 4.12c illustrates an alternative training condition, in which the robot generates random arm movements in the presence of a second, randomly moving robot. An important consequence of learning in this condition is that the observed range of efferent-afferent delays is significantly smaller than when the robot observes its movements in isolation. Thus, this finding suggests that the capacity for self-perception is facilitated by self-observation in a noisy environment that includes multiple sources of sensory data.

While the Stoytchev model uses a relatively simple feature set (i.e., motion of the colored markers), other models have expanded the set of visual features so that more complex spatiotemporal relationships can be detected and exploited. Examples of these approaches include computing visual salience, rhythmic or periodic movement, and similarities in physical appearance (e.g., Fitzpatrick and Arsenio 2004; Kaipa, Bongard, and Meltzoff 2010; Michel, Gold, and Scassellati 2004; Sturm, Plagemann, and Burgard 2008). An important limitation of most of these models, however, is that they rely on a moment-to-moment match between internal states (e.g., motor commands) and external sensory data (e.g., visual motion). To address this issue, Gold and Scassellati (2009) propose a Bayesian modeling framework, in which belief estimates are

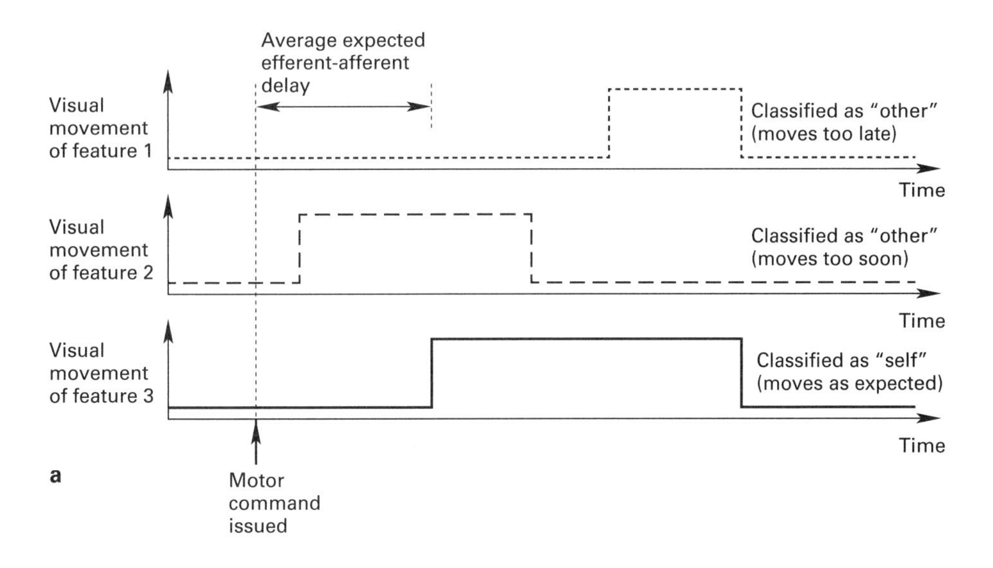

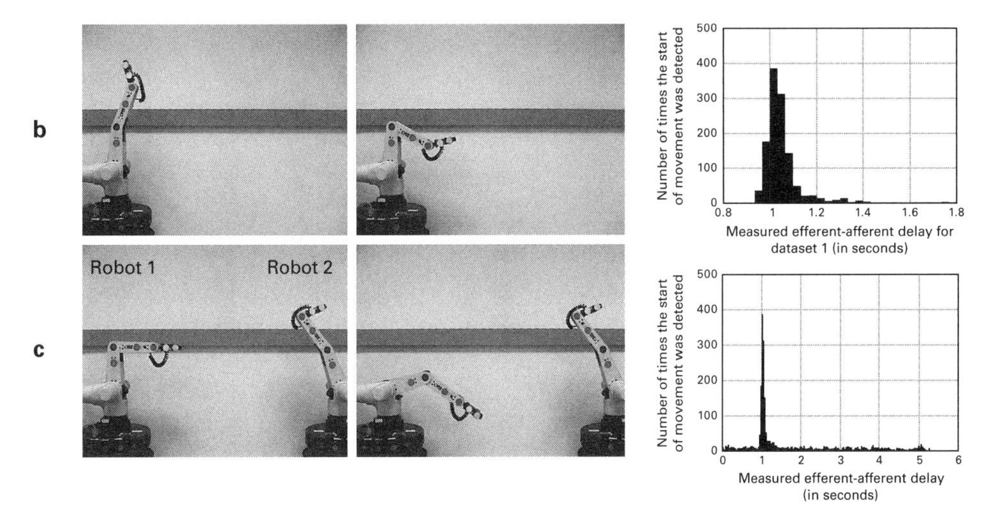

Figure 4.12
Illustration of the algorithm for self-perception (a), and examples of motor babbling and temporal latencies while alone (b), and with a second robot (c). From Stoytchev 2011. Reprinted with permission from Cambridge University Press.

computed and maintained over time for three dynamic belief models. The advantage of this approach is that it not only incorporates the entire observation history of the robot in its belief estimates, but it also allows those estimates to vary in time as new sensory data are acquired. During training, the self-perception model first uses motion cues to segment the visual input into discrete regions, and then tracks the movement of these regions over time. The first Bayesian model estimates the probability that an object corresponds to "self" (i.e., that its motion is correlated with self-produced motor signals), while the second and third models compute estimates for "animate" (other) and "inanimate," respectively. These belief estimates are established during a four-minute training period, in which the robot observes its arm moving, and are then tested by placing the robot in front of a mirror. During testing, the robot not only correctly identifies its reflection, but also differentiates its own movement in the mirror from the movements of a nearby experimenter.

Moreover, Asada et al. (1999) showed not only moment-to-moment matching, but also state vector estimation in the reinforcement-learning scheme. This leads to the automatic classification of three kinds of categories of self, passive agent (static object), and active agent (others), by applying a system-identification method. The capability to autonomously distinguish between self and the other passive/active agents can be utilized in real robot applications, as in the passer and shooter roles in RoboCup scenarios.

4.5 Developmentally Inspired Models of Object Perception

In the domain of object perception, two fundamental questions have been investigated thus far by researchers from the developmental robotics perspective. First, how can a robot learn to parse or segment the visual scene into discrete objects? One strategy for answering this question is to use motor skill development—particularly reaching, grasping, and object exploration—as a bootstrap for the development of object perception (e.g., Fitzpatrick and Arsenio 2004; Fitzpatrick et al. 2008; Natale et al. 2005). For example, Natale and collaborators (ibid.) describe a humanoid robot that learns to associate visual object features (e.g., colored surfaces or "blobs") with proprioceptive information that is acquired during grasping and exploration of the object. Proprioceptive information includes not only the configuration of the hand during grasping, but also other properties of the object such as weight. After a brief series of interactions with an object, the robot is able to visually segment the object when it is placed within a natural scene, and can also recognize it from multiple angles. Zhang and Lee (2006) propose an alternative approach that focuses on the role of *optic flow* as a perceptual cue for object segmentation. In particular, the Zhang and Lee model is presented with seminatural scenes that include moving objects. The model first detects optic flow, and then uses this feature to identify bounded regions that specify candidate objects; color and shape filters are then applied, to segment the bounded regions into objects. In a series of systematic comparisons, Zhang and Lee (ibid.) show that their model outperforms a conventional model that uses edge detection as the segmentation mechanism.

A second and perhaps more challenging question is: how can a robot learn to integrate multiple surfaces of an object that are separated by an occluder (i.e., perceptual completion)? The Zhang and Lee (2006) model addresses this question by constructing a stored representation for each object that it learns to recognize, and then applying a hand-built fill-in mechanism to identify the partially occluded object. However, while effective at a functional level, this approach is not able to explain how the fill-in mechanism itself develops. There are several other models of the development of perceptual completion that avoid specifying the fill-in mechanism as an "innate" or hand-built capacity (e.g., Franz and Triesch 2010; Mareschal and Johnson 2002; Schlesinger, Amso, and Johnson 2007). Each of these models captures one or more critical aspects of the developmental pattern observed in infants between birth and age four months when assessed with the unity-perception task (see box 4.1). We briefly highlight here the Schlesinger, Amso, and Johnson (2007) model, first because it simulates both perceptual processing and eye movements, and second because it has been tested on the same displays used with young infants (e.g., Amso and Johnson 2006).

The Schlesinger, Amso, and Johnson (2007) model uses an adapted version of the saliency-map paradigm proposed by Itti and Koch (2000). Processing in the model is illustrated in figures 4.13a–d: an input image (from the unity-perception task) is presented to the model (a); four synchronous visual filters extract intensity, motion, color, and oriented edges from the input image (b–c); and the image maps are pooled into an integrated salience map (d). Eye movements are then generated by selecting locations with large values from the salience map stochastically.

As we noted in box 4.1, three-month-old infants can be categorized as perceivers or nonperceivers, as a function of whether they prefer viewing the solid-rod or broken-rod test displays. In addition, recall that Amso and Johnson (2006) found that perceivers and nonperceivers generated different scanning patterns when they viewed the occluded-rod display. In particular, as figure 4.13e illustrates, perceivers fixated the rod (i.e., "rod scans") more often than nonperceivers. In contrast, there was no difference between the two groups in other fixation measures, including gaze shifts between the upper and lower halves of the display (i.e., "vertical scans"). These behavioral findings were replicated in the Schlesinger, Amso, and Johnson (2007) model, by hand-tuning a parameter that modulates the amount of competition that occurs between features within each feature map (figure 4.13c). In particular, an increase in this parameter value corresponds to increased competition within the posterior parietal cortex, and provides further evidence for Amso and Johnson's (2006) hypothesis that the development of perceptual completion is due to progressive improvements in oculomotor skill and visual selective attention. It is important to note, though, that these simulations'

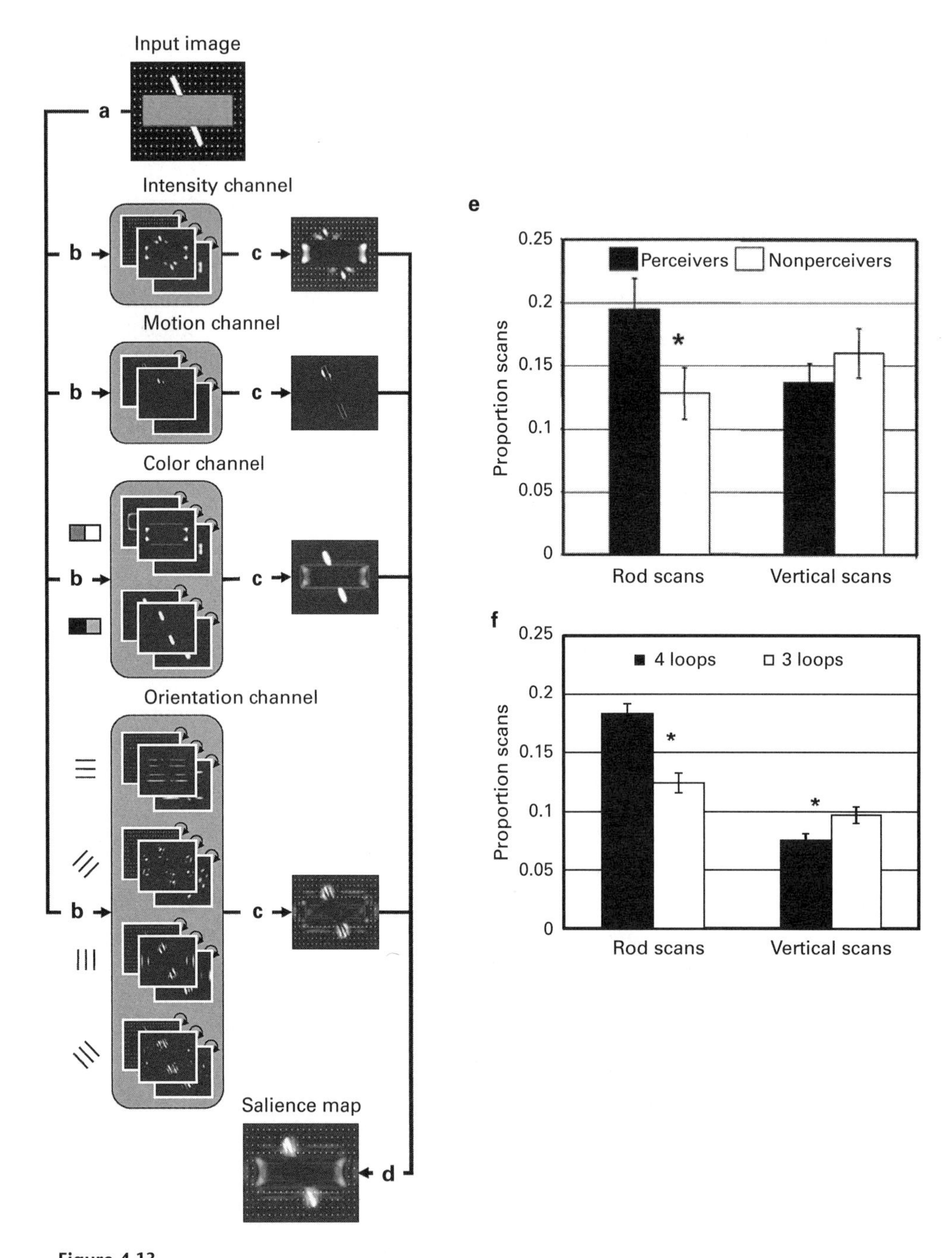

Figure 4.13
Schematic diagram of the Schlesinger, Amso, and Johnson (2007) model of perceptual completion (a–d), and a comparison of the performance of three-month-olds (e) and the model (f).

findings leave open the question of whether the predicted changes in parietal cortex and the corresponding perceptual improvements are due to maturation, experience, or both (Schlesinger, Amso, and Johnson 2012).

4.6 Affordances: Perceptually Guided Action

While the perception of affordances in robots is an active area of research, much of the work to date has focused on using affordances in a conventional cognitive robotics framework as a strategy for improving performance (see Sahin et al. 2007 for a comprehensive review). Nevertheless, several researchers have investigated the perception of affordances from a developmental perspective. This includes models that explore the role of affordances in prediction learning, motivation-based learning, and imitation (e.g., Cos-Aguilera, Cañamero, and Hayes 2003; Fritz et al. 2006; Montesano et al. 2008).

We highlight here two examples of studies that complement the type of work done with infants presented earlier (i.e., tool-use and locomotion/navigation). First, Stoytchev (2005, 2008) describes a robot that learns to perceive the affordances of several tools, which can be grasped and moved in different ways to retrieve a goal object. Figure 4.14a illustrates the robot's workspace, including a goal object (an orange hockey puck), several rigid, stick-like tools, and a square target location. During the preliminary stage of training, the robot first explores each tool in a trial-and-error manner by moving it in a variety of directions, while observing the effect on the hockey puck. The data acquired during exploration are used to populate an *affordance table*, which relates each of the tools to a particular action, a particular effect on the puck, and most important, a particular visual feature of the tool. Figure 4.14b provides a schematic illustration of a portion of the affordance table, for the T-shaped tool. For each action, the arrows indicate how the corresponding region of the tool affects the movement of the puck. After training the robot, Stoytchev evaluated its tool-use performance by presenting the robot with a tool, placing the puck on the table, and then using a search algorithm through the affordance table to identify a series of actions that moved the puck to the target location. Over a series of four goal locations and five tools, the robot succeeded on 86 percent of the test trials.

This work has two major implications for the study of affordance perception in human infants. First, the model provides a compelling demonstration for how affordances can be discovered through object exploration and trial-and-error learning. Thus, it may offer a key insight into how infants solve comparable problems, and in particular, which perceptual features they detect and exploit. Second, by periodically testing the model during the exploration phase, the model can also generate a developmental trajectory that can be directly compared to the pattern of infant development (e.g., Schlesinger and Langer 1999; van Leeuwen, Smitsman, and van Leeuwen 1994).

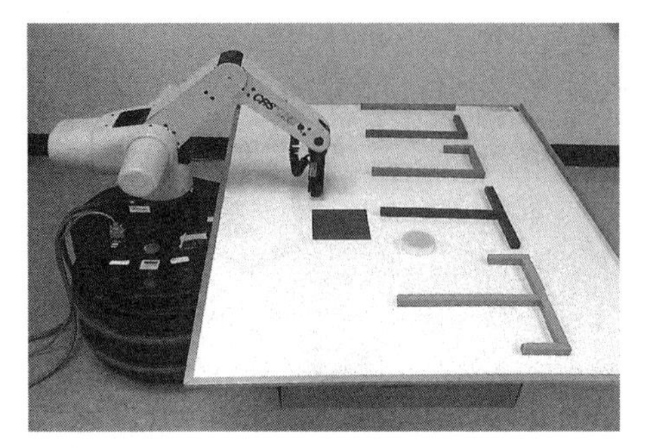

a

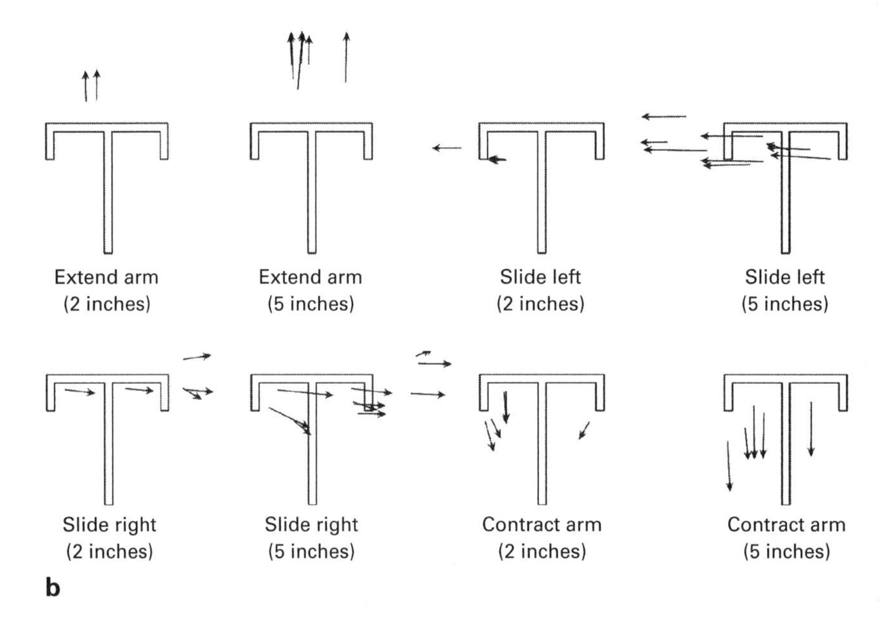

b

Figure 4.14
Learning tool-use affordances through object exploration illustrating the robot's workspace (a) and the portion of the affordance table for the T-shaped tool (b). From Stoytchev 2005 and 2008. Figure courtesy of Alexander Stoytchev. Reprinted with permission from IEEE.

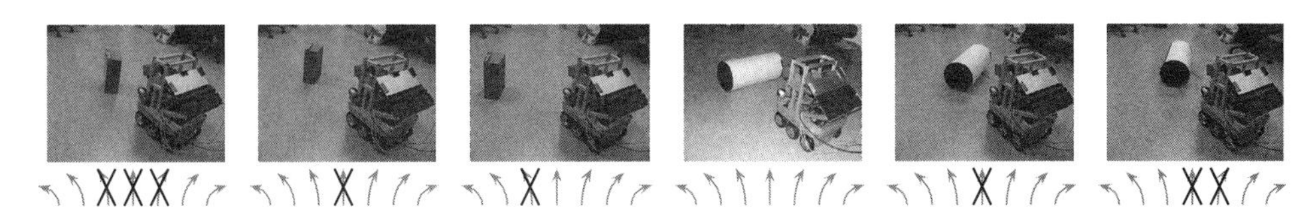

Figure 4.15
Learning to perceive "traversability" affordances in a mobile robot. From Ugur et al. 2007. Figure courtesy of Erol Sahin. Reprinted with permission from IEEE.

A second area of research on affordance perception in robots focuses on navigation and obstacle avoidance. For example, Sahin et al. (2007; Ugur et al. 2007) propose a model for learning affordances in a mobile robot. Like the Stoytchev model, the Sahin et al. model robot first learns to relate its movements through the environment with changes in its visual sensors (obtained from a 3D range scanner). Affordance categories emerge through a classifying process, in which a subset of all visual features is weighted as a function of the movement outcome. For example, figure 4.15 illustrates the robot's performance in the real environment, after training in simulation, in response to a series of novel objects. In each case, the robot perceives the affordance of "traversability" when an object is placed in its path, and it successfully associates with its ongoing perception the set of movements that will enable it to move around the obstacle. It should be noted that although models like this have not yet been tested on the same experimental paradigms used to study infant locomotion and affordance perception (e.g., visual cliff, crawling and walking on slopes, etc.), there are no reasons in principle to prevent this.

4.7 Conclusions

In this chapter, we began by first introducing the two primary paradigms used to study visual perception in infants—preferential looking and habituation-dishabituation—and describing the rationale and assumptions of each method. The fact that these two methods are widely used raises an important question: should developmental robotics researchers also use the same paradigms to assess the performance of their models? For the time being, such models are relatively uncommon (e.g., Chen and Weng 2004; Lovett and Scassellati 2004). Instead, most work to date on simulating the development of visual perception has tended to employ generic tasks or skills, such as reaching and grasping or navigation, rather than replicating specific behavioral paradigms or experiments. Hopefully this gap will continue to narrow in future years.

Our brief survey of face perception highlighted the empirical finding that infants rapidly learn not only to detect faces, but also to differentiate between familiar and novel faces. Systematic study of infants' responses to both natural and artificial face-like stimuli provides support for a two-stage pattern of development, in which an innate or early-emerging mechanism promotes orienting to face-like stimuli, while a subsequent mechanism enables learning of specific facial features. We also noted that face perception continues to develop well into the first year, including changes in the ability to differentiate male and female faces, as well as members of the same or different race groups.

Spatial perception develops more slowly than face perception, in part, due to the fact that spatial relations such as depth and distance are closely tied to movement *through* space. Thus, the onset of crawling and walking are also associated with important changes in how infants perceive space. In addition, infants have an egocentric bias in how they perceive space: relations between objects and locations and landmarks are initially encoded in relation to self-position. Between ages nine and twelve months, however, infants gradually transition to an allocentric perception of space, in which spatial relations are viewed objectively.

Like the perception of space, self-perception also benefits from the ability to move. Thus, a fundamental cue for perceiving the self is spatiotemporal contingency between one's own movements, and the seen motion of another object or body in the environment. Indeed, infants become increasingly skilled at detecting their own movements when placed in front of a mirror (or when viewing themselves on a video display, etc.).

We also noted some of the important developmental achievements that occur in the perception of objects during infancy. First, there is some evidence that neonates are able to transfer sensory information about an object from one modality to another. Second, between ages one and three months, infants acquire the ability to link the surfaces of an object together into a coherent percept of a solid object. Over the next several months, infants become increasingly skilled at perceiving objects that are partially or fully occluded.

The final topic we surveyed was the development of affordance perception, which spans both visual development and motor-skill acquisition. One area of affordances that has been relatively well studied with infants is the ability to use simple tools like supports, strings, and hooks; use of tools such as these begins as early as age six months, and continues to improve during the second year. A second type of affordance perception develops in parallel with the ability to crawl and walk, as infants discover the visual cues that signal the relative safety or danger of slopes, stairs, and other surfaces in their environment.

Developmental models of face perception, for the most part, have focused on the computational mechanisms that support detection of, and discrimination between, faces. In particular, we described two models that rely on prenatal experience as a critical influence. For example, Fuke, Ogino, and Asada (2007) proposed that infants develop a multimodal face representation through the process of exploring their own face with their hands. Alternatively, Bednar and Mikkulainen (2003) describe a neural pattern generator that stimulates the visual pathway during prenatal development, resulting in a postnatal ability to process face information. In addition, we also described computational modeling work on the other-race effect, which examines the biasing effect on face perception of viewing members of a racial/ethnic group more frequently than nonmembers.

Our review of spatial perception focused more narrowly on the transition from an egocentric to allocentric reference frame. In particular, we presented the Hiraki, Sashima, and Phillips (1998) model, which uses a mobile robot to simulate the experimental paradigm investigated by Acredolo, Adams, and Goodwyn (1984). There are two important features of the model worth noting. First, Hiraki, Sashima, and Phillips (1998) evaluated the model using a procedure that was analogous to the method employed by Accredolo and colleagues. Second, they also explicitly tested and confirmed Accredolo's hypothesis, which proposed that self-produced locomotion promotes the shift to allocentric perception of space.

We next described a series of modeling and robotics studies on the development of self-perception. Perhaps not surprisingly, the dominant research strategy for much of this work is for the robot to move parts of its body, and to then discover which sensory events in the environment correlate with self-produced movement (which are then labeled as "self"). In general, these models have been largely successful in accounting for the capacity for self-perception, which confirms the role of spatiotemporal contingency as a fundamental cue. An open challenge for future work, however, is to expand this account so that it can help explain the full trajectory of self-perception development from early infancy into the second year.

Like other developmental models of visual perception reviewed in this chapter, models of object perception seek to capture key features of the developmental process observed in human infants. The Zhang and Lee (2006) model, for example, exploits optic flow as a perceptual cue for object segmentation. A potential weakness of this model, however, is that it may treat a partially occluded object as two distinct objects rather than one. The Schlesinger, Amso, and Johnson (2007) model offers a solution to this problem by demonstrating that an increase in spatial competition produces increased attention to surfaces of an object that are separated by occlusion.

The final topic that we surveyed was robotic models of affordance perception. For example, Stoytchev (2008) describes a robot that learns through trial and error to select a hook-like tool, which it then uses to retrieve a goal object. An important feature of the model is that while it was not specifically assessed with a tool-use paradigm corresponding to an infant study, doing so would be a relatively straightforward extension of the model. A second area of work focuses on movement through a cluttered environment, and in particular, learning to perceive whether a space between two objects is wide enough to allow passage.

Additional Reading

Gibson, J. J. *The Ecological Approach to Visual Perception.* Hillsdale, NJ: Lawrence Erlbaum Associates, 1986.

Gibson's text is an essential read for students of perception, whether they study humans, animals, or machines. His book introduces the reader to the *ecological theory of perception,* which resonates with two fundamental themes in developmental robotics: (1) that the structure of the physical environment is perceived through agent-environment interaction, and (2) that the agent does not passively "absorb" the details of the environment, but rather, the agent learns through active exploration. A key concept introduced by Gibson in this text is the notion of *affordance.*

Fitzpatrick, P., A. Needham, L. Natale, and G. Metta. "Shared Challenges in Object Perception for Robots and Infants." *Infant and Child Development* 17 (1) (Jan.–Feb. 2008): 7–24.

Fitzpatrick and colleagues' seminal paper not only provides an eloquent and convincing argument for the mutual benefits of studying infants and robots, but more specifically, also systematically sketches a set of common goals for investigating perceptual development. As a case study, the paper introduces the problem of visual object segmentation (also know as "segregation"), and weaves together a discussion of ideas and research methods that bridge human and machine-learning approaches.

5 Motor-Skill Acquisition

For human infants, mastering two basic sets of motor skills—manipulation (e.g., reaching and grasping) and locomotion (e.g., crawling and walking)—provides an essential foundation for many of the behaviors and abilities that develop during the first two years of postnatal life. In addition, for parents, the emergence of these core motor skills is dramatic and rapid: for example, at age three months, their child learns to roll over, at six months she is sitting up, by eight months she has started to crawl, and on her first birthday, she takes her first steps. As we discussed in chapter 3, these developmental milestones appear to be intrinsically motivated, that is, rather than serving some immediate need or goal, they appear to be driven by the desire to improve the skill itself (e.g., Baldassarre 2011; von Hofsten 2007). While infants may benefit from observing their parents or siblings, family members typically provide no direct instruction, feedback, or assistance to the infant during motor-skill acquisition. And yet, these skills consistently develop in most infants on a predictable schedule (e.g., Gesell 1945).

In this chapter, we survey the core motor skills that infants develop in the first two years, and contrast this developmental pattern with the data obtained from models and robotics experiments that simulate learning of the same skills. In particular we note, as we illustrated in chapter 2, that recent advances in the design and construction of child-sized robots allow developmental robotics researchers to study the rich spectrum of physical behaviors that infants and children produce. Thus, robot platforms provide a unique and valuable tool for investigating motor-skill acquisition: because they are designed to mimic the size and shape of human children, and in some cases also capture physical processes at the skeletal and muscle levels, the study of humanoid robots like iCub, NAO, and CB^2 may reveal fundamental principles underlying the emergence of these skills in both natural and artificial organisms.

It is important to note that developmental robotics not only differs from conventional robotics in terms of the size and strength of the robot platforms, but also in terms of the modeling philosophy itself. Thus, a traditional approach often found in "adult" humanoid robotics is to first estimate the robot's current position (using vision, joint-angle sensors, etc.), and then to compute the necessary change in joint angles and joint torques or forces that will produce the desired movement or end position, or both. In other words, this approach focuses on solving the *inverse-kinematic* and *inverse-dynamic problems* (e.g., Hollerbach 1990). In contrast, as we will highlight, an alternative strategy employed in developmental robotics is instead to learn a mapping between spatial locations and joint positions and joint forces through the production of a wide range of exploratory movements. The key difference here is that the developmental approach typically focuses on *learning* of motor skill by trial and error, rather than by computing a desired movement trajectory in advance.

As an example of the kind of research question that developmental robotics is well suited to address, consider the following developmental pattern. The production of smooth, skilled reaching movements is a long-term achievement that takes human children two to three years to master (e.g., Konczak and Dichgans 1997). Along the way, infants often pass through a period of development in which their movements are comparatively jerky and appear somewhat reflex-like (e.g., von Hofsten and Rönnqvist 1993; McGraw 1945; Thelen and Ulrich 1991). Shortly after birth, for instance, neonates generate spontaneous hand and arm movements that are visually elicited (e.g., Ennouri and Bloch 1996; von Hofsten 1982). While these "prereaching" movements are directed toward nearby objects, infants rarely make contact with the target. Over the next two months, prereaches decline in frequency. At age three months, a more robust and organized set of reaching behaviors begins to emerge, which includes several important and new properties (e.g., the hand or grasp preshape, corrective movements, etc.; Berthier 2011; von Hofsten 1984). A similar pattern is also observed during the development of walking: very young infants produce a highly stereotyped stepping behavior (i.e., the stepping reflex), which disappears by age three months, but can be elicited in older infants by placing them in water or supporting them on a treadmill (e.g., Thelen and Ulrich 1991). Like the development of reaching, infants begin to crawl and subsequently walk a few months after the reflex-like stepping behavior has disappeared.

To date, there are no computational models that account for this U-shaped developmental pattern. However, an important concept that plays a critical role in many models of motor-skill acquisition, and that may help explain the U-shaped pattern, is the phenomenon of *motor babbling* (e.g., Bullock, Grossberg, and Guenther 1993; Caligiore et al. 2008; Kuperstein 1991). Analogous to preverbal babbling in human infants, motor babbling is a more general phenomenon, in which infants learn to control their bodies by actively generating a wide variety of movements in a trial-and-error fashion. Thus, the precocious or early movements that infants produce (such as prereaches) may also be understood as a form of motor babbling. What remains unexplained, however, is why these movements diminish and later reemerge in a more mature form. One possible explanation is that the "decline" phase represents the transition from ballistic, stereotyped movements to visually guided action, which requires integrating multiple sensory inputs (e.g., vision and proprioception; Savastano and Nolfi 2012; Schlesinger, Parisi, and Langer 2000).

In the next section, we describe the major milestones observed in infants during the development of manipulation and locomotion. In particular, we highlight four fundamental skills: reaching, grasping, crawling, and walking. We focus on these skills for two reasons. First, not only are they essential survival skills for the species, but they also have profound influence on both perceptual and cognitive development (e.g., Kermoian and Campos 1988). Second, manipulation and locomotion are also comparatively well-studied skills within robotics, and in particular, from the developmental robotics perspective (e.g., Asada et al. 2009; Shadmehr and Wise 2004). After reviewing the pattern of human development, therefore, the remainder of the chapter surveys the array of models that have been proposed for simulating motor-skill acquisition, and the major findings provided by this work.

5.1 Motor-Skill Acquisition in Human Infants

In chapter 4, we noted Greenough and Black's (1999) concept of *experience-expectant* development, that is, a pattern of development in which a skill or ability reliably and consistently emerges across all members of the species. This pattern contrasts with *experience-dependent* development, in which a particular skill or ability only emerges under a set of very specific conditions. Both of these developmental patterns are observed during motor-skill acquisition in human infants. On the experience-expectant side are behaviors such as reaching and walking, which are—barring developmental pathology or pathological rearing conditions—universal skills that all children acquire, regardless of cultural-historical context, geographical region, native language, and so on. Other early skills that fall in the same category are oculomotor control, trunk and neck control, and nursing. In contrast, experience-dependent skills typically develop at a later age, and require explicit instruction, such as swimming, playing musical instruments (e.g., violin or piano), and drawing. These skills also vary widely across cultural, historical, and geographical contexts.

In this section, we focus on the development of experience-expectant abilities, and in particular, on four motor skills that emerge in infancy: reaching, grasping, crawling, and walking.

5.1.1 Manipulation: Reaching

To help provide a context for our discussion of the development of reaching, table 5.1 summarizes the major milestones in reaching and grasping that occur during the first two years. As we noted in the introduction, the earliest example of reaching behavior is observed in newborn infants, who produce brief forward extensions of their hand toward nearby objects (e.g., Bower, Broughton, and Moore 1970; Ennouri and Bloch

Table 5.1
Timescale and major milestones for reaching and grasping development in human infants (adapted from Gerber, Wilks, and Erdie-Lalena 2010)

Age (months)	Competence
0–2 months	Grasp reflex Prereaching appears and increases in frequency
2–3 months	Prereaching declines in frequency
3–4 months	Onset of reaching Hands held predominately open
4–6 months	Hand preshape emerges Palmar/power grasp
6–8 months	Radial-palmar grasp Scissors grasp Hand preshape predominates
8–12 months	Radial-digital grasp Pincer/precision grasp Online reach corrections
12–24 months	Adult-like reaching kinematics Prospective (prereach) arm control

1996; Trevarthen 1975; von Hofsten 1984). Figure 5.1 illustrates the apparatus used by von Hofsten (1984) to study the development of prereaching movements in young infants. Von Hofsten (ibid.) reported an increase in object-oriented forward extensions of the infant's hand between birth and two months. The behavior then appears to decline between seven and ten weeks; interestingly, as prereaches become less frequent, infants spend increasingly more time fixating the target object. Between ten and thirteen weeks, reaching movements become more frequent again, this time co-occurring with fixations of the object. In addition, at this age the tendency to open the hand during the *transport phase* (i.e., movement of the hand toward the target) also increases (e.g., Field 1977; von Hofsten 1984; White, Castle, and Held 1964). Taken together, these findings support the idea that early prereaches are spontaneous, synergistic movements, which become temporarily suppressed as intentional, visually guided reaching begins to emerge between three and four months.

The U-shaped pattern of reaching development from birth to age three months not only implicates the role of vision as a major influence, but also raises an important and interesting question: do infants learn to control the movement of their hands by visually guiding them toward the target (e.g., Bushnell 1985; Clifton et al. 1993; McCarty et al. 2001a)? Clifton et al. (1993) addressed this question by following a group of infants longitudinally between ages six and twenty-five weeks, and found that the onset of reaching for objects in an illuminated room, versus objects that glowed (or made a

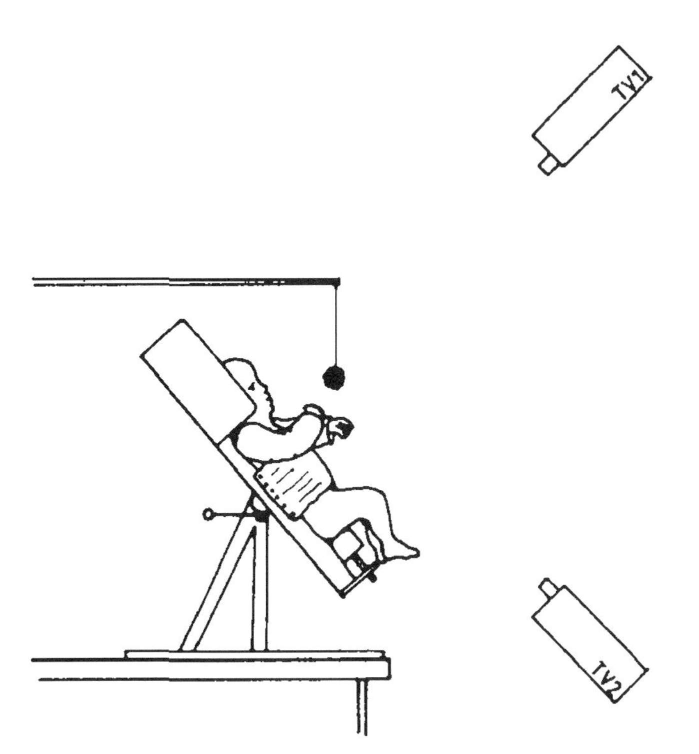

Figure 5.1
Experimental setup used by von Hofsten (1984) to study prereaching in young infants. Copyright © 1984 by the American Psychological Association. Reproduced with permission.

sound) in the dark, did not differ statistically. In both cases, the average age of the first successful reach was twelve weeks in both conditions, and the average age of the first successful grasp followed at fifteen weeks. These data suggest, at least at the onset of reaching, that localization of the target and proprioceptive input from the hand and arm are sufficiently well coordinated. It remains an open question, however, if this coordination is achieved in the weeks prior to reach onset through visual feedback of the hand.

After the onset of reaching around age four months, there are several subsequent improvements. For example, between ages four and twelve months, infants become more effective at using visual information from the target to update their reaching movements. When presented with targets that appear to abruptly change location during a reach, five-month-olds typically reach toward the original target location, while nine-month-olds adjust their trajectories mid-reach (Ashmead et al. 1993). In addition, guidance of the hand toward the target is not only updated online using vision of the target, but also with stored information about the target location. For example, as early as age five months, infants will continue to reach for an object even if it is darkened during mid-reach (McCarty and Ashmead 1999). Infants at this age can also learn to adapt their reaching movements while wearing displacing prisms, which shift the apparent target location (e.g., McDonnell and Abraham 1979).

In addition, there are a number of changes in the kinematic properties of infants' reaching movements that continue to occur after the onset of reaching. One important feature is that infants' reaches become progressively straighter and smoother with age (e.g., Berthier and Keen 2005). This pattern is due in part to the fact that infants begin reaching by holding the elbow relatively fixed, while rotating the shoulder, which results in comparatively rounded or curved reaching trajectories (e.g., Berthier 1996). During the second year, infants begin to coordinate elbow and shoulder rotation (e.g., Konczak and Dichgans 1997). Another feature is that infants' hand-speed profiles become more adult-like; in particular, the peak movement speed shifts "backward" in time, closer to the onset of the reach (e.g., Berthier and Keen 2005). This shift reflects a tendency toward producing one large, rapid initial movement, followed by a subsequent series of smaller, slower corrective movements as the hand nears the target.

5.1.2 Manipulation: Grasping

While there is a three- or four-month gap between the earliest form of reaching (i.e., prereaching in neonates) and the onset of voluntary grasping, the two behaviors overlap considerably, not only in real time but also in their respective developmental periods. Thus, soon after the onset of visually controlled reaching movements, the palmar grasp also emerges and begins to develop. We highlight here two important patterns that occur during the development of grasping.

First, infants' grasping behaviors follow a consistent and predictable developmental pattern between ages six and twelve months (see figure 5.2; Erhardt 1994; Gerber, Wilks, and Erdie-Lalena 2010). The earliest voluntary grasp is the *palmar grasp* (or *power grasp*), in which infants approach an object with an open hand, and once contact is made with the palm, enclose the object with their fingers. The palmar grasp then transitions into the *scissors grasp* at eight months, where the four fingers form a unit that opposes the thumb. At age nine months, only two fingers (the index and middle fingers) oppose the thumb, creating the *radial-digital grasp*. This is followed at ten months

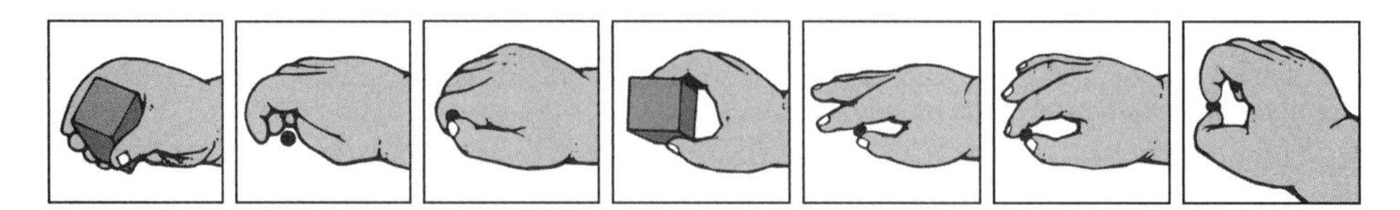

Figure 5.2
Development of grasping between six and twelve months (Erhardt 1994).

by an early form of the *pincer grasp* (or *precision grasp*), characterized by use of the thumb and index finger; by age twelve months, infants have mastered this technique, and can now use the *mature pincer grasp* to pick up small objects, such as pieces of food.

In addition to the role of grasping as a fine-motor skill, its development also reflects a second important function: planning and prospective action. In chapter 3, we touched on this issue in our discussion of prediction-based intrinsic motivation, as well as in chapter 4, where we described the development of the perception of affordances. In the context of grasping, this issue has been explored by investigating the development of *hand* (or *grasp*) *preshaping*. The hand preshape is an orienting of the hands and fingers during the transport phase (i.e., movement of arm and hand toward the target), as a function of the size, shape, and position of the target object.

The emergence of the hand preshape lags several weeks behind the onset of reaching: until approximately age four months, infants typically reach with their hand in a stereotyped, open configuration, regardless of the size and shape of the target object (e.g., Witherington 2005). The earliest form of preshaping is orienting of the hand to match the orientation of the target, which begins to emerge at age four and a half months (e.g., horizontally vs. vertically oriented rods; von Hofsten and Fazel-Zandy 1984; Lockman, Ashmead, and Bushnell 1984). Orienting of the hand to match the target object orientation continues to improve over the next several months. For example, by nine months infants will correctly orient their hand, given a preview of the target but no visual information once the reach is initiated (e.g., McCarty et al. 2001a).

A second, more complex dimension of hand preshape is correctly positioning the fingers prior to contacting the object. Newell et al. (1989) investigated this skill by comparing the grasps of infants between ages four and eight months as they reached toward a small cube and three differently sized cups. Interestingly, infants at all ages distinguished between the objects by generating size- and shape-specific grasp configurations (e.g., small cube versus large cup). However, the timing of the hand shape varied with age: in particular, the youngest infants relied almost exclusively on the strategy of shaping the hand *after* making contact with the object. With age, infants progressively increased the frequency of shaping their hand *prior* to contact. McCarty, Clifton, and Collard (1999) report a similar developmental pattern in infants aged nine to nineteen months, on a more complex task that involves grasping objects with handles (e.g., a spoon with applesauce) oriented in different directions. We describe this study in detail in box 5.1.

5.1.3 Locomotion: Crawling

Table 5.2 presents the major stages in the development of crawling and walking. While infants do not begin to locomote independently until approximately age six or seven months (i.e., "creeping"), there are several prior achievements that precede the development of crawling that help make it possible. For example, infants typically begin to

Box 5.1
Prospective Grasp Control: The "Applesauce" Study

Overview

McCarty, Clifton, and Collard (1999) describe an elegant and simple method for studying planning and prospective action in infants: a spoon is loaded with applesauce, and placed in front of the child, on a wooden stand (see figure below). On some trials, the spoon is oriented with the handle on the left, while on other trials the handle is on the right. As a grasping task, the spoon can be gripped in three ways: (1) the handle can be grasped with the ipsilateral hand (i.e., the one on the same side as the handle), producing a radial grip; (2) the handle can be grasped with the contralateral hand, producing an ulnar grip; or (3) the hand on the same side as the goal can grasp the "bowl" of the spoon directly, producing a goal-end grip (see top, middle, and lower hand-spoon diagrams, respectively). The radial grip is the most efficient, allowing the goal-end to be directly transported to the mouth. In contrast, the ulnar grip requires either transfer of the handle to the opposite hand before transport to the mouth, or an awkward orientation of the grasping hand and arm before eating. McCarty, Clifton, and Collard (ibid.) suggest that if infants want to retrieve the applesauce as efficiently as possible, they should (1) show a bias toward producing a radial grip, and therefore, (2) prospectively select the ipsilateral hand prior to reaching for the spoon.

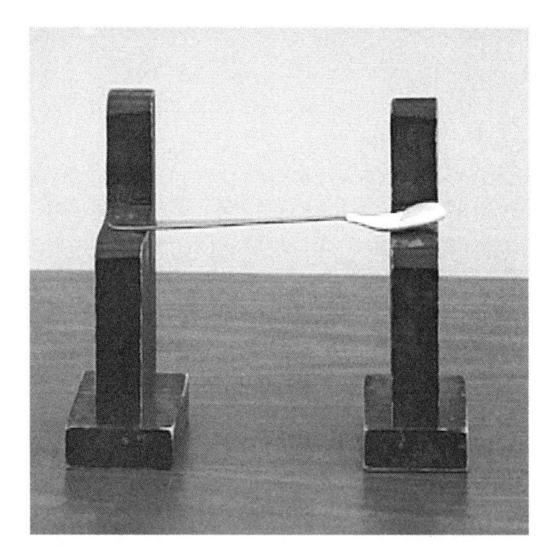

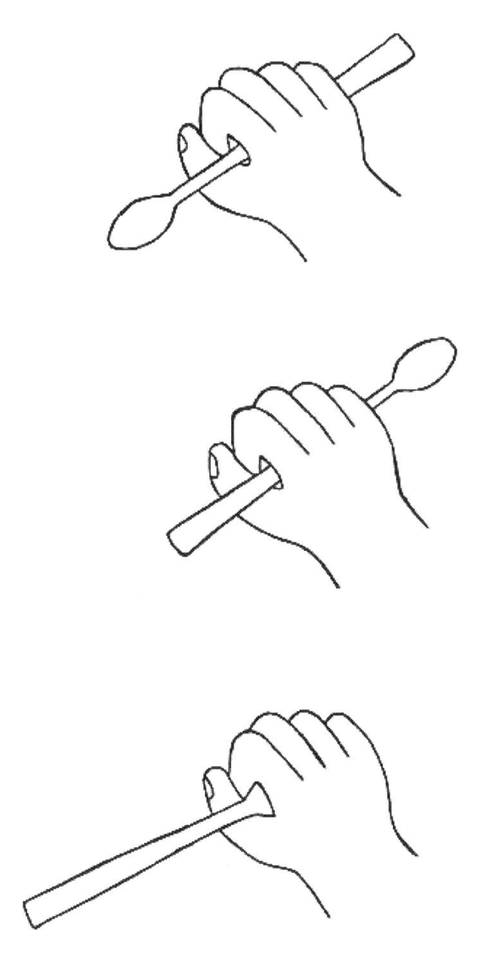

Procedure

Infants at ages nine, fourteen, and nineteen months participated in the study. Prior to the applesauce task, each infant's hand preference was assessed by placing a series of toys at midline, and noting which hand was used to grasp each object. Infants were then presented with a series of test trials, including first a set of toys mounted on handles (e.g., a rattle) and then the applesauce-loaded spoon. The orientation of the handle was alternated across trials.

Results

The pattern of results across the two handle conditions (i.e., toys vs. applesauce) was comparable. We present here the findings from the applesauce condition. After each infant's hand preference was determined, the test trials were divided into two categories: *easy trials* are those in which the handle of the spoon was on the same side as the infant's dominant hand, while *difficult trials* are those in which the handle of the spoon was on the opposite side. The line chart below presents the proportion of radial, ulnar, and goal-end grips for the three age groups, during easy and difficult trials, respectively (upper and lower plots). On easy trials, the proportion of radial grips was significantly higher for older infants; thus, even when the handle was oriented in a direction that facilitated the radial grip, younger infants produced it less often than older infants. On the difficult trials, a similar pattern emerged. However, as the lower plot illustrates, nine-month-olds were equally likely to use all three grips when the handle was on the opposite side of their dominant hand. The tendency to produce a radial grip on difficult trials (i.e., with the nondominant hand) increased with age, reaching almost 90 percent by age nineteen months.

Easy Trials

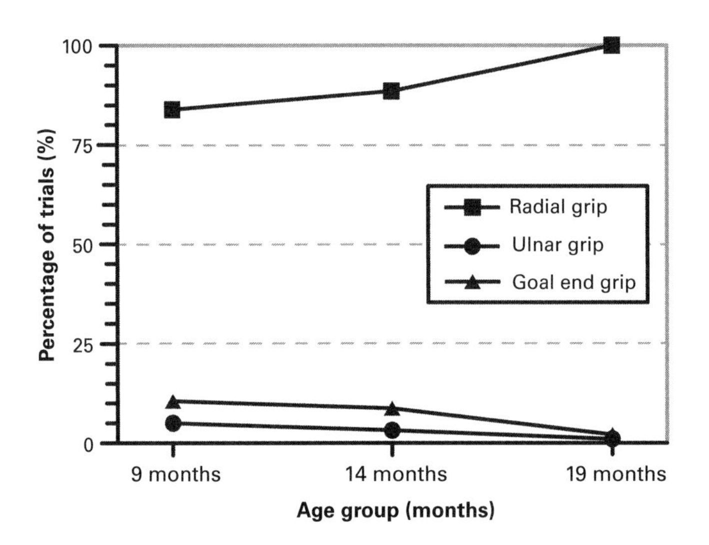

In a follow-up study, McCarty, Clifton, and Collard (2001b) used a similar paradigm to study how infants grasped (and subsequently used) tools with handles, such as a hairbrush, and how prospective grasping behavior varied as a function of whether the goal action was directed toward the self or toward an external goal. As in their previous study, McCarty and colleagues (ibid.) found that radial grips increased in frequency with age. In addition, they also found across all ages that radial grips were more likely in the context of a self-directed action than an externally directed action.

Difficult Trials

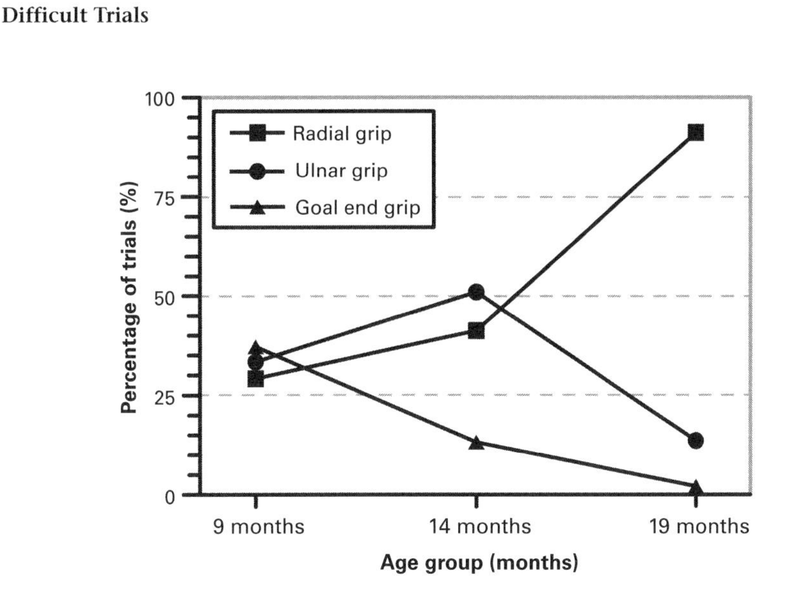

Source: Figures copyright © 1999 by the American Psychological Association. Reproduced with permission. Courtesy of Michael McCarty.

roll over on their side at three months, from front to back at four months, and from back to front at five months (Gerber, Wilks, and Erdie-Lalena 2010). Most infants can also sit without support at six months. These developments in trunk control indicate not only significant gains in gross motor skill, but also increased strength in the same muscle groups that are used during locomotion.

In the month prior to the onset of crawling, infants often produce a number of behaviors that fall into the category of "precrawling," including alternating between prone and sitting positions, rocking on the hands and knees, and pivoting or rotating in the prone position (e.g., Adolph, Vereijken, and Denny 1998; Goldfield 1989;

Table 5.2
Major stages of locomotion development during the first year (adapted from Vereijken and Adolph 1999)

Age (months)	Stages of locomotion development
0–6 months	Immobility
7 months	Belly crawling ("creeping")
8 months	Hands-and-knees crawling
9 months	Sideways cruising
10 months	Frontward cruising
12 months	Independent walking

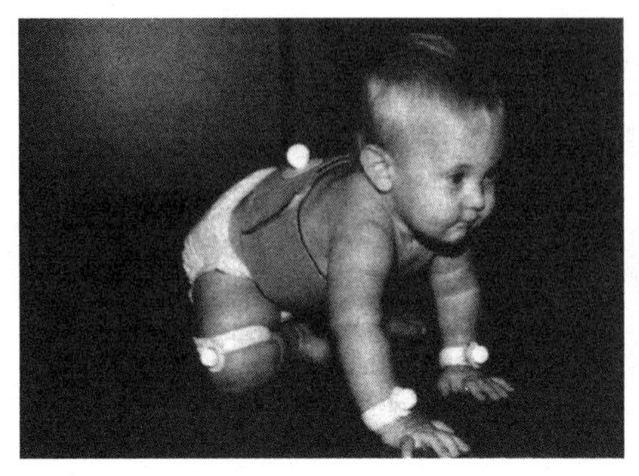
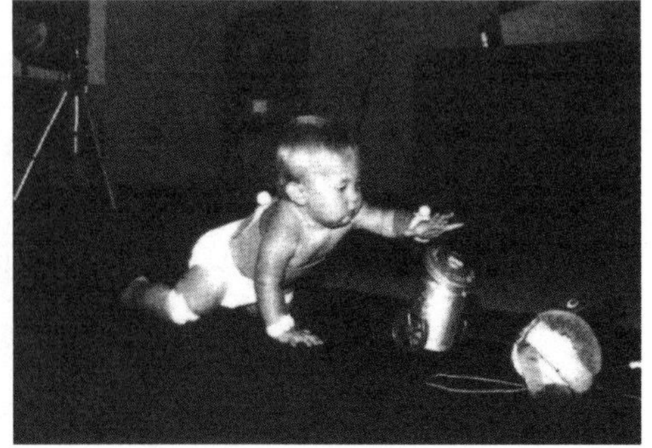

Figure 5.3
An infant participant demonstrating hands-and-knees crawling in the Freedland and Bertenthal (1994) study.

Vereijken and Adolph 1999). Adolph, Vereijken, and Denny (1998) also identified a precursor form of crawling at age seven months that they described as "belly crawling" or creeping, in which the infant remains prone on his stomach and pushes with his legs. While experience with belly crawling did not influence the average onset age of hands-and-knees crawling, Adolph, Vereijken, and Denny (ibid.) also reported that at onset, belly-crawlers were more efficient at hands-and-knees crawling, and also more consistent in "diagonal" alternation of the limbs (e.g., coordinated movement of the left leg and right arm).

Figure 5.3 illustrates an example of hands-and-knees crawling in a young infant. The emergence of the diagonal pattern during the development of hands-and-knees crawling, around age eight months, is important for several reasons. First, it reflects a relatively skilled level of timing and coordination of the arms and legs (e.g., Freedland and Bertenthal 1994). Second, it also affords the infant a flexible strategy for locomotion while also maintaining balance, especially in contrast to other crawling patterns, such as ipsilateral limb movements (e.g., Freedland and Bertenthal 1994). Finally, the diagonal crawling pattern is important because it illustrates *asymmetry*: in particular, theorists such as Gesell (1946) proposed that infants move from one stage of motor skill development to the next by "overcoming" or "breaking" the symmetric organization imposed by the biomechanical system. Support for this claim, in the context of learning to reach and crawl, is provided by Goldfield (1989), who found that the emergence of a hand preference in infants—as assessed by a reaching task—was highly associated with the onset of crawling.

5.1.4 Locomotion: Walking

As we noted at start of the chapter, both reaching and walking initially appear as reflex-like behaviors in newborn infants. In the case of walking, neonates produce a stepping reflex that is elicited by supporting the infant in an upright position, and then placing the infant's feet on a horizontal surface (e.g., Bril and Breniere 1992; Zelazo, Zelazo, and Kolb 1972). With this contextual support, young infants produce a well-organized pattern of alternating stepping movements with their legs. By age three months, however, this response no longer occurs.

What role, if any, does this early stepping behavior play in the process of learning to walk? One proposal is that the "loss" of the stepping reflex is due to the maturational shift of neural control from the spinal cord and brain stem, to the cortex (e.g., McGraw 1941). In other words, learning to walk involves suppressing or inhibiting the stepping reflex, while gradually developing voluntary control over the same muscle groups. However, in a series of studies, Thelen (1986; Thelen, Ulrich, and Niles 1987; Thelen and Ulrich 1991) found that between ages two and nine months, even when infants do not generate stepping movements on a stationary surface, they continue to produce the stepping pattern when placed on a moving treadmill. In addition, at this age treadmill stepping is not purely reflex-like, but instead relatively flexible and well-coordinated. For example, when the treadmill is divided into two parallel belts, each moving at different speeds, seven-month-olds maintain a regular gait by modulating the timing of the stepping movements with each leg (Thelen, Ulrich, and Niles 1987). Taken together, these results are consistent with the idea that neonatal leg movements are the result of a *central pattern generator* (CPG) at the spinal cord level, which is responsible for the basic coordination and timing of the leg movements, and that it is not suppressed, but instead gradually integrated with several other emerging abilities, including voluntary control of the legs and postural control of the trunk and upper body (Thelen and Ulrich 1991). As we will highlight, CPGs play a central role in developmental robotic studies as a neural mechanism that supports crawling and walking.

Thelen and Ulrich (1991) propose, thus, that the early stepping pattern does not disappear, but instead continues to develop and improve (e.g., during supine kicking of the legs), though it is not expressed as supported or independent walking until other necessary skills are in place. In particular, a rate-limiting factor (i.e., control parameter) that may constrain the emergence of independent walking is postural control (e.g., Bril and Breniere 1992; Clark and Phillips 1993; Thelen 1986), which begins to develop shortly after birth and follows a cephalocaudal (i.e., top-to-bottom) developmental pattern through the first year. From zero to three months, infants are first able to lift their heads (while prone on their stomach), then their head and chest, and finally, their upper body while using their arms for support (Johnson and Blasco 1997). Between three and six months, as we noted earlier, infants then learn to roll over, and maintain a seated, upright posture. Between six and nine months, infants are able to transition from the prone to seated position on their own, and they can also pull themselves up to a standing position. At age ten months they are competent at cruising (i.e., supported walking), while at eleven months they can stand alone without support. Finally, near their first birthday, most infants take their first steps and begin to develop independent walking.

While postural control (especially strength and balance) helps make the first steps possible, early walking behavior is far from adult-like (e.g., Bril and Breniere 1992; Clark and Phillips 1993; Vereijken and Adolph 1999). There are two important changes that occur during the second year. First, infants increasingly synchronize and coordinate the movement of their arms and legs. For example, at the onset of walking, rotations of the upper and lower leg segments (i.e., thigh and shank) are loosely coupled, and the phase relation between the two is irregular; within three months of walking, however, thigh and shank rotations are highly correlated and the phase relation resembles the pattern produced by adults (Clark and Phillips 1993). Similarly, early walkers extend and lift their arms, which helps create "ballast" but also precludes use of the arms during the swing phase. With experience, however, skilled walkers lower their arms, and swing them in phase with leg and hip rotation (Thelen and Ulrich 1991).

Infants also explore other forms of interlimb coordination as they learn to crawl, cruise, and walk. An interesting example is bouncing, which develops near the emergence of crawling but differs by recruiting a simultaneous (rather than alternating) pattern of leg movements. Goldfield, Kay, and Warren (1993) investigated the development of this skill by placing infants in a spring-mounted harness (i.e., the "Jolly Jumper"), and conducting a kinematic analysis of infants' bouncing behavior. Testing sessions were repeated over six consecutive weeks. Goldfield, Kay, and Warren (ibid.) hypothesized three phases of learning: (1) an initial phase focused on exploring the relation between kicking movements and bouncing (i.e., "assembly"), (2) a second phase, focused on "tuning" the timing and force of the component behaviors, and (3) a final phase, in which an optimal bouncing pattern emerges. They defined optimality as a stable pattern of bouncing, at or near the resonant period of the spring-mass system, with low variability in the period and high amplitude. Kinematic analyses provided clear support for each of the learning phases. In particular, a stable pattern emerged near age eight months, in which infants timed their kicking movements to coincide with the lowest point in the oscillation.

In addition to coordination within and between limbs, a second important dimension of walking that emerges during the second year is the ability to dynamically maintain balance while moving. Thus, early walkers tend to rely on a "stiff leg" strategy, which helps preserve upright posture, but fails to recruit available degrees of freedom (DOFs) in the legs and hips (e.g., Clark and Phillips 1993). With experience, however, older infants learn to "relax" these joints and incorporate their rotation into the stepping movement (e.g., Vereijen and Adolph 1999). Note that this is the same qualitative pattern that is observed in infants during the development of reaching, as early movements are characterized as stiff and rigid, while subsequent movements are smoother, more fluid, and recruit a higher number of joint DOFs. Indeed, the strategy of freezing and then freeing DOFs also plays a central role in motor-skill models, which we will highlight (e.g., Berthouze and Lungarella 2004; Lee, Meng, and Chao 2007; Schlesinger, Parisi, and Langer 2000). As a result of this developmental pattern, there are several improvements in walking that emerge during the second year, including (1) longer steps, (2) reduced distance between the feet during steps, (3) pointing the feet more forward, and (4) straighter paths while walking (e.g., Bril and Breniere 1992; Vereijken and Adolph 1999).

5.2 Reaching Robots

We describe here two classes of developmental robotics models of reaching. In the first set are *developmentally inspired* models, which exploit known properties and principles of motor-skill acquisition in human infants. In the second set, meanwhile, are models which are not only inspired by reaching development in human infants, but also seek to reproduce the developmental pattern in an artificial agent or robot.

A fundamental goal of developmentally inspired models of reaching is to implement a cognitive architecture, learning algorithm, or physical design that borrows key features (e.g., physical, neurophysiological, etc.) from human infants, and to demonstrate that these features constrain or simplify the problem of learning to reach in a fundamental way (e.g., Kuperstein 1988, 1991; Schlesinger, Parisi, and Langer 2000; Sporns and Edelman 1993; Vos and Scheepstra 1993). In addition, these models also help reveal and identify the underlying neural mechanisms that shape motor development. For example, Schlesinger, Parisi, and Langer (2000) highlight Bernstein's *DOF problem* (Bernstein 1967), that is, the fact that biomechanical systems have a large and redundant number of DOFs, including joints, muscles, neurons, and so on, which from a control perspective means there are an unlimited number of ways to produce a given movement trajectory. Schlesinger, Parisi, and Langer (2000) investigate this issue by using a genetic algorithm (GA) as a proxy for trial-and-error search (i.e., "hill climbing") in a population of artificial neural networks, which controls the movements of a one-DOF eye and a three-DOF arm. As we noted in the previous section, one strategy for solving the DOF problem at the joint level is to lock or "freeze" redundant joints, which reduces the dimensionality of the resulting joint space that must be explored. Schlesinger, Parisi, and Langer (ibid.) demonstrate that the freezing strategy need not be programmed into the model, but that it can in fact emerge "for free" as a result of the learning process: in particular, the model quickly learns to freeze the shoulder joint, and to reach by rotating the body axis and elbow joint.

A related problem involves learning a function that associates visual input of the target object with a motor program that moves the endpoint to the target location. An early and influential approach to this problem is Kuperstein's INFANT model (1988, 1991), which was implemented with a dual-camera vision system and a multijoint arm.

INFANT divides the process of coordinating vision and movement into two phases. During the first phase, the arm is driven to a series of random postures, while the hand grasps an object. At the end of each posture, the vision system records the resulting view of the grasped object. A multilayer neural network is then trained to produce the motor signal that corresponds to the given visual input; in particular, the prior (randomly generated) motor signal is used as the teaching pattern, against which the computed motor signal is compared. As we noted at the beginning of the chapter, this training strategy—in which the simulated infant generates its own training data through random movement—illustrates the phenomenon of *motor babbling*. During the second phase, visual input of the target at a new location is presented to the neural controller, which then drives the arm to the learned posture and results in reaching to the target location. More recently, Caligiore et al. (2008) have proposed a related approach that also bootstraps motor babbling, but in this case to solve a more difficult reaching problem (i.e., reaching around obstacles). A key feature of the Caligiore model is the use of CPGs, which contribute to the production of cyclic movements that, together with motor babbling, help solve the problem of obstacle avoidance during reaching.

Two potential criticisms of INFANT are that, first, an extended period of visuomotor training is required before the model begins to reach, and second, it benefits from a feedback rule that explicitly corrects movement errors. More recent approaches have been proposed to address these issues (e.g., Berthier 1996; Berthier, Rosenstein, and Barto 2005; Sporns and Edelman 1993). For example, Berthier (1996) proposed a reinforcement-learning (RL) model that simulates movement of the infant's hand in a 2D plane. In a subsequent version (Berthier, Rosenstein, and Barto 2005), the model was updated to include dynamic force control of the arm, as well as a 3D workspace. In particular, the muscles controlling rotation of the shoulder and elbow joints are modeled as linear springs. Rather than a supervised learning signal, the model is provided with a scalar reward signal (i.e., estimated time to the target), which does not explicitly specify how the arm should be moved. The model therefore learns by exploratory movements, which are generated through the addition of Gaussian noise to the output motor signal.

The Berthier model produces a number of important findings. First, the model captures the *speed-accuracy trade-off*: movements to smaller targets have longer durations than those to larger targets. Second, performance varies as a function of the magnitude of the Gaussian noise added to the movement signal. Interestingly, and perhaps counterintuitively, the model performs more accurately for *larger levels of noise*, than for smaller levels. Berthier, Rosenstein, and Barto (2005) interpret this result as supporting the idea that incorporating stochasticity into the motor signal promotes the exploratory process, and facilitates learning. Finally, the model also reproduces several important kinematic features of infants' reaches, including (1) the number of reaching "submovements," (2) the shape of the speed profile, and (3) the tendency for the largest speed peak to occur early in the movement (i.e., near reach onset; see Berthier and Keen 2005).

The models described thus far succeed in demonstrating that a core set of developmentally inspired principles are sufficient for learning to reach. There are a number of additional models that extend these findings by implementing the model within a physical robot, which develops the ability to reach in a manner that is analogous to the experience of human infants. An example of this approach is described by Lee and colleagues (e.g., Hulse et al. 2010; Law et al. 2011), who focus on the problem of coordinating visual input and arm postures and movements within a common spatial reference frame. Figure 5.4a illustrates the general architecture proposed to solve this problem. Visual data are acquired through a dual-camera input, which is then mapped from a retinotopic coordinate system to an active-vision gaze system (see figure 5.4b). At the same time, a second map coordinates sensory data from the arm with the gaze-space system. Thus, the gaze-space system not only provides a common reference frame for visual and proprioceptive (i.e., arm position) data, but also offers the means for either modality to direct the movement of the other, that is, to shift gaze toward the arm endpoint position, or alternatively, to move the arm toward the current gaze position.

An important feature of the model is the mechanism used to generate training data. In contrast to the strategies described earlier (e.g., motor babbling and Gaussian noise), the Lee model leverages *visual exploratory behavior* (i.e., scanning and search) as a source of visual activity, which not only serves to drive learning of the retina-gaze space mapping, but also provides a motivation for producing and improving reaching movements (i.e., reaching toward fixated objects). A series of analyses illustrates that the model quickly learns to coordinate vision and arm movements, and also to recalibrate the visuomotor mapping when one of the sensory inputs is shifted in space (e.g., the camera head is translated thirty centimeters; Hulse, McBride, and Lee 2010). The ability of the model to rapidly adjust to a shift in the camera position is particularly noteworthy, and may help explain how infants adapt their reaches while wearing displacing prisms (McDonnell and Abraham 1979).

Like the Lee model, Metta, Sandini, and Konczak (1999) propose a learning strategy that also relies on gaze position: once gaze is directed toward a visual target and the target is "foveated," the position of the eyes provides a proprioceptive cue that can be used to specify the location of the target in gaze space. The primary task of the Metta model, then, is to learn the mapping that coordinates the position of the eyes with movement of the arm to the fixated location. To accomplish this, the model relies on the *asymmetric tonic neck reflex* (ATNR). Sometimes referred to as the "fencer's pose," the ATNR is a synergistic movement of the infants' head and arm: when a young infant turns his or her head to the side, the arm on the corresponding side is raised and straightened. The Metta model exploits this mechanism in a four-DOF robotic platform, by directing gaze

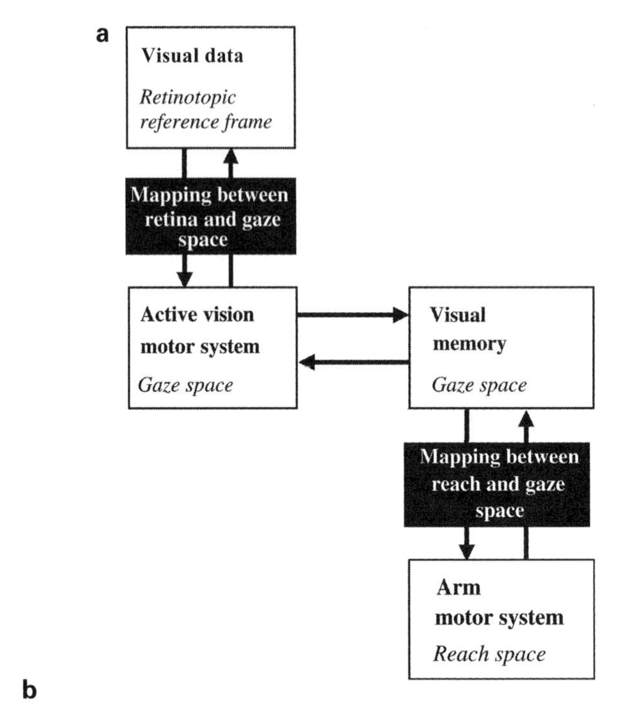

b

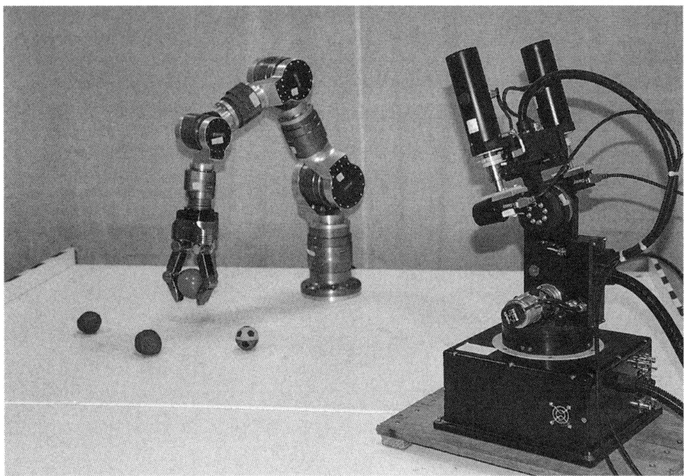

Figure 5.4
Diagram of the Hulse et al. (2010) architecture (a), and the active-vision and robot arm system (b). Reprinted with permission from IEEE.

to the hand once it enters the visual field (i.e., "hand regard"). Thus, when the hand is fixated, the model quickly learns to calibrate the sensory maps that link eye and arm positions. Indeed, Metta, Sandini, and Konczak (ibid.) demonstrate that after roughly five minutes of experience with hand-eye movement, the robot is able to accurately move its arm to a visually fixated location. A limitation of this method, however, is that the model does not reach accurately to locations that are infrequently (or never) "visited" by the hand. Natale et al. (2007; see also Nori et al. 2007) addressed this problem, while also extending the robot platform to a twenty-two-DOF humanoid upper torso, by employing a motor babbling strategy that drives the hand to a wider range of locations. After calibrating the hand-eye map, the robot is then able to reach to both familiar and novel locations.

As we noted at the beginning of the chapter, the iCub robot is an ideal platform for simulating the development of reaching in infants (e.g., Metta et al. 2010; see figure 5.5a). Indeed, a number of researchers have begun to use the platform as a tool for studying motor-skill learning, and in particular, to design and test models of reaching. However, not all of this work fits within the developmental robotics perspective; some studies, for example, are intended to address more general questions that span cognitive/humanoid robotics and machine learning, such as computational strategies for dynamic motion control (e.g., Mohan et al. 2009; Pattacini et al. 2010; Reinhart and Steil 2009).

An example of recent work on the iCub that explicitly adopts the developmental-robotic perspective is the model proposed by Savastano and Nolfi (2012), who simulate the development of reaching in a fourteen-DOF simulation of the iCub. As figure 5.5a illustrates, the iCub is trained and tested in a manner that corresponds to the method used by von Hofsten (1984; see figure 1), in which the infant is supported in an upright position while a nearby object is presented within reach. Like Schlesinger, Parisi, and Langer (2000), the Savastano model uses a GA to train the connection weights in a neural network that controls body, head, and arm movements (figure 5.5b; solid lines denote pretrained, fixed connections, in other words, orienting and grasping "reflexes," while dashed lines indicate modifiable connections). Two unique features of the model are (1) that it is initially trained with low-acuity visual input, which gradually improves over time; and (2) a secondary pathway (i.e., "internal" neurons)—representing cortical control of movement—remains inactive until the midpoint of training. Several major results are reported. First, while the model is designed to generate prereaching movements, as visual acuity improves, the percentage of these movements declines, replicating the developmental pattern reported by von Hofsten (1984) and others. Second, with experience, reaches also become progressively straighter (e.g., Berthier and Keen 2006). Finally, and perhaps most interesting, if the limitations on acuity and the secondary pathway are lifted at the start of training, overall reaching performance is lower in the model. This provides further support for the idea that early constraints on

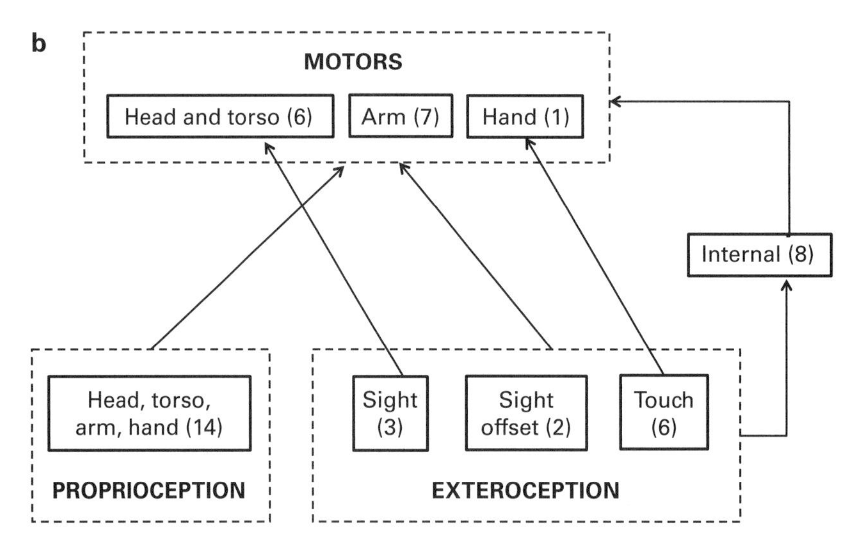

Figure 5.5
The iCub robot simulator (a) and the neurorobotic controller (b) investigated by Savastano and Nolfi (2012). Figures courtesy of Stefano Nolfi.

motor-skill development can facilitate long-term performance levels (e.g., Schlesinger, Parisi and Langer 2000).

5.3 Grasping Robots

As we noted in section 5.1.2, reaching and grasping are skills that overlap in both developmental and real time. In addition, grasping can also be characterized as "reaching with the fingers" to specific locations on the target object (e.g., Smeets and Brenner 1999). It is therefore not surprising that some of the same ideas used to study reaching are also exploited in models of grasping. An example of this is the model proposed by Caligiore et al. (2008) that we noted in the previous section, which examines the utility of motor babbling as a developmental mechanism for learning to reach around obstacles. In a similar vein, Caligiore et al. (ibid.) propose that motor babbling can also provide a bootstrap for learning to grasp. In particular, they use the iCub simulator to study a dynamic reaching model robot, which develops the ability to grasp in the following way: (1) either a small or large object is placed in the iCub's hand; (2) the preprogrammed grasping reflex causes the hand to perform an "involuntary" palmar grasp (see figure 5.2); (3) a motor-babbling module drives the arm to a random posture; and once the posture is achieved, (4) the iCub fixates the object. As figure 5.6a illustrates, position and shape information from the target are propagated through two parallel networks, which associate the corresponding sensory inputs with proprioceptive inputs from the arm and hand (i.e., arm posture and grasp configurations, respectively). Through Hebbian learning, the model is then able to drive the arm to the seen location of an object, and "re-establish" the grasp configuration. Figure 5.6b illustrates the performance of the model after training; the ends of the reaching movements are presented for the small (left) and large (right) objects, placed at twelve locations in the workspace. Thin lines represent successful reaches, while thick lines represent successful reaches and grasps. Caligiore et al. (2008) note that the overall grasp success of the model is relatively low (2.8 percent and 11.1 percent for the small and large objects, respectively), which is due in part to the comparatively basic mechanisms used to drive learning. Nevertheless, the fact that the model is more successful at grasping the large object is consistent with the developmental pattern that the power grasp (i.e., voluntary palmar grasp) emerges before the precision grasp (Erhardt 1994).

A related approach is proposed by Natale, Metta, and Sandini (2005a), who use the Babybot platform to study reaching and grasping (see figure 5.7a). Comparable to the strategy used by Natale et al. (2007), the Babybot first learns to fixate its hand, which is driven to a range of locations through motor babbling. Once a variety of arm postures are associated with gaze positions, the Babybot then learns to grasp seen objects. Figure 5.7b illustrates the process: (1) an object is placed in Babybot's hand, and a preprogrammed palmar grasp is executed (images 1–2); (2) the object is brought to the center

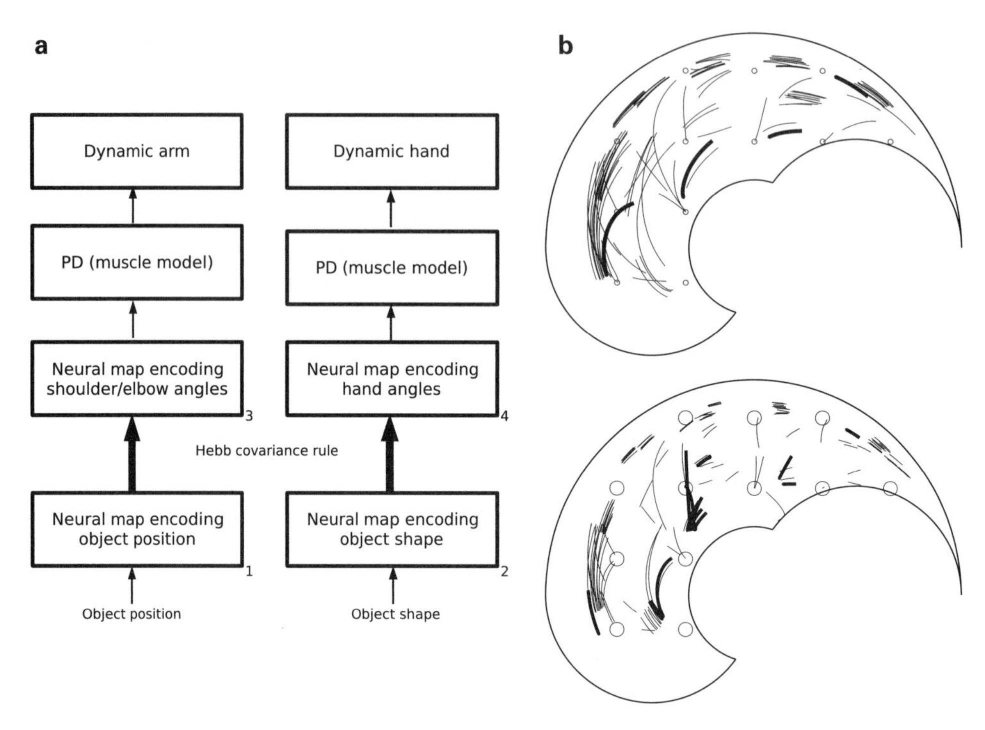

Figure 5.6
Diagram of the model architecture investigated by Caligiore et al. (2008) (a), and grasping performance of the model (left = small object, right = large object) (b). Figures courtesy of Daniele Caligiore.

of the visual field (image 3); (3) the object is returned to the workspace, and the Babybot uses a preprogrammed object-recognition module to search for it (images 4–6); and (4) once localized, the previously acquired reaching behavior is used to guide the hand toward the object and grasp it (images 7–9). While the grasping behavior described by Natale, Metta, and Sandini (2005a) is not systematically evaluated—and like Caligiore et al. (2008), the skill represents a relatively early stage of grasping development—the model makes an important contribution by illustrating how the components that are exploited or developed while learning to reach (e.g., motor babbling, visual search, etc.) also provide a foundation for learning to grasp.

While the Caligiore and Natale models capture the early emergence of grasping behavior, Oztop, Bradley, and Arbib (2004) describe a model that develops a comparatively large repertoire of grasp configurations. Figure 5.8a illustrates the seventeen-DOF arm/hand platform, as well as examples of both power grips (top panel) and precision grips (middle and bottom panels) that are produced by the model after training. During

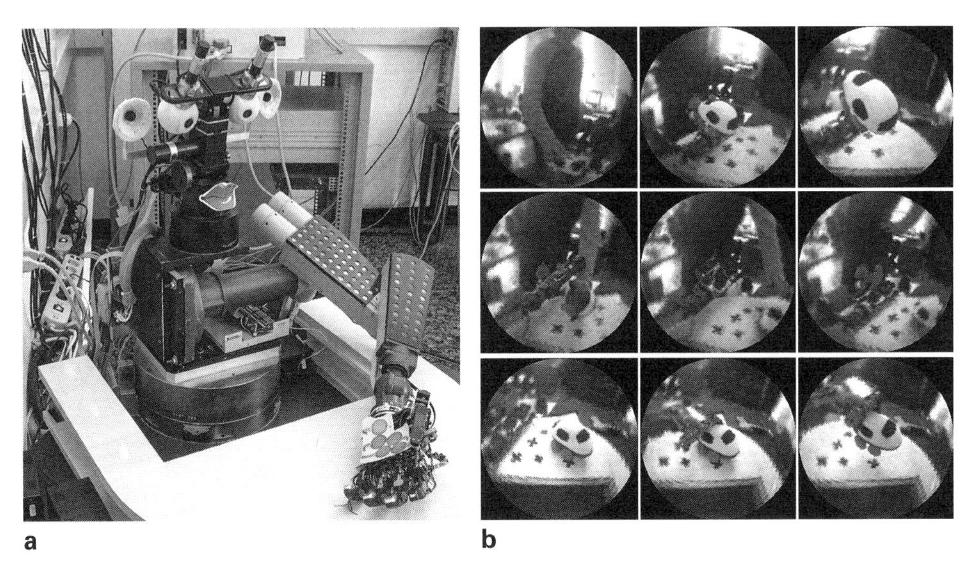

Figure 5.7
The Babybot robot designed by Natale, Metta, and Sandini (2005). View of a grasping sequence (a) recorded from the Babybot's left camera (b). Figures courtesy of Lorenzo Natale.

training, the model is presented with input information specifying the location and orientation of the target object, which propagates to a layer that activates a corresponding arm and hand configuration; this layer then drives movement of the arm and hand toward the object, resulting in a reach and grasp. A reinforcement-learning (RL) algorithm then updates connections in the model between the input, arm/hand configuration, and movement layers. Figure 5.8b presents the results of an experiment that replicates Lockman, Ashmead, and Bushnell (1984), in which five- and nine-month-olds reached for an oriented cylinder. The first panel presents the data from the human study, which found that during the reaching movement, nine-month-olds (dotted line) were more successful at orienting their hand correctly toward the vertical dowel than five-month-olds (solid line). Oztop, Bradley, and Arbib (2004) hypothesized that young infants were less successful due to poor use of the visual information provided by the object. They then simulated five-month-olds by providing the model with only target location information, while nine-month-olds were simulated by providing both location and orientation information. In figure 5.8b, the second panel presents the simulation data, which corresponded to the pattern observed in human infants and provided support for Oztop, Bradley, and Arbib's (ibid.) hypothesis.

As we described in box 5.1, a relatively advanced form of grasping behavior involves reaching for an object that is then manipulated, such as an applesauce-loaded spoon. In box 5.2, we present the Wheeler, Fagg, and Grupen (2002) version of the "applesauce

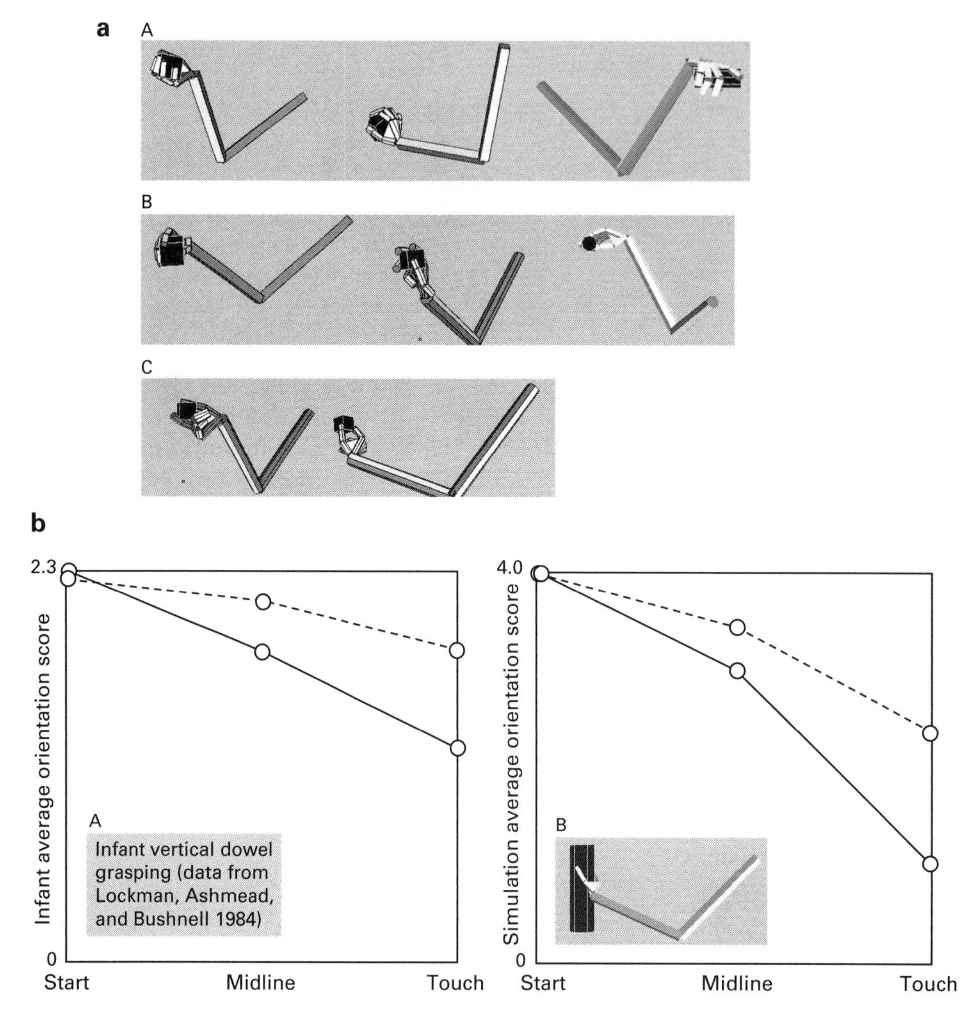

Figure 5.8

Diagram of the model architecture investigated by Oztop, Bradley, and Arbib (2004) (a), and grasping performance of the model (left = small object, right = large object) (b). Reprinted with permission from Springer.

Box 5.2
Prospective Grasp Control: The "Pick-and-Place" Robot

Overview

The McCarty, Clifton, and Collard (1999) "applesauce" study not only highlights how reaching and grasping can be used to index the development of planning and prospective action, but it also demonstrates how this capacity develops during infancy. There are a number of important questions raised by this work, including: what is the developmental mechanism that promotes the shift toward flexible use of the grasping hand? What kinds of experiences are part of this shift, and how does the development of a hand preference influence the process?

In order to address these questions, Wheeler, Fagg, and Grupen (2002) designed an upper-torso robot with two arms and a binocular vision system ("Dexter," figure below). Analogous to the applesauce study, Dexter is presented with a pick-and-place task in which an object (mounted to a handle) must be grasped and placed in a receptacle. Like the spoon, the object can be grasped at either end; only grasping the handle, however, permits insertion in the receptacle. With this paradigm, Wheeler, Fagg, and Grupen (ibid.) investigated whether learning by trial and error would produce a learning trajectory that mirrored the developmental pattern in infants. In addition, they also systematically manipulated a side bias in Dexter, in order to study the effects of hand preference.

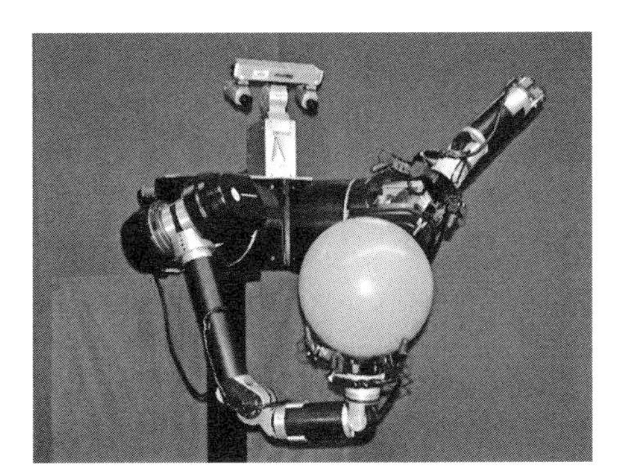

Procedure

The figure below illustrates the cognitive architecture designed and studied by Wheeler, Fagg, and Grupen (2002). A set of high level controllers are built in, which select from six basic actions: (1/2) grasp with left or right hand, respectively, (3/4) swap from left to right hand, or vice versa, (5/6) insert the object held in the left or right hand, respectively, into the receptacle. A precoded feature generator encodes Dexter's current state, which is used to generate an action. Exploration is achieved through occasional selection of random actions

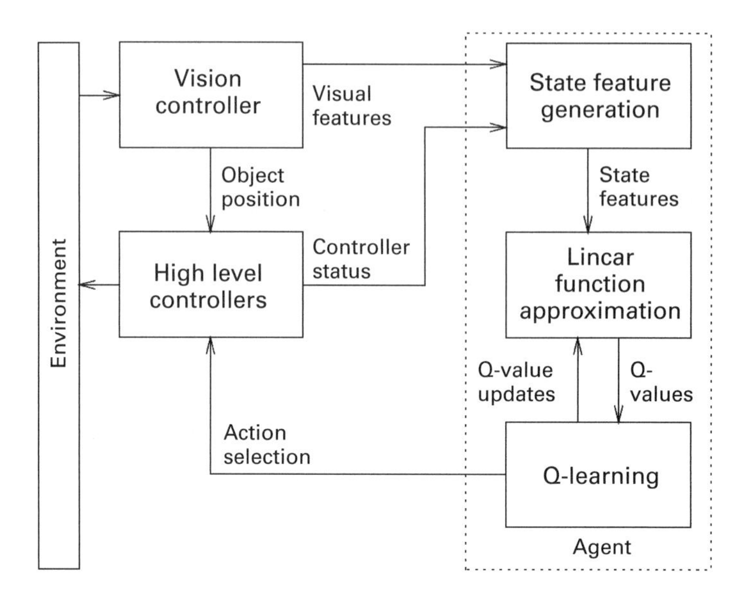

(i.e., ε-greedy). Successful insertion of the object in the receptacle results in a reward of 1, while failed trials result in a 0. In order to simulate the development of a hand preference, Dexter was first pretrained for 400 trials with the object oriented in the same direction. For the remaining 600 trials, the object was then oriented randomly across trials.

Results

As in the infant study, easy and difficult trials were analyzed separately. Easy trials were those in which the object was oriented in the same direction as pretraining trials, while difficult trials were in the opposite direction. The solid curve in the chart below presents the proportion of optimal trials, that is, in which Dexter grasped the handle with the arm that was on the same side as the handle. The upper panel illustrates that Dexter reached near-optimal performance during the pretraining phase, but when orientation of the object varied across trials (starting at 400 trials), performance rapidly dropped, before slowly approaching optimality again.

Interestingly, this finding mirrors the result observed with infants that even when the spoon is oriented in the direction that favors the dominant hand, infants occasionally reach with the nondominant hand, which is a suboptimal solution. In the case of the Wheeler, Fagg, and Grupen (2002) model, this behavior is the result of the exploration policy, which enables Dexter to learn how to respond to difficult trials, by reaching with the nondominant hand. As the figure below illustrates, total corrections (i.e., the analog of ulnar and goal-end grips) initially increases for both easy and difficult trials (upper and lower figures, respectively), but then gradually declines as Dexter learns to use the handle orientation as a visual cue for selecting the appropriate arm to use.

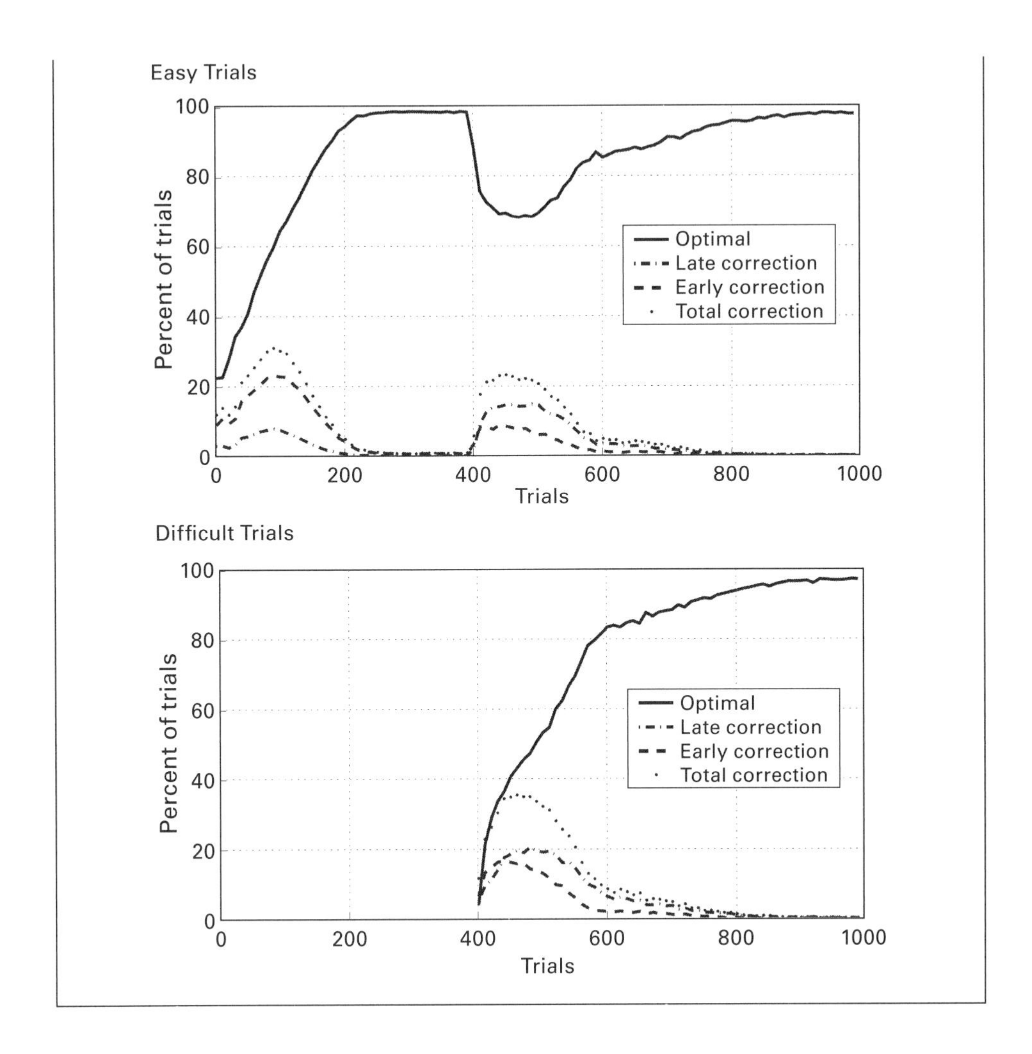

task," in which a humanoid robot learns to reach for, grasp, and place a tool-like object in a container.

5.4 Crawling Robots

As we noted in section 5.1.4, CPGs play a fundamental role in both the theories and computational models of locomotion (e.g., Arena 2000; Ijspeert 2008; Wu et al. 2009). Indeed, all of the models that we describe in this section leverage the CPG mechanism as a core element. As an example, figure 5.9a illustrates the CPG model of locomotion proposed by Kuniyoshi and Sangawa (2006). Their model represents the CGP as a neuron in the medulla (within the brainstem) that generates an oscillatory signal; its

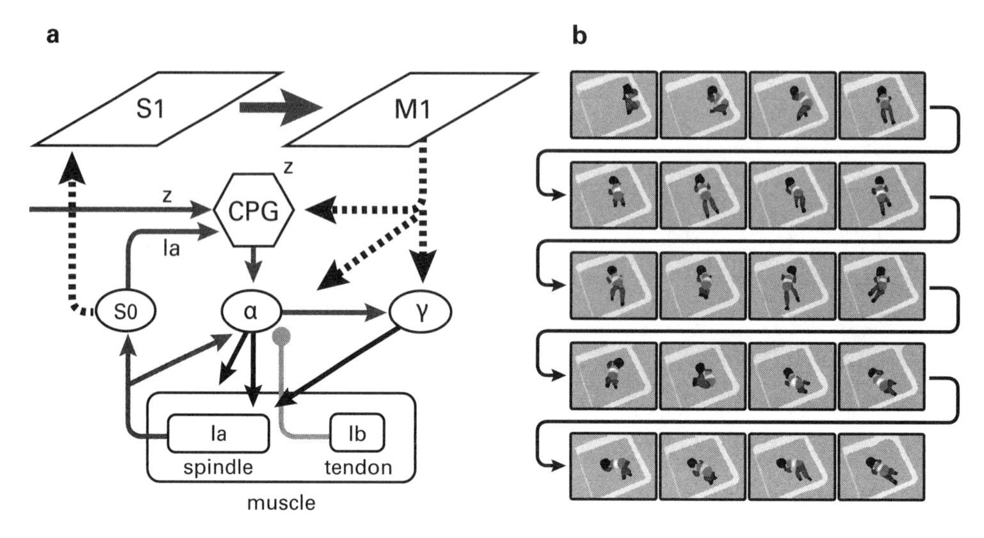

Figure 5.9
Diagram of the CPG-circuit proposed by Kuniyoshi and Sangawa (2006) (a), and crawling behavior of the simulated infant (b). Figures courtesy of Professor Yasuo Kuniyoshi, University of Tokyo. Reprinted with permission from Springer.

output stimulates the activity of muscle spindles, which produce body movements. At the same time, spinal afferent signals (S0) travel to the primary somatosensory area (S1), which produces activity in the primary motor area (M1). The sensorimotor loop is closed by connections from M1 back to the CPG, which allow patterned activity from M1 to modulate the output of the CPG. Figure 5.9b is a snapshot of a simulated infant with 19 joint segments and 198 muscles that is controlled by a pair of circuits like the one illustrated in figure 5.9a (i.e., one circuit for the left and right hemispheres/body sides). Two important results emerge from the model. First, the simulated infant often produces "disorganized" movements, which may provide an adaptive exploratory function (i.e., spontaneous motor babbling). Second, coherent, synergistic movements also occur: for example, at the start of the illustrated sequence, the infant rolls from face up to face down. In addition, after rolling over, the infant generates a crawling-like behavior. It is important to note that these coordinated behaviors are not preprogrammed, but are instead the result of linkages and synchronous oscillations between the CPGs that briefly occur.

While the model proposed by Righetti and Ijspeert (2006a, 2006b) also focuses on CPGs, they employ a unique approach by first gathering kinematic data from human infants, which they use to inform and constrain their model. In particular, a motion capture system is used to identify and record the positions of infants' limbs and joints while crawling. A key finding from these kinematic analyses is that skilled infant

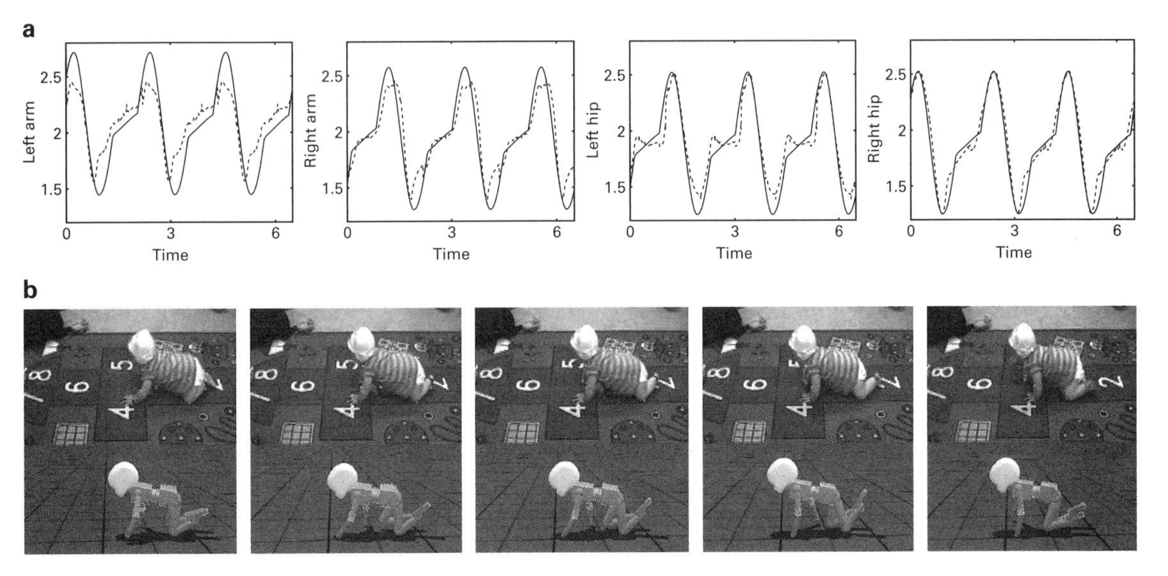

Figure 5.10
Observed shoulder and hip trajectories in a crawling human infant (solid line) and the corresponding trajectories produced by Righetti and Ijspeert's (2006) crawling model (dashed line) (a), and crawling examples in a human infant and the simulated iCub robot (b). Figures courtesy of Ludovic Righetti.

crawlers produce a "trot-like" gait, in which the diagonally opposed leg and arm move in phase, while moving a half period out of phase with the opposite pair of limbs (i.e., confirming the diagonal-alternation pattern we described in section 5.1.3). However, a key difference between the trot-like gait of infant crawlers and a classic trot is that infants spend 70 percent of the movement cycle in the stance phase. In order to capture this temporal asymmetry in the stance and swing phases, Righetti and Ijspeert (2006a) model the CPGs as spring-like oscillatory systems that switch between two stiffness parameters as a function of the phase (i.e., direction of oscillation). Figure 5.10a presents a comparison of the trajectories of the shoulder and hip joints (i.e., arm and leg, respectively) of the infants (solid line) and the corresponding oscillations produced by the model (dashed line). In addition, Righetti and Ijspeert (ibid.) also evaluate the model by using it to control the crawling behavior of a simulated iCub robot. As figure 5.10b illustrates, the crawling pattern produced by the robot (figure 5.10b, lower panel) corresponds to the gait pattern produced by a human infant (figure 5.10b, upper panel). Subsequent work on the iCub platform has provided a number of important advances, including the ability to combine periodic crawling movement with shorter, ballistic actions, such as reach-like hand movements (e.g., Degallier, Righetti, and Ijspeert 2007; Degallier et al. 2008).

In addition to work on the iCub, another humanoid robot platform that has been used to study crawling development is the NAO (see chapter 2). Li et al. (Li et al. 2011; Li, Lowe, and Ziemke 2013) raise an important challenge: they argue that a successful model of locomotion should be general and robust enough that it can be "ported" from one platform to another with minimal modification. In order to test this proposal, they implement the CPG architecture designed by Righetti for the iCub platform (e.g., Righetti and Ijspeert 2006a) on the NAO. However, as Li et al. (2011) note, a key difference between the two platforms is that NAO has larger feet. In order to address this issue, the legs are spread wider apart (than in the iCub) to assist with forward movement. Figure 5.11a illustrates the four-cell CPG architecture, which includes inhibitory connections between limb pairs and excitatory connections across "diagonal" limbs (e.g., right arm and left leg); the inhibitory connections enforce a half-period phase shift between limbs, while the excitatory connections produce in-phase movements of the corresponding limbs. As expected, crawling successfully generalizes to the NAO

a

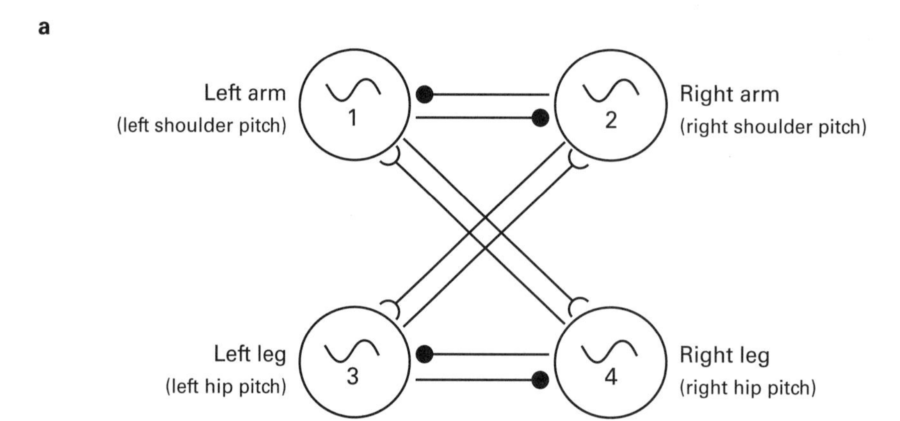

b

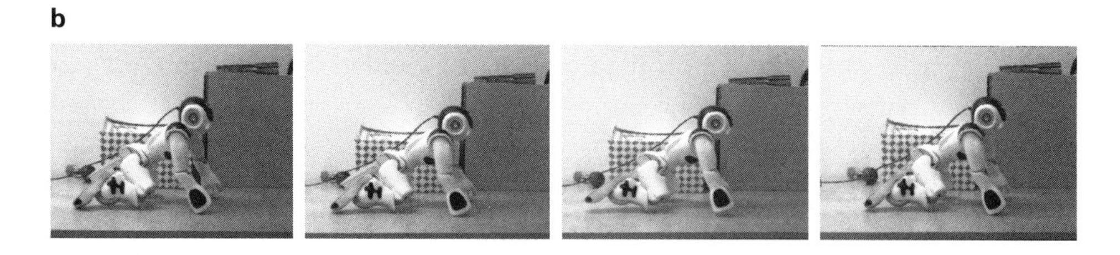

Figure 5.11
CPG architecture employed by Li et al. (Li et al. 2011; Li, Lowe, and Ziemke 2013) to produce crawling (a), and an illustration of crawling behavior on the NAO robot platform (b). Figures courtesy of Li Cai.

platform. Figure 5.11b provides an illustration of crawling behavior on the NAO robot, which resembles the pattern produced by iCub (see figure 5.10b). In order to provide further support for the idea that the same underlying CPG mechanism can produce multiple forms of locomotion, Li, Lowe, and Ziemke (2013) have recently extended their model with a six-cell CPG architecture that produces an early form of upright, bipedal walking behavior.

5.4 Walking Robots

While developmental robotics researchers have not yet captured the complete transition from crawling to independent walking, there are several models of walking that have provided insight into how this process occurs. As we noted earlier in the chapter, one mechanism that may play a crucial role is successive changes in the joint or neuromuscular DOFs that are utilized during body movement. In particular, Taga (2006) proposes a CPG model in which various components of the system are systematically fixed or released (i.e., freezing and freeing of DOFs). Figure 5.12 illustrates the model: PC represents the posture-control system, and RG is the rhythm generator, which produces periodic signals. In neonates, the PC system is nonfunctional, while the oscillators in the RG are linked via excitatory connections. At this stage, the system produces reflex-like stepping movements when the simulated infant is supported in an upright position. Subsequent developments (e.g., standing, independent walking, etc.) are accounted for in the model through a process of tuning excitatory and inhibitory connections within the RG and between the RG and PC.

Lu et al. (2012) describe a model that focuses specifically on supported walking. In particular, a wheeled platform is designed for the iCub robot, which provides both upper-body trunk support and balance. An interesting constraint imposed by the "walker" is that because it is rigidly attached to the iCub, the robot cannot move in the vertical dimension. In other words, while walking, iCub must find a movement strategy that keeps the waist and upper body at a fixed height from the ground. The solution proposed to solve this problem is the "bent-leg" gait, in which iCub keeps both of its knees bent while stepping. A series of tests of the bent-leg gait in the iCub simulator compared a range of heights for the walker, resulting in a stable solution where the maximum bend in iCub's legs is approximately 45 degrees. Lu et al. (ibid.) conclude by demonstrating that the solution acquired in simulation successfully transfers to the real robot platform.

Hase and Yamazaki (1998) take a similar approach, but also implement a unique feature: rather than representing the supporting forces as a rigid body that holds the infant at a fixed height, they instead model support as a set of tuneable forces. Figure 5.13a presents a diagram of the simulated infant, which has the physical properties (i.e., size and muscle capacity) of a twelve-month-old. As the figure illustrates,

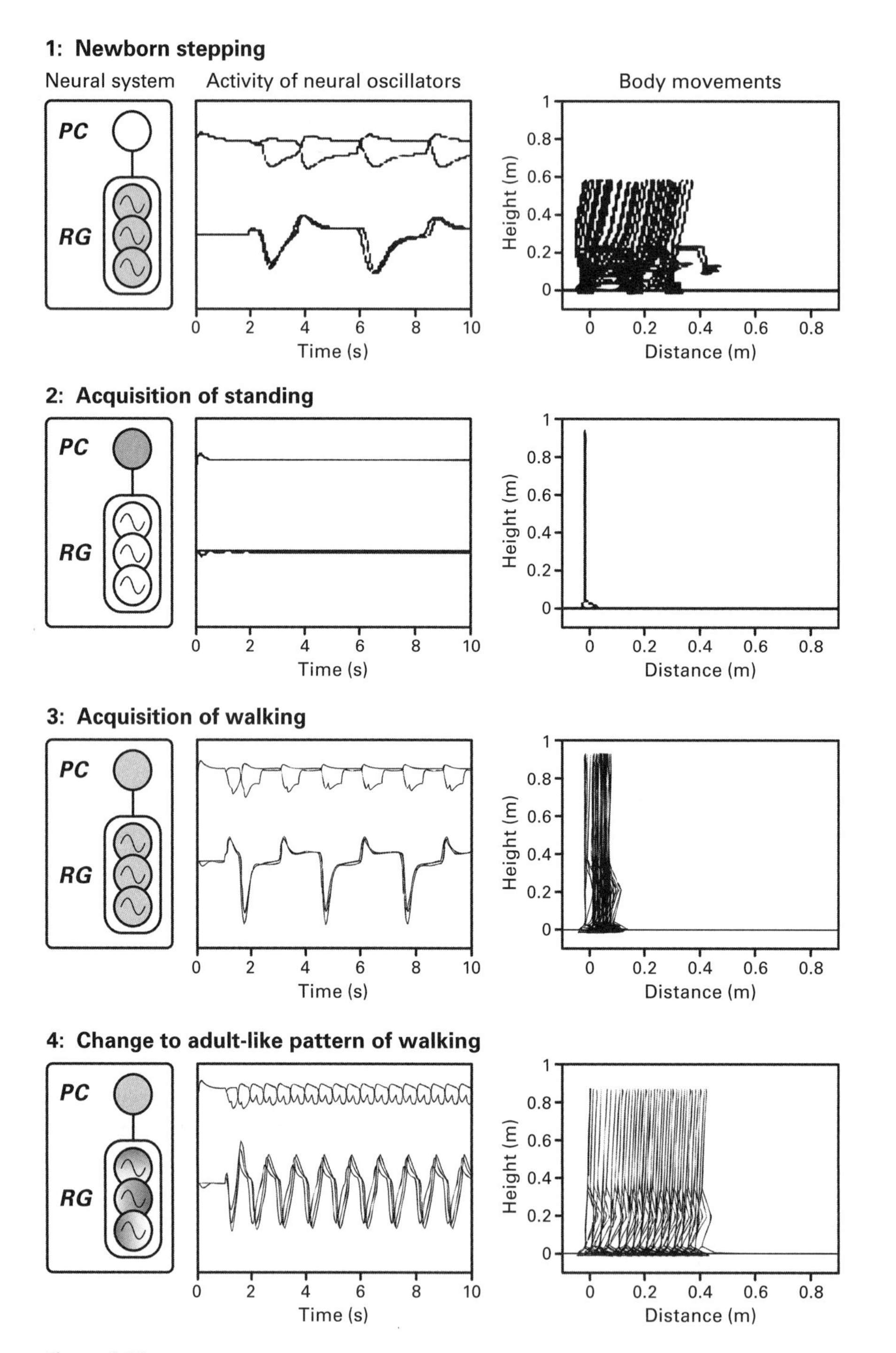

Figure 5.12
The CPG model proposed by Taga (2006) to account for stages of walking by freezing and freeing DOFs.

supporting forces are modeled as spring-damper systems that operate on the shoulder and hip joints and that apply upward force to the corresponding joint when it falls below a height threshold (i.e., the restoring force is graded). Movement of the model is controlled by a CPG architecture, and the major system parameters are tuned by a GA with the goal of optimizing four performance measures: (1) minimal use of supporting forces, (2) target step length (i.e., 22 cm, as determined by empirical measure), (3) minimal energy use, and (4) minimal muscle stress (or fatigue). Figure 5.13b presents the findings from the simulated model. Between 0 and 5,000 search steps (i.e., simulated developmental time), the infant relies on supported walking (dark line), while consuming low energy levels (thin line) and generating minimal muscle stress (dashed line). Between 5,000 and 10,000 search steps, however, the transition to independent walking begins, and by 10,000 steps the infant can walk without external support. In particular, it is interesting to note that independent walking comes at a price: energy and fatigue levels initially rise at the onset of independent walking, but then continue to decline afterward.

We conclude this section by returning to a skill that we briefly described in section 5.1.4, that is, bouncing, which involves coordination of the same joints and muscles as those recruited for crawling and walking, but in a qualitatively different pattern. Because bouncing is a rhythmic behavior, like crawling and walking it can also be captured and described mathematically with CPGs. Lungarella and Berthouze (2003, 2004) investigate the bouncing behavior produced by a twelve-DOF humanoid robot that is suspended in a harness with springs (see figure 5.14a). Like other models of locomotion, Lungarella and Berthouze also use a network of CPGs to control the motor activity of the robot, which produces kicking movements (i.e., flexions and extensions of the knee joint). Figure 5.14b illustrates the system in which the CPG network is embedded, which includes sensory input, motor activity, as well as interaction with the skeletomuscular system and the external environment (i.e., ground and harness-spring system). An important finding from the model results from a comparison of bouncing with versus bouncing without sensory feedback: with feedback, the robot is more successful at maintaining a stable bouncing regime than without feedback. Thus, sensory feedback is used by the model to modulate the output of the CPG network, resulting in a self-sustaining limit-cycle pattern. It is important to note, however, that this result depends on hand-tuning of the model. As figure 5.14b suggests, a motivational or value system can be implemented within the model that drives exploration of the parameter space and selects parameter configurations that produce "desirable" patterns (e.g., stable bouncing, maximal bounce height, etc.).

5.5 Conclusions

We began the chapter by proposing that manipulation and locomotion are two fundamental skills that develop in early infancy. They are essential because they provide the

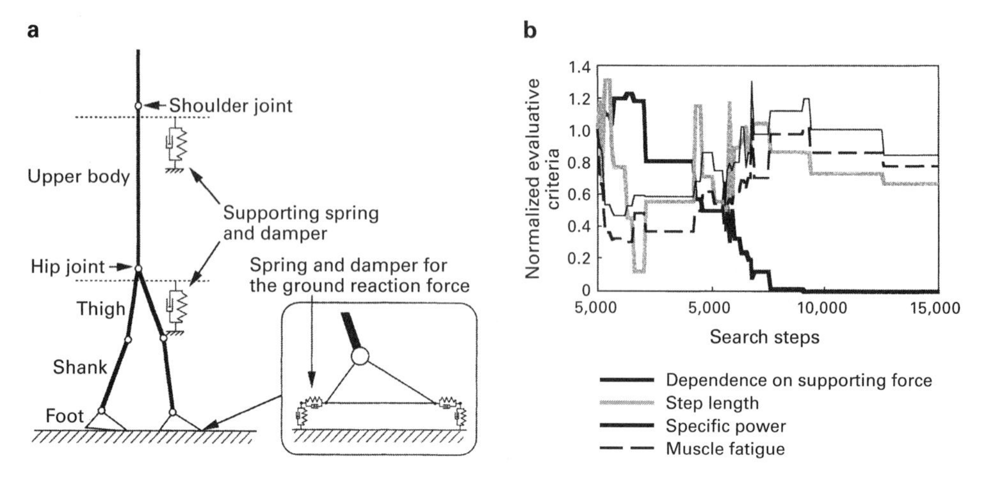

Figure 5.13
Simulated twelve-month-old studied by Hase and Yamazaki (1998) (a), and performance of the model during training (b). Figures reprinted with permission of The Anthropological Society of Nippon.

means for probing, exploring, and interacting with both the physical and social world, and they therefore create a "motor" that drives not only physical, but also perceptual, cognitive, social, and linguistic development. While our brief survey of motor-skill development in human infants focused on reaching, grasping, crawling, and walking, we also noted there are a number of additional motor skills—some that are necessary for survival and appear universally, and others that are not survival-related and vary widely across cultures and historical time periods—which have not yet been modeled or studied from the developmental robotic perspective. We suspect that as the state of the art continues to improve, and the field moves toward a more domain-general approach (e.g., Weng et al. 2001), researchers will seek to capture a wider spectrum of motor-skill development in their models.

The review of reaching and grasping in infants identified a number of important themes, some of which have been incorporated into computational models of development. First, both skills initially appear as simple, reflex-like behaviors, and eventually become transformed into volitional actions that are adapted to the ongoing goals of the infant. A unique aspect of reaching development is that its initial form (i.e., prereaching) first declines in frequency, and subsequently reemerges as an intentional or goal-directed behavior. Second, both reaching and grasping demonstrate a developmental trend toward *differentiation*: early behaviors are somewhat ballistic, open-loop behaviors that once triggered, unfold in a stereotyped fashion. In contrast, more mature reaching and grasping movements are flexible and varied as a function of the

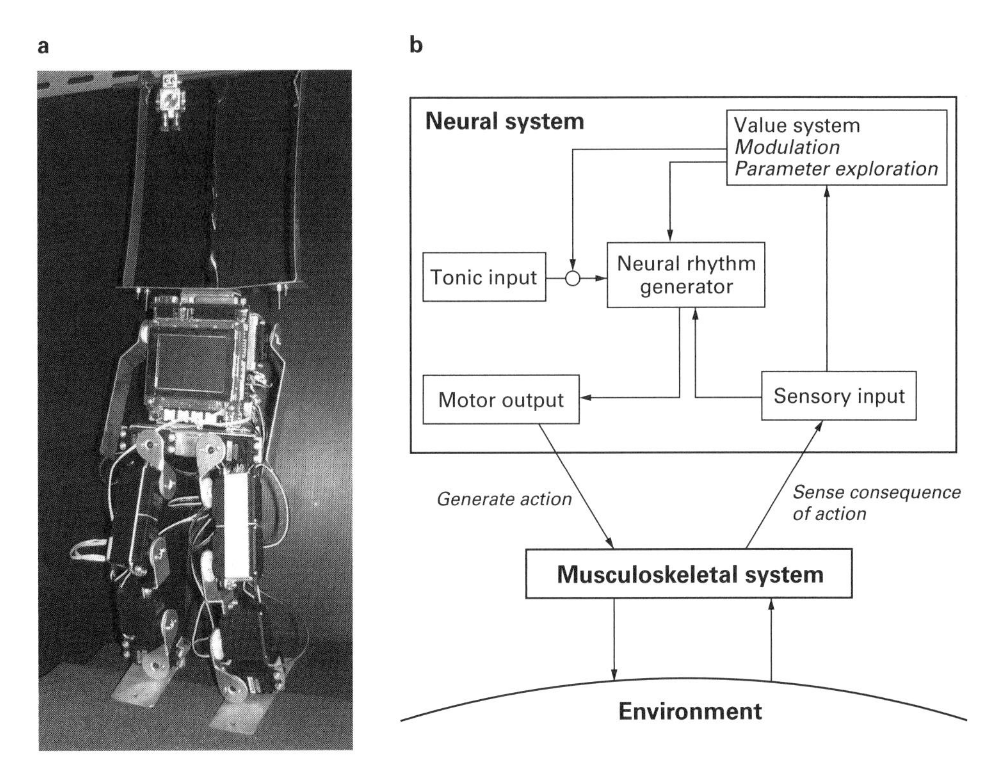

Figure 5.14
The "bouncing" robot designed by Lungarella and Berthouze (2004) (a), and the neural controller used to produce bouncing behavior (b). Figures courtesy of Luc Berthouse.

task constraints (e.g., location and size of the target object). Figure 5.2, which illustrates the changes in infants' grasping ability from ages six to twelve months, provides a clear example of this pattern. A third and related aspect is not only an increasing use of visual information during reaching and grasping, but also the use of closed-loop movement strategies, in which flexible adjustments or corrective movements are produced in the face of changing task demands.

We also described the development of crawling and walking, the two earliest forms of locomotion. Like manipulation, self-produced locomotion also has long-ranging consequences for development. For example, in chapter 4 we discussed how the emergence of crawling transforms infants' perception and experience of space (i.e., the onset of the fear of heights on the visual cliff). An important theme in the development of crawling is the idea of posture control as a rate-limiting factor. In particular, infants must first gain the strength and coordination necessary to produce the component actions (e.g., lift and support the head, alternate leg and arm movements) before they begin to crawl. In addition, we noted that infants explore a diverse range of movement patterns, such as creeping and rotating, before the canonical form of hands-and-knees crawling appears. Within two months of crawling, infants then quickly transition to pulling themselves upright, and using stable surfaces in their environment to provide balance as they master supported walking. After two months of supported walking, the typical infant then lets go of the couch or coffee table, and begins independent walking. An important concept highlighted in the development of walking was the central pattern generator or CPG, which we subsequently noted plays a central role in developmental models of locomotion.

Our review of research from the developmental robotics perspective on motor-skill acquisition focused on the same four abilities. First, we described a series of models that incorporate the properties, strategies, or constraints found in human infants, and demonstrate that these features help simplify or facilitate the task of learning to reach. A ubiquitous example is the use of motor babbling, which enables the simulated infant or robotic agent to simultaneously experience—and thus learn to correlate or coordinate——a wide range of arm postures and visual inputs. Another important issue addressed by developmental models is the DOF problem. A common computational strategy for addressing this problem, which mirrors the strategy exploited by human infants—is to "freeze" a subset of the joints available for generating a reaching movement, such as the elbow, and to subsequently "free" the joints after the reduced movement space has been explored.

Despite the availability of several child-sized robot platforms, the use of these robots for studying the development of reaching is still at an early stage; as we noted, for example, much of the work to date on the iCub platform has concentrated on more general issues, while only a few studies have specifically modeled the developmental process in infants. A similar pattern emerges from the review on developmental models of grasping. Nevertheless, there are several important findings from this work. First, as with reaching, a fundamental challenge is to relate visual information from the target object into not simply a corresponding grasp configuration, but to complicate matters, into a series of finger movements that result in the desired configuration. While there are only a few models thus far that explicitly adopt a developmental approach to this problem, they each suggest a potential solution that is not only available to human infants, but also succeeds in either simulation or on a real robot platform. Second, a key ingredient in the learning process, across each of the models we described, is variability in movement and a learning mechanism that benefits from trial-and-error learning.

Finally, there are also several developmental robotic models of crawling and walking, some of which take advantage of available robot platforms like the iCub and NAO. It is worth noting that this work is also at a relatively early stage, and that the full spectrum of locomotion behaviors has not yet been captured by existing models. In particular, most models have focused on a subset of locomotion skills, such as hands-and-knees crawling, rather than attempting to account for the entire timespan from the onset of crawling to the mastery of independent walking. Analogous to the use of motor babbling in models of reaching and grasping, a common element in most models of infant locomotion is the CPG. A recurring theme in this work is that development consists of solving two problems: first, appropriately linking excitatory and inhibitory connections across CPG units or neurons, in order to optimally coordinate the movement of corresponding limbs, and second, learning to use input from sensory systems (e.g., visual, vestibular, and proprioceptive) to modulate the output of the CPG neurons.

Additional Reading

Thelen, E., and B. D. Ulrich. "Hidden Skills: A Dynamic Systems Analysis of Treadmill Stepping during the First Year." *Monographs of the Society for Research in Child Development* 56 (1) (1991): 1–98.

Thelen and Urich's comprehensive study of the development of stepping behavior illustrates a number of key phenomena, including U-shaped development, the stability of motor patterns in the face of perturbation, and the emergence of new forms of sensorimotor coordination. While the theoretical perspective tilts toward dynamic systems theory, many of the core ideas—like embodied cognition and learning through exploratory movement—resonate clearly with ongoing work in developmental robotics.

Asada, M., K. Hosoda, Y. Kuniyoshi, H. Ishiguro, T. Inui, Y. Yoshikawa, M. Ogino, and C. Yoshida. "Cognitive Developmental Robotics: A Survey." *IEEE Transactions on Autonomous Mental Development* 1, no. 1 (May 2009): 12–34.
Metta, G., L. Natale, F. Nori, G. Sandini, D. Vernon, L. Fadiga, C. von Hofsten, K. Rosander, J. Santos-Victor, A. Bernardino, and L. Montesano. "The iCub Humanoid Robot: An Open-Systems Platform for Research in Cognitive Development." *Neural Networks* 23 (2010): 1125–1134.

Asada and Metta and their respective research teams have authored two excellent surveys of developmental robotics. While each article goes beyond the topic of motor-skill acquisition, both provide outstanding introductions to the issue, as well as numerous examples of state-of-the-art research. A unique feature of Asada's overview is the discussion of fetal movements and their role in postnatal motor development. Metta's article, meanwhile, highlights the use of the iCub platform, and in particular, it provides a detailed discussion of how the physical structure of the robot is designed to provide a testbed for exploring critical questions, such as how vision and hand movements become coordinated.

6 Social Robots

The child psychology studies on motivational, visual, and motor development in children, and the numerous developmental robotics models of intrinsic motivation and sensorimotor learning, focus on the acquisition of individual capabilities. However, one of the fundamental characteristics of human beings (as well as of other highly social animal species) is the fact that the human infant, from birth, is embedded in the social context of parents, caregivers, and siblings, and naturally reacts to this social presence and has an instinct to cooperate with others (Tomasello 2009). Evidence exists that newborn babies have an instinct and capability to imitate the behavior of others from the day they are born. After just thirty-six minutes old, the newborn can imitate complex facial expressions such as happiness and surprise (Field et al. 1983; Meltzoff and Moore 1983; Meltzoff 1988). As the child cannot walk until nearly her first birthday, she is dependent on the continual care, presence, and interaction of her parents and caregivers. This further reinforces the social bonding between infants and their parents, family members, and caregivers. Moreover, social interaction and learning is a fundamental mechanism for the development of empathic and emotional skills, as well as for communication and language.

The infant's social development depends on the gradual acquisition, refinement, and enrichment of various social interaction skills. Eye contact and joint attention support the establishment of the emotional bonding with the caregivers, as well as the cognitive capabilities to create a shared context of interaction. For example, the infant first learns to establish eye contact with the adult, and later develops an ability to follow the adult's gaze to look at objects placed within the child's own field of vision, but later also to look for an object outside her view. In addition to eye contact, the child also gradually learns to respond to, and then produce, pointing gestures initially to draw the adult's attention and request an object such as food or toy (imperative pointing), and then just to draw attention toward an object (declarative pointing). Imitation capabilities, as in the face imitation studies with one-day-old newborns, also go through various developmental stages, with qualitative changes of imitation strategies, from simple body babbling and body part imitation, to the inferring of goals and intentions of the demonstrator when repeating an action (Meltzoff 1995). The acquisition of advanced joint attention and imitation skills creates the bases for the further development of cooperation and altruistic behavior. Children learn to collaborate with adults and other peers to achieve shared plans. Finally, the parallel development of all these social competences and skills leads to the acquisition of a complex ability to correctly attribute beliefs and goals to other people, and the emergence of a "theory of mind" that supports the children's social interaction until they reach adulthood.

In the sections that follow we will first look at the developmental psychology studies and theories on the acquisition of joint attention, imitation, cooperation, and theory of mind. These sections are followed by the analysis of current developmental robotics models mirroring developmental shifts in the emergence of these social capabilities in robotic agents. Such skills are essential to support a fluid interaction between robots and humans, as mechanisms like gaze following and imitation are essential to allow robots to understand and predict the goals of the humans. At the same time, the social cues and feedback from the robots can be used by human agents to adapt their behavior to the perceived robot's sensorimotor and cognitive potential.

6.1 Children's Social Development

6.1.1 Joint Attention

Joint attention is based on the ability to recognize the face of another agent and their position, to identify the direction of their gaze and then to simultaneously look toward the same object gazed at by the partner. However, this goes well beyond a simple perceptual act. In fact, as Tomasello et al. (2005) and Kaplan and Hafner (2006b) stress (see also Fasel et al. 2002), within the context of social interaction between two intentional agents, joint attention must be seen as a coupling between two intentional actions. In a developmental context, this implies that a child looks at the same target object gazed at by the parent with the shared intention to perform an action on the object or discourse about properties (e.g., name, visual or functional features) of the object. Therefore in child development joint attention plays the fundamental function to support the acquisition of social and cooperative behavior.

In chapter 4 (sections 4.1.2 and 4.2) we looked at the child's (and robot's) capability to recognize faces. So if we start with the assumption that a child has a capability, and preference, to recognize and look at faces, we should then look at the developmental stages in gaze following to support intentional joint attention. Butterworth (1991) has investigated the early stages of gaze following, and has identified four key developmental stages (figure 6.1): (1) *Sensitivity Stage*, around six months, when the infant can discriminate between the left or right side of the caregiver's gazing direction; (2) *Ecological Stage*, at nine months, which is based on the strategy of scanning along the line of gaze for salient objects; (3) *Geometrical Stage*, at twelve months, when the infant

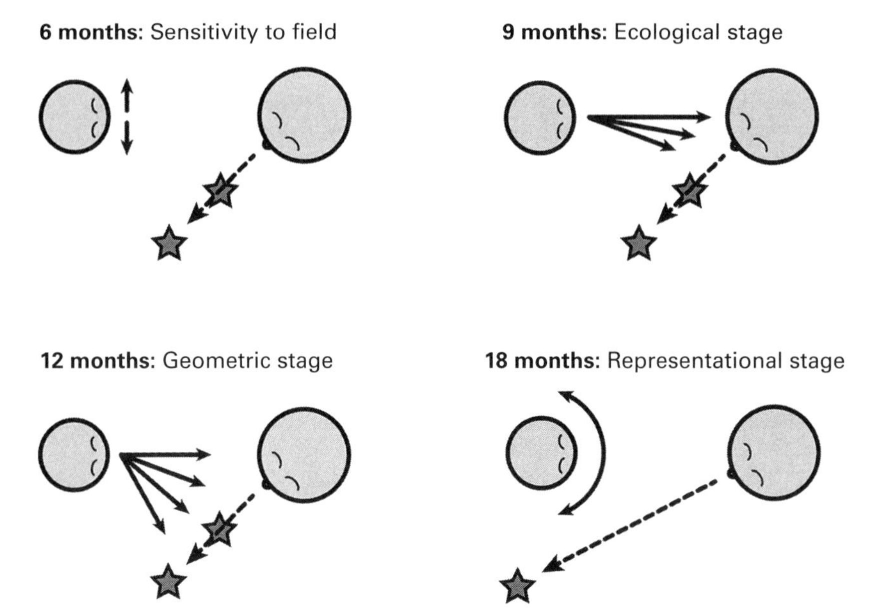

Figure 6.1
Developmental progression of gaze following. Adapted from Scassellati 2002.

can recognize the orientation angle of the caregiver's eyes to localize the distal target; (4) *Representational Stage*, reached at around eighteen months, when the child can orient outside the field of view, to gaze precisely at the target object that the caregiver is viewing.

Scassellati (1999) uses these four stages to identify the contribution of gaze following capabilities in agents for a theory of mind in humanoid robots. Kaplan and Hafner (2006b) have also proposed an illustration of different joint attention strategies, which include some of the key developmental stages summarized in table 6.1 and based on the developmental literature on gaze and pointing. The four strategies are as follows: (1) *mutual gaze*, when both agents are attending to each other's eyes simultaneously; (2) *gaze following*, when one agent looks at the object, and the other watches the first robot's eyes to detect where the partner is looking; (3) *imperative pointing*, when the first agent points to the object regardless of whether the other agent is attending to it; and (4) *declarative pointing*, when the first robot points to the object, while the second is also looking at it, thus achieving shared attention. These are represented in the four attentional strategies of the SONY AIBO robot (see section 6.2 below for a full discussion).

Kaplan and Hafner (2006b) have also proposed a developmental timescale for shared attention and social skills based on four key prerequisites: (1) attention detection, (2) attention manipulation, (3) social coordination, and (4) intentional understanding.

Table 6.1
Developmental timescale for joint attention skills (adapted from Kaplan and Hafner 2006b)

Age (months)	Attention detection	Attention manipulation	Social coordination	Intentional understanding
0–3 months	Mutual gaze through eye contact	Mutual gaze through maintaining eye contact	Protoconversation, simple turn taking mediated by caregiver	Early identification with other persons
6 months	Sensitivity only to the left/right side that the caregiver is gazing at	Simple forms of attention manipulation	Shared routines: caregiver-child conversational games	Animate-Inanimate distinction; physical/social causality distinction
9 months	Gaze angle detection, fixation of first salient object	Imperative pointing to ask for object/food	Joint activities, imitative games of caregiver's movement	First goal-directed behavior
12 months	Gaze angle detection, fixation of any salient object	Declarative pointing, draw attention with gestures	Joint activities and imitative games for goal sharing	Goal understanding, behavior understood as goal directed
18 months	Gaze following toward object outside field of view	First predications with words and gestures	Coordination of joint action plans	Intentional understanding, same goal for different action plans

These cognitive capabilities, whose developmental milestones are summarized in table 6.1, constitute the prerequisite for full achievement of shared attention between two agents. *Attention detection* refers to the capability of an individual to perceive and track the attentional behavior of other agents. In infants this starts as a simple capability of mutual gaze through detecting of the eyes of a social partner and establishing eye contact (first three months), and reaches a stage where the agent can look at an object outside the field of view by following the gaze of the other agent (eighteen months). *Attention manipulation* involves a more active capability to influence and direct the attentional behavior of others. In addition to mutual gaze, attention manipulation is achieved around nine months with imperative pointing (e.g., the infant points at food when she is hungry), at twelve months with declarative pointing (e.g., gesture to draw attention to an object), up to the eighteen-month stage when she can use predications based on word-gesture combinations, and later with complex linguistic interactions.

The capacity of *social coordination* allows two agents to engage in coordinated interaction. In the very early stages of development the infant can take turns through the mediation of the caregiver. Later the child can perform joint activities such as imitation games of caregiver's actions (at nine months) and imitative games for goal sharing (twelve months), up to social coordination for action plans (see section 6.1.3 for more discussion on social coordination). Finally, *intentional understanding* refers to the development of a capability of the agent to understand that she, as well as other agents, can have intentions and goals. Developmentally, this first manifests with the identification of the physical presence of other people (zero–three months) and with the distinction between animate and inanimate agents (six months). This then proceeds to the understanding of the goals of action and prediction of the behavior of other agents in terms of joint action plans for common goals (eighteen months).

This detailed developmental timescale and milestones of social development in human infants have also provided a roadmap for developmental robotics research on shared attention (Kaplan and Hafner 2006b).

6.1.2 Imitation

The consistent evidence of the existence of imitation capabilities in neonates, even in the first hours after birth (Field et al. 1983; Meltzoff and Moore 1983), provides a bridge between the nature and nurture views of sensorimotor, social, and cognitive development (Meltzoff 2007). In addition to supporting the development of motor skills, imitation is a fundamental capability of social agents. As infants have been shown to be able to imitate facial gestures as early as forty-two minutes following birth (Meltzoff and Moore 1983, 1989), this provides support for the existence of a basic mechanism for comparing states across modalities, and for the instinct to imitate other agents.

Given the wide uses of terms such as "imitation," "emulation," and "mimicry" in the developmental and comparative psychological literature, Call and Carpenter (2002) have proposed an operational definition of imitation that differentiates imitation from other forms of copying behavior such as mimicry and emulation. When considering the situation of two social agents where one (the imitator) copies the other agent (the demonstrator), we need to distinguish three sources of information: the goal, the action, and the result. For example, if the demonstrator shows the action of opening a wrapped gift box, this implies several hierarchically organized goals (unwrap the box, open the box, get the present), hierarchical actions (unfold or tear the wrapping paper, open the box, take the present) and hierarchical results (the box is unwrapped, the box is open, I hold the present). Call and Carpenter propose to use the term "imitation" only for the case when the imitator copies all the three sets of goals, actions, and results. "Emulation," in contrast, implies the copying of the goal and the results, but not the exact copying of the action itself (e.g., I use scissors to get the present out of the box). And "mimicry" involves the copying of the action and results, but not the goal (I unwrap and open the box, but leave the present inside). The distinction among these different types of copying behavior, and how the three sources of information are involved in each, allows us to better differentiate various imitation capabilities in animals and humans, and between different developmental stages of imitation strategies in infants. Moreover, this can better inform the modeling of imitation in robotics experiments (Call and Carpenter 2002).

Meltzoff and colleagues have identified four stages in the development of imitation abilities: (1) body babbling, (2) imitating body movements, (3) imitating actions on objects, and (4) inferring intentions (Rao, Shon and Meltzoff 2007). With *body babbling*, the infant randomly (by trial and error) produces body movements, and this allows her to learn about the sensorimotor consequences (proprioceptive and visual states) of her own actions and gradually develop a body schema ("act space"). In stage 2, the infant can *imitate body movements*. Studies with infants in their first month of age show that neonates can imitate facial acts (e.g., tongue protrusion) that they have never seen themselves perform, and twelve- to twenty-one-day-old infants can identify body parts and imitate differential action patterns with the same body part. This is referred to as "organ identification" in child psychology (Melzoff and Moore 1997) or "self-identification" in the robotics literature (Gold and Scassellati 2009). These studies demonstrate that newborns have an innate observation-execution mechanism and representational structure that allows infants to defer imitation and to correct their responses without any feedback. Stage 3 regards the *imitation of actions on objects*. Infants between one-year to one-and-a-half-years old can imitate not only movements of face and limbs, but also actions on objects in a variety of contexts. In stage 4, *inferring intentions*, the child can understand the underlying goal and intention behind the demonstrator's behavior. In an experiment by Meltzoff (2007), eighteen-month-old children were shown unsuccessful acts involving a demonstrator trying but failing to achieve his goal. Because children are able to imitate what the adult intended to achieve, and are not just copying the behavior, this indicates that children can understand the intentions of others. These incremental stages of the development of imitation skills show how the understanding and inference of the goal and intention of the others are crucial in adult imitation capabilities. And referring again to Carpenter and Call's classification of imitation terminology, proper imitation requires the full capability to understand the goals of demonstrated actions, as in stage 4, inferring intentions. As for stages 1 to 3, further investigations are needed to distinguish the stage when the child goes from simple emulation of action to full intentional imitation.

Meltzoff and Moore (1997) have proposed a developmental model that can explain the imitation of both facial and manual actions, called the Active Intermodal Matching (AIM) model (figure 6.2). This model consists of three main subsystems: (1) perceptual system, (2) motor acts system, and (3) supramodal representational system. The *perceptual system* specializes for visual perception of the observed facial acts, or motor behavior. The *motor acts system* involves the imitator's active copying of the demonstrated act, supported by the essential feature of the proprioceptive information. The proprioceptive feedback permits the core matching to target between the visual input and the infant's own acts, and serves as a basis for correction. What is crucial here is the

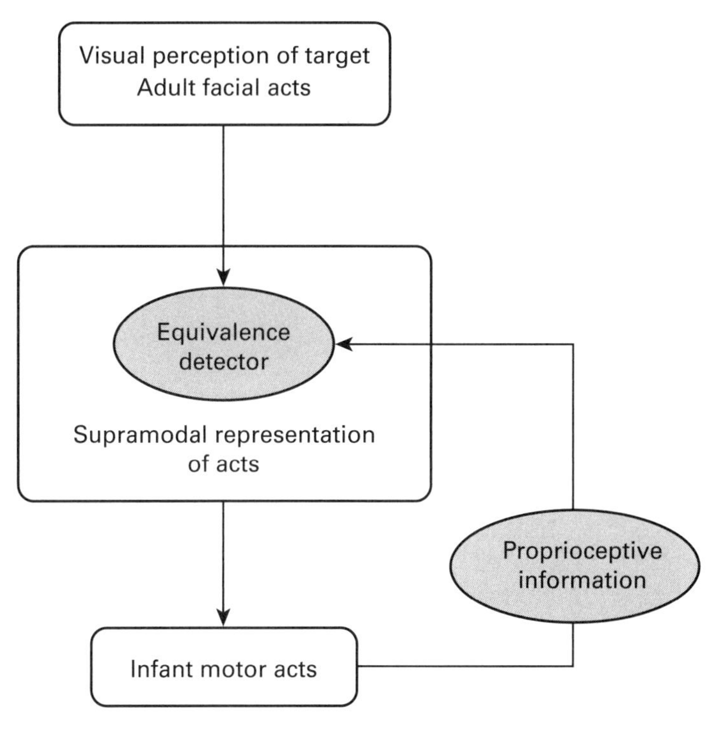

Figure 6.2
The Active Intermodal Matching model for imitation of Meltzoff and Moore 1997.

role of the *supramodal representational system*, which provides a common (supramodal) framework to encode, and detect the equivalence, between the perceived and the produced behavior. Given the benefits of the AIM model to provide a functional, operational description of the sensorimotor and the representation mechanisms involved in imitation, it has inspired the design of developmental robotics studies of imitation in mobile and humanoid robots (Demiris and Meltzoff 2008).

Significant progress has also been achieved in the last decade on theories and experimental evidence on the neural bases of imitation. In particular, the discovery of "mirror neurons" by Rizzolatti and colleagues (Rizzolatti, Fogassi, and Gallese 2001) has contributed to the renewed interest in this field. The mirror neurons, first studied in the monkeys' premotor cortex (area F5), are neurons that fire both when the animal performs an action, and when it observes the same action being performed by another animal or a human experimenter. These neurons—which are involved in both the perception and production of an action—are obvious candidates for supporting tasks that require the matching of the visual and motor systems, as in the AIM model. Moreover, it has been demonstrated that in monkeys a mirror neuron fires only when the goal of the action is the same, as in the goal-based definition of imitation. However, as monkeys have a mirror system too primitive to code the details of the observed action, they cannot therefore replicate the observed action (Rizzolatti and Craighero 2004).

The existence of the mirror neuron system in humans, and its specific involvement in imitation tasks, has recently received experimental evidence (Iacoboni et al. 1999; Fadiga et al. 1995). This follows early theoretical stances on the anatomic and evolutionary similarities between the monkey's F5 area and the humans' Broca area, and the role that these play in motor and language imitation and in the evolution of human language (Rizzolatti and Arbib 1998).

6.1.3 Cooperation and Shared Plans

Cooperating with other people to achieve a common goal is another fundamental milestone in the development of social skills. Cooperation requires the capability of building and using shared plans and shared intentions between two or more interacting individuals. This can also explain forms of altruistic behavior observed in children, and in some cases also observed in nonhuman primates.

The study of the development of social cooperation capabilities in children has been closely associated with comparative psychology experiments carried out with both human children and nonhuman primates, such as chimpanzees. One of the key findings in the developmental comparative psychology of cooperation is that human children appear to have a unique ability and motivation to share goals and intentions with others. In nonhuman primates, however, this capability is absent. Tomasello and colleagues (Tomasello et al. 2005; Tomasello 2009) have proposed that human children's ability to share intentions emerges from the interaction of two distinct capabilities: (1) intention reading and (2) motivation to share intentions. *Intention reading* refers to the ability to infer the intentions of other people through observation of their behavior. This implies the more general ability to understand that others are intentional goal-directed agents. The *motivation to share intentions* implies that once we identify the intention of the other agent, this is then shared in a cooperative way. Tomasello has demonstrated in numerous experiments that both nonhuman and human primates are skilled at reading the intentions of others through observation of their actions and gaze direction. However, only human children possess the additional capability to altruistically share intentions.

One seminal study that has clearly demonstrated this difference between human children and other primates is the work by Warneken, Chen, and Tomasello (2006). They carried our comparative experiments with eighteen- to-twenty-four-month-old children and with young chimpanzees of thirty-three and fifty-one months of age. In this study they used two goal-oriented problem-solving tasks and two social games that required cooperation (see box 6.1 for more details). They also manipulated the condition of complementary versus parallel roles. In activities requiring complementary roles, the child and the adult have to perform different actions to achieve the same problem-solving or ludic goal. In parallel role tasks, the two partners must perform similar actions in parallel to achieve the joint task. For example, in the double-tube game task, the experimenter (cooperating adult) starts by inserting a wooden block in the upper hole of one of the two tubes. The block slides down the inclined tube, and finally has to be caught by the child, or chimpanzee, holding a tin cup at the other end of the tube. This task requires complementary roles, that is, one agent has to insert the block, and the second agent uses the container to catch it. In the trampoline game, both agents instead perform the same task: tilting their half of the trampoline border to make the wooden block jump.

Box 6.1

Cooperation Skills in Humans and Chimpanzees (Warneken, Chen, and Tomasello 2006)

Overview

This study compares cooperation behavior in human children and young chimpanzees to identify the existence of a form of cooperative activity involving shared intentionality, and to establish if this is a uniquely human capability. Four different cooperative tasks were used (two joint problem-solving activities and two social games), with either complementary or parallel roles with a human adult partner. Complementary role tasks require the child and the adult to perform different actions, while parallel role tasks require the two to perform the same action in parallel. In the second part of each experiment, the adult participant is instructed to stop the cooperative behavior, for example, by walking away or not responding to the child's (or chimpanzee's) requests. This condition aims at testing the capacity of the participants to reengage the human adult in the cooperative task.

Participants

In the four experiments, the group of young human participants consisted of sixteen eighteen-month-old children and sixteen twenty-four-month-old children. The two age groups were chosen to test the hypothesis that older children can better adjust their cooperative behavior to that of the adult, and that older children can better regulate the activity during interruption periods.

The animal group consisted of three young chimpanzees. Two were fifty-one-month-old female chimpanzees (Annet and Alexandra) and one was a thirty-three-month-old male chimpanzee (Alex), all housed together at the Leipzig Zoo. The materials used in the four experiments were subject only to minor adjustments (e.g., change in material and dimension, and food reward instead of toys) and directly matched the four cooperative tasks used with the human children.

Tasks

	Complementary role	*Parallel role*
Problem solving	Task 1—Elevator The goal of this task is to retrieve an object placed inside a vertically movable cylinder. A single person cannot solve this, as it requires two complementary actions on the opposite sides of the apparatus. One person must first push the cylinder up and hold it in place (role 1), and only then the other person can access the object through the opening of the cylinder from the opposite side (role 2).  	Task 2—Tube with handles The goal of this task is to retrieve a toy placed inside a long tube with two handles. A single person cannot solve this, as the length of the tube makes it impossible for a person to grasp and pull both handles at the same time. This tube can only be opened by two persons playing simultaneously the two parallel roles of pulling at each end until the tube opens and releases the toy. 

Social game

Task 3—Double tube

The game requires one person to insert a wooden block in either of two tubes and let the other person catch it from the end of the correct tube. It requires one participant to insert the block in the top opening of the tube (role 1), and the other person to catch the block in a tin cup at the bottom end of the tube (role 2).

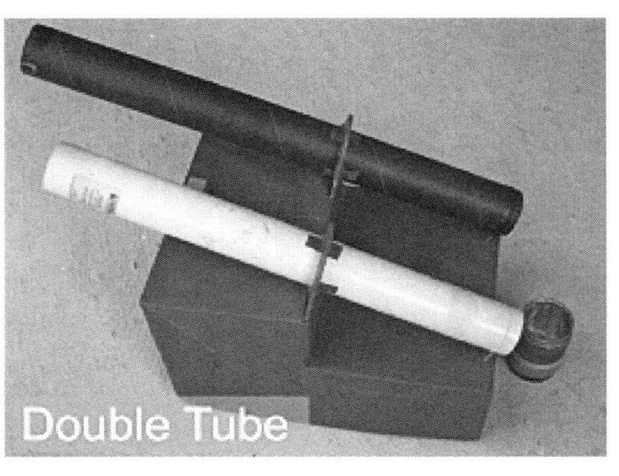

Task 4—Trampoline

The game requires two people to make a wooden block jump on the trampoline by holding the rim on opposite sides. Because the trampoline is made of two C-shaped hoses connected with flexible joints, covered with cloth, it requires parallel roles of synchronously shaking the hose ring.

Results Summary

- When the cooperation task is very simple, human children can achieve coordination by eighteen to twenty-four months of age; chimpanzees too can engage in simple cooperation.
- Children spontaneously cooperate and are motivated not just by the goal but also by the cooperation itself.
- With interruption conditions, all children attempted to reengage the partner, but no chimpanzee made a communicative attempt to reengage the partner.

Using these experimental tasks, Warneken, Chen, and Tomasello (2006) first studied cooperation in normal conditions, when both agents consistently cooperate in the game. They also observed the children's and chimpanzees' behavior in disrupted cooperative settings, such as when the adult experimenter stops performing the collaborative task by refusing to use the trampoline or to insert the block in the rube.

Results of the experiments show that in normal cooperative settings all the children enthusiastically participate both in goal-directed, joint problem-solving tasks and in social games. They engage in cooperation not only for the aim of achieving the joint goal, but also for the sake of cooperation itself. Chimpanzees can also cooperate is simple tasks. But in the condition when the adult experimenter stops cooperating, results show that children spontaneously and repetitively attempt to reengage the adult, while chimpanzees appear disinterested in non-goal-directed social games. The animals are only focused on obtaining the food goals, independently of cooperation. Warneken, Chen, and Tomasello (2006) thus conclude that there is a very early and unique human motivation and capacity for actively engaging in cooperative activities.

In a follow-up study, Warneken, Chen, and Tomasello (2006) demonstrated that eighteen-month-old infants spontaneously and altruistically help adults in a variety of situations, for example, when the adult is struggling to achieve the goal (instrumental help) and even if the child receives no direct benefit from his action. Children were tested with ten different situations in which an adult was having trouble performing a task, manipulating the difficulty across four conditions: (1) an out-of-reach object, (2) an object obstructed by an obstacle, (3) achieving the wrong goal when this can be corrected by the child, and (4) using the wrong means that the child can correct. Results show clear evidence of altruistic behavior in the children, with twenty-two of the twenty-four infants helping in at least one of the tasks. They also carried out a comparable experiment with three young chimpanzees using the same four types of tasks. All three chimpanzees consistently helped in the task with out-of-reach objects, even if the target object was not food. However, they did not help the human in the types of tasks with obstacles, wrong results, or wrong means. The chimpanzee's altruistic behavior with the reaching task is explained by the fact that in this task the goal is easier to detect than in the other conditions. This suggests that both children and chimpanzees are capable of altruistic behavior, but that they differ in their ability to interpret the other's need for help.

The results of these studies on cooperative behavior are consistent with Carpenter, Tomasello, and Striano's (2005) role reversal analysis. They observe the emergence of a "triadic" cooperative strategy in role reversal, which involves the child, the collaborative adult/peer partner, and an object. This is for example the case in which the adult holds out a container so the child can place toys into it. In subsequent interactions, reversing the roles, the child can then hold out the basket to the adult so that he can put toys into it. This also implies that the child has the capacity to read the adult's intended goal and to perform, on his behalf, the action the adult would have wanted to perform. The role reversal capacity corresponds to a bird's-eye view, or third-person representation of the interaction, that allows the child to take either role in the cooperation task. These child studies have inspired the design of developmental robotics models implementing different cooperation strategies (Dominey and Warneken 2011—see section 6.4).

6.1.4 Theory of Mind

The parallel development of social learning capabilities, including skills such as eye-gaze detection, face recognition, observation and imitation of other people's actions, and cooperation gradually leads to the acquisition of a complex ability to correctly attribute beliefs and goals to other people. This is normally referred to as the theory of mind (ToM), which is the capacity to understand the actions and expressions of others and to attribute mental states and intentionality to the other social agents. The understanding of the cognitive mechanisms leading to ToM development is crucial for the design of social robots capable of understanding the intention and mental stages of human agents and of other robots. Here we will follow Scassellati's (2002) analysis of two of the most influential ToM hypotheses, proposed by Leslie (1994) and by Baron-Cohen (1995), and their influence in social developmental robotics. In addition, Breazeal et al. (2005) have proposed a ToM based on simulation theory and Meltzoff's AIM model that has been applied to developmental robotics. Moreover, work on theory of mind in nonhuman primates also provides useful insights on mechanisms involved in the (evolutionary) development of the theory of mind (Call and Tomasello 2008).

Leslie's (1994) ToM is based on the core concept of the attribution of causality to objects and individuals during the perception of events. Leslie distinguishes three classes of events according to the causal structure involved: (1) mechanical agency, (2) actional agency, and (3) attitudinal agency. *Mechanical agency* refers to the rules of mechanics and physical interaction between objects. *Actional agency* describes events in terms of the agent's intent and goals and their manifestation in actions. *Attitudinal agency* explains events in terms of the attitudes and beliefs of agents.

Leslie (1994) claims that the human species has evolved three independent, domain-specific cognitive modules to deal with each type of agency, and that these gradually emerge during development. The theory of body module (ToBY) handles mechanical agency, for the understanding of the physical and mechanical properties of the objects and the interaction events they enter into. This reflects the infant's sensitivity to the spatiotemporal properties of object-interaction events. Classical examples of mechanical agency phenomena are Leslie's experiment on infants' perception of causality between two animated blocks. He manipulates conditions such as direct launching (where one moving block hits a second static object and this immediately starts to move), delayed reaction (same collision, but delayed movement of the second object),

Table 6.2
Developmental timescale for theories of mind by Leslie and by Baron-Cohen

Age (months)	Leslie	Baron-Cohen
0–3 months	Sensitivity to spatiotemporal properties of events for theory of body (innate?)	Intentionality detector for self-propelled agents vs. inanimate objects (innate?)
6 months	Detecting of actions and their goals (through eye gaze) for ToM-System1	Eye direction detector
9 months		Appearance of shared attention mechanism
18 months	Initial development of attitudinal agency for ToM-System2	Initial development of ToM mechanisms
48 months	Fully developed ToM-System2; meta-representations	Full development of ToM mechanisms

launching without collision, collision with no launching, and static objects. Children with fully developed ToBY, at six months of age and later, can perceive causality of interaction in the first condition, and not in the others. Leslie suggests that this might be an innate ability, as mechanical agency is seen in the very first months of life (table 6.2).

The second module is called theory of mind system 1 (ToM-S1), and is used for actional agency to explain events in terms of goals and actions. This is manifested through eye-gaze behavior, and leads to the child's identification of the actions and their goals. This capacity starts to appear at six months of age. The third module is called theory of mind system 2 (ToM-S2), and is used for attitudinal agency for the representations of other people's beliefs that might differ from our own knowledge or from the observable world. Children develop and use meta-representations, where the truth properties of a statement depend on mental states, instead of the observable world. This module starts to develop at eighteen months, until its full development at around forty-eight months.

Baron-Cohen's theory of mind, called the "mindreading system," is based on four cognitive capabilities: (1) intentionality detector, (2) eye direction detector, (3) shared attention mechanism, and (4) theory of mind mechanism. *Intentionality detector* (ID) is specialized for the perception of stimuli in the visual, auditory, and tactile senses that have self-propelled motion. As such it distinguishes between animate entities (i.e., agents) versus nonanimate entities (i.e., objects). This leads to the understanding of concepts such as "approach" and "avoidance," and for representations such as "he wants food," "he goes away." The intentionality detector appears to be an innate capability in infants (table 6.2). The *eye direction detector* (EDD) specializes in face perception and the detection of eye-like stimuli. Baron-Cohen identifies various functions for the EDD. These include the detection of eye gaze, the detection of the target of eye gaze (looking at an object or at another person), and the interpretation of gaze direction as a perceptual state (the other agent is looking at me). The eye detection capability is available in the first nine months of age. Both ID and EDD produce dyadic representations, with one agent and one object or another agent, as in the "he wants food" and "the other agent is looking at me" cases.

The *shared attention mechanism* (SAM) combines dyadic representations to form triadic concepts. For example two dyadic percepts "I see something" and "you want food" can be combined into the "I see (you want food)" representation. In particular, combinations of ID and EDD representations allow the infant to interpret the gaze of others as intentions. SAM develops between nine and eighteen months of age. Finally, with the *theory of mind mechanism* (ToMM) the triadic representations are converted into meta-representations (as in Leslie's ToM-S2) through the understanding and representation of the mental states and beliefs of other agents. ToMM permits the construction of representations of the form "Mary believes (I am hungry)," as well as representation of knowledge that might not correspond to true states in the world, such as "Mary believes (dogs can speak)." The advanced ToMM starts to appear around eighteen months, until it fully develops by forty-eight months. An important advantage of such a theory is that it is possible to identify ontogenetic impairments in the development of the four capabilities, which can in turn explain various autism spectrum disorders (ASDs).

In addition to Leslie and Baron-Cohen's explanations of the mechanisms involved in the development of theory of mind in infants, other theoretical stances on ToM focus on imitation and simulation theory (Breazeal et al. 2005; Davies and Stone 1995). Simulation theory claims that we can make predictions about the behaviors and mental states of others by simulating another person's actions and sensory state. We can use our own imitation skills and cognitive mechanisms to recreate how we would think, feel, and act in the other agent's situation, thus inferring the emotions, beliefs, goals, and actions of the others. Simulation theory is also consistent with the embodied cognition approach (Barsalou 2008) and the role of mirror neurons in linking action perception with action execution (Rizzolatti, Fogassi, and Gallese 2001).

In addition to the theory of mind development in humans, there has been an important debate on the evolutionary origins of the theory of mind, and the existence of theory of mind capabilities in animals. One of the main contributions in this field has come from Call and Tomasello's (2008) investigation of the capacity to infer the goals and intentions of other agents in nonhuman primates such as chimpanzees. They review extensive evidence in support of the hypothesis that at least chimpanzees understand both the goals and intentions of other agents (humans and chimpanzees), as well as the perception and knowledge of others. Primates use these social capabilities to produce intentional action. No evidence is available, however, to support the fact that primates can understand false belief, as in the meta-representation of Leslie's and Baron-Cohen's theories, or to support the view that primates can appreciate that other agents might use mental representations of the world to drive their actions and that do not correspond to the observable reality.

The operationalization of various cognitive mechanisms involved in the gradual development of the theory of mind, as in Leslie's and Baron-Cohen's theories, and evolutionarily in animal studies, provides a useful framework for the modeling of theory of mind in cognitive robots (Scassellati 2002) and for the use of robots in the treatment of autism disorder (François, Dautenhahn, and Polani 2009a; François, Powell, and Dautenhahn 2009b; Tapus, Matarić, and Scassellati 2007).

6.2 Joint Attention in Robots

The study of joint attention in robots has been a key focus in developmental robotics, for its link between social skills and human robot interaction and communication. The developmental classification of attentional and social skills proposed by Kaplan and Hafner (2006b) provides a useful framework for the review of developmental robotics models of joint attention. They use the example of the two AIBO robots facing each other to show different types of gaze sharing and pointing gestures. Figure 6.3a–b

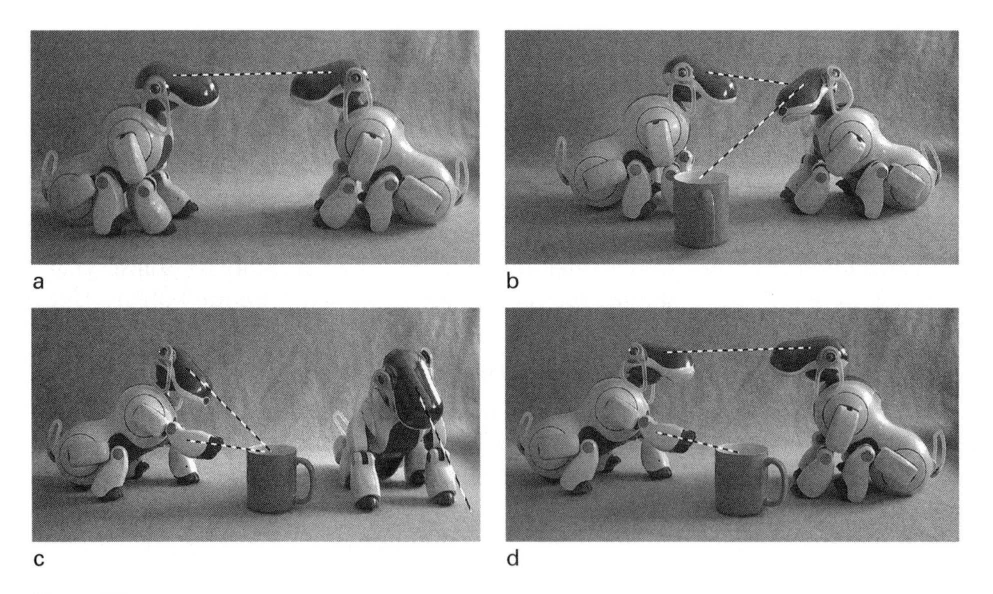

Figure 6.3
Different joint attention strategies: (a) mutual gaze, (b) gaze following, (c) imperative pointing, (d) declarative pointing (from Kaplan and Hafner 2006b). Figure courtesy of Verena Hafner. Reprinted with permission from John Benjamins.

illustrates two attention-detection strategies of the robots, namely (1) mutual gaze when the two robots establish mutual eye contact, and (2) gaze following, when the partner looks at the object gazed at by the first agent. In figure 6.3c–d, two attention manipulation behaviors are shown, i.e., (3) imperative pointing to request an object or food even when the other agent is not initially looking at it, and (4) declarative pointing to put emphasis and create shared attention on an object focus of the interaction.

A developmental robotics investigation into the learning recognition of pointing gestures was originally carried out by Hafner and Kaplan (2005) with the AIBO robots, and more recently by Hafner and Schillaci (2011) with the NAO. In the original Hafner and Kaplan (2005) study, two Sony AIBOs sit on the floor and face each other, with an object in the middle. The "adult" robot is equipped with a capacity to recognize the object location and point at it. The "child" robot can learn to recognize the partner's pointing gesture. It first looks at the pointing gesture of the adult, guesses the direction of the object, and turns its head to look at it. A supervision signal is given to check if the learner looks at the correct direction, and this is used to update the neural control architecture. To train the robot's neural controller, the learner robot's camera captures 2,300 images of the partner's pointing gesture (half for left and half for right pointing) with a variety of backgrounds, lighting conditions, and distances between the robots. Each image is divided into a left and right side, and then processed to extract two levels of brightness and the horizontal and vertical edges through Sobel filtering. Further operators are then applied to identify the vertical and horizontal centers of mass of the image pixels. The selected features intuitively correspond to the detection of the vertical shift of brightness when the robot lifts its arm, to the increase of horizontal edges and to the decrease of vertical edges on the side of the pointing. Through a pruning method based on greedy hill climbing, three key features are selected out of the combination of the various features and operators, leading to a pointing recognition performance of over 95 percent.

To train the robot to recognize the left/right direction of the pointing gesture, Hafner and Kaplan use a multilayer-perceptron with three input neurons for the selected visual features, three neurons in the hidden layer, and two outputs for the left/right pointing direction. The backpropagation algorithm used to train the multilayer perceptron is considered comparable to a reward-based system, when a binary left/right decision is involved. Hafner and Kaplan (2005) argue that this capacity to learn to recognize the direction of pointing is at the basis of manipulative attention capability, which can further develop to include imperative and declarative pointing behavior. Other robotics models have focused on the developmental emergence of shared gaze, as in Nagai and collaborators (Nagai et al. 2003; Nagai, Hosoda, and Asada 2003) model of the gradual extension of the gaze-followable area. Nagai and colleagues follow Butterworth's (1991; Butterworth and Jarrett 1991) developmental framework based on the incremental acquisition of ecological (the infant looks at an interesting object regardless of the caregiver's gaze direction), geometric (joint attention only when the

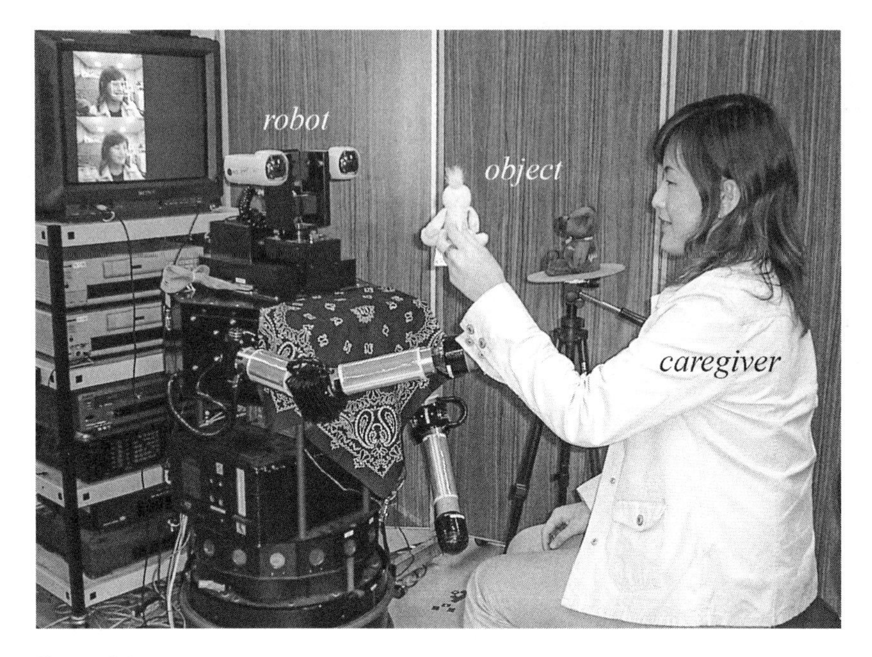

Figure 6.4
Experimental setup for Nagai, Hosoda, and Asada (2003) experiment. Figure courtesy of Yukie Nagai.

object is in the infant's field of view), and representational (the infant can find a salient object outside its own field of view) gaze strategies.

The experimental setup consists of a robot head with two cameras, which rotate on the pan and the tilt axes, and a human caregiver with various salient objects (figure 6.4). In each trial, the objects are randomly placed, and the caregiver looks at one of them. She changes the object to be gazed at every trial. The robot first has to look at the caregiver by detecting her face through template matching and extracting a face image. It then locates the salient, bright-colored objects by using thresholds in color space.

Through the cognitive architecture as in figure 6.5, the robot uses its camera image and the angle of the camera position as inputs, to produce in output a motor command to rotate the camera eyes. The architecture includes the visual attention module, which uses the salient feature detectors (color, edge, motion, and face detectors) and the visual feedback controller to move the head toward the salient object in the robot's view. The self-evaluation module has a learning module based on a feedforward neural network and the internal evaluator. The internal evaluator gauges the success of the gaze behavior (i.e., whether there is an object at the center of the image, regardless of the success or failure of joint attention) and the neural network learns the sensorimotor

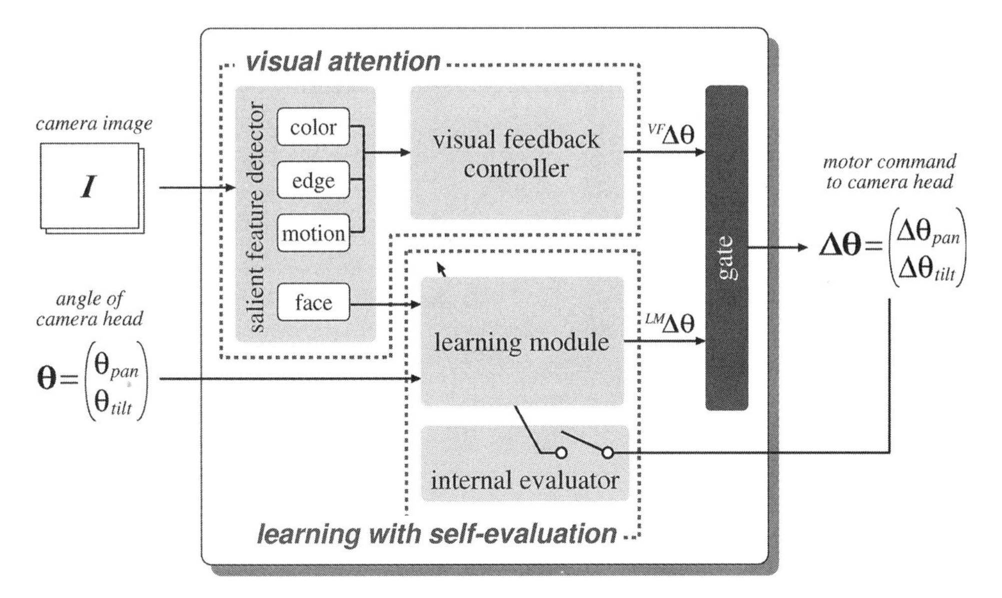

Figure 6.5
Cognitive control architecture for shared gaze in the Nagai, Hosoda, and Asada (2003) experiment. Figure courtesy of Yukie Nagai.

coordination between face image and current head position, and the desired motor rotation signal. The gate module selects between outputs from the visual feedback controller and the learning module. This uses a selection rate that is designed to choose mainly the output of the attention module at the beginning of the training, and then, as learning advances, gradually select the learning module's output. The selection rate uses a sigmoid function to model the nonlinear developmental trajectory between bottom-up visual attention and top-down learned behavior.

To train the robotic agent, when a face is detected, the robot first looks at the caregiver and captures the image. The filtered image and the camera angles are used in input to the visual attention module. If the attention module detects one salient object, the robot produces the motor rotation signal to look at this object. In parallel, the learning module uses the head image and camera angles to produce its own motor rotation output. The gate module uses the selection rate to choose either the attention module's or the learning module's motor output signal, and the camera is then rotated to gaze at the object. If the robot has successfully gazed at one object, the learning module uses the motor rotation output as training signal for the feedforward neural network. Successful gaze is defined as the action to center the camera on any object in the scene regardless of whether this is the one gazed at by the caregiver. However, the feedforward network is able to learn the correlation between joint gazed objects due to

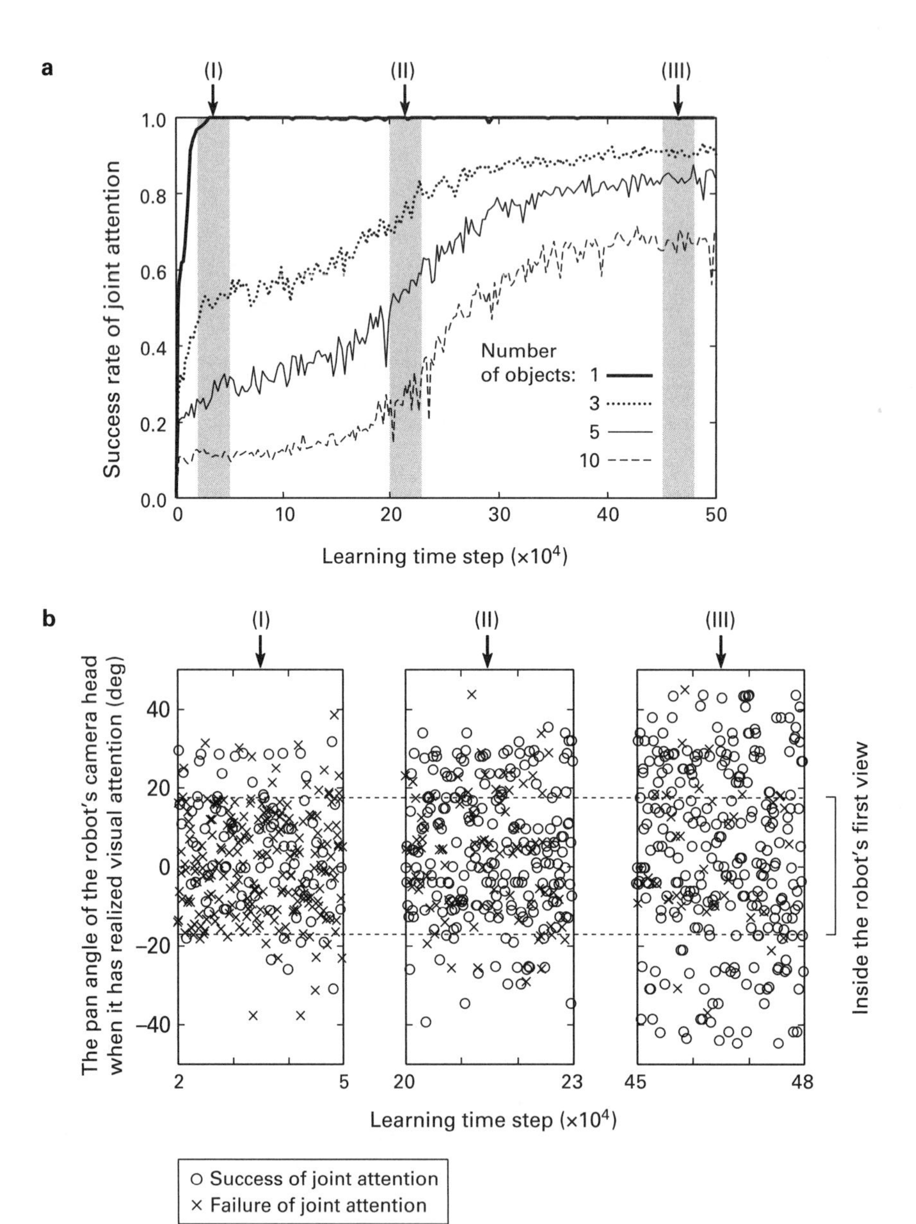

Figure 6.6
(a) Learning performance as success rate during the Nagai, Hosoda, and Asada (2003) experiment. (b) Data on the three stages, showing the emergence of ecological (I), geometrical (II) and representational (III) joint attention strategies. Figure courtesy of Yukie Nagai.

the systematic correlation between the image of the face's eye direction and the object position (when the wrong object is looked at, the position of the object that the robot has gazed at does not uniquely correspond to the image of the caregiver's face). This mechanism has the advantage that the robot can develop the ability of joint attention without any task evaluation or direct feedback from the caregiver. And as the gate module gradually allows the learning module to take over control of gazing direction, this produces the incremental improvement of the shared gaze behavior.

Nagai et al. (2003) and Nagai, Hosoda, and Asada (2003) carried out a series of experiments with this robot head varying the number of objects from one to ten. When only one object is used, success is easy and the robot achieves 100 percent success. When the experimenter uses five objects, in the first stage of learning joint attention is only possible for 20 percent, in other words, due to random choice of one of the five objects. However, at the end of the training, the success rate reaches 85 percent, well above chance, when the learning module output has achieved low training error and it has taken over the bottom-up visual attention preference for the most salient visual object. With ten objects the task is quite hard, and the training only reaches a performance of just above 60 percent success.

The most important result from a developmental perspective are the analyses of the emergence of Butterworth's (1991) three stages of gaze and shared attention—I: ecological, II: geometrical, and III: representational. Figure 6.6a shows the increase of the rate of successful joint attention occurrences during the whole experiment, with a highlight of the three stages selected for further analysis in figure 6.6b. Figure 6.6b shows the number of successes (symbol o) and failures (symbol ×) of joint attention events at three different developmental stages (sample analyses on the case with five objects). At the beginning of training (stage I), the robot mostly looks at objects located within the robot's view, and can only achieve joint attention at a chance level. This is due to the predominance of choices by the bottom-up visual attention module, as per the gate module's sigmoid selection rate. In the intermediate steps of training (stage II), the agent is capable of achieving joint attention in the great majority of cases when the object is within the image, and at the same time the robot increases the gazing at location outside the eye's view. Finally (stage III), the robot achieves joint attention in almost all trials and positions.

How does the robot achieve the capability to look at the target (caregiver's gazing direction) object outside the robot's own field of view, as in stage III? During training, the robot's learning module gradually acquires the capability to detect the sensorimotor correlation between the eye position of the caregiver's face image and a direction of camera rotation. This association is learned especially during successful joint gazes when the object is visible as in stages I and II. However, even when no objects are visible, the robot tends to produce a motor rotation signal consistent with the caregiver's eye direction. The robot's head rotation toward the direction of gaze will gradually lead to the visualization of the target object on the border of the image. Once the object is detected in the image, the robot will be able to identify its location and gaze directly on its center.

The Nagai et al. (2003; Nagai, Asada, and Hosoda 2006) study is an elegant example of a developmental robotics model directly investigating known developmental stages, as the joint attention stages proposed by Butterworth (1991). Moreover, it shows how the changes between these qualitative stages are the results of gradual changes in the robot's neural architecture. This is due to the subsymbolic and distributed nature of the robot's neural controller, that is, trained through small changes to the network parameter (weights). However, a gradual accumulation of changes can result in nonlinear learning phenomena, as in the well-known connectionist models of learning past tense (Plunkett and Marchman 1996) or the vocabulary spurt (Mayor and Plunkett 2010) and general U-shape modeling in robots (Morse et al. 2011) (see also chapter 8). Numerous other developmental robotics models of joint attention have been proposed. A few have a focus on human-robot interaction and how a robot can achieve and support joint attention with a human. For example, Imai, Ono, and Ishiguro (2003) carry out experiments on pointing for joint attention. Their Robovie robot (Ishiguro et al. 2001) is able to attract the attention of a human participant by pointing at an object and establishing mutual gaze. Kozima and Yano (2001) used the Infanoid baby robot, and modeled the robot's capacity to track human faces and objects, to point to and reach for objects, and to alternate the gaze between faces and objects. Jasso, Triesch, and Deak (2008) and Thomaz, Berlin, and Breazeal (2005) modeled social referencing, meaning, the phenomenon when an infant is presented with a novel object and consults the adult's facial expression before reacting to this object. If the adult's facial expression is positive (e.g., smile), the infant touches and interacts with the object, while she avoids it when the adult's face shows a negative reaction. Jasso, Triesch, and Deak (ibid.) use a reward-driven modeling framework, based on temporal-difference reinforcement learning, for social referencing in simulated agents. Thomaz, Berlin, and Breazeal (ibid.) use the humanoid robot Leonardo to model social referencing dependent on both visual input and an internal affective appraisal of the object.

Developmental models of joint attention are also relevant to the investigation of normal and atypical development, given the importance of shared attention and social interaction in disorders such as autism spectrum disorders. For example, in this field, Triesch et al. (2006; see also Carlson and Triesch 2004) have modeled healthy, autistic, and Williams syndrome infants by changing characteristics of their preferences for gaze shift in simulated agents. They propose a computational model of the emergence of gaze-following skills in infant-caregiver interactions, as the infant's learning of his caregiver's gaze direction allows him to predict the locations of salient objects. The modeling of shared attention for normal and atypical development is also relevant to the robotics models of the theory of mind, as in section 6.5.

6.3 Imitation

The study of social learning and imitation has been one of the main topics of research in cognitive robotics and human-robot interaction (e.g., Demiris and Meltzoff 2008; Breazeal and Scassellati 2002; Nehaniv and Dautenhahn 2007; Schaal 1999; Wolpert and Kawato 1998). Robots provide a useful tool to investigate the developmental stages of imitation as they require the explicit operationalization of the various components and mechanisms necessary to achieve. To imitate, a robot must be able to possess the following skills: (a) motivation to observe and imitate the others, (b) perception of movement, and (c) conversion of the observed actions into their own body schema (correspondence problem) (Breazeal and Scassellati 2002; Hafner and Kaplan 2008; Kaplan and Hafner 2006a).

For studies on the evolutionary and developmental origins of the motivation to imitate, most developmental robotics models start from the assumption that the robot is equipped with an innate motivation (instinct) to observe and imitate others. Although numerous comparative psychology and neuroscience studies have looked at the origins of imitation capabilities in animals and humans (e.g., Ferrari et al. 2006; Nadel and Butterworth 1999) and through the involvement of the mirror neuron systems (Rizzolatti, Fogassi, and Gallese 2001; Ito and Tani 2004), few computational models exist that have explicitly explored the origins of the imitation capabilities. For example, Borenstein and Ruppin (2005) used evolutionary robotics to successfully evolve adaptive agents capable of imitative learning, through the emergence of a neural mirror device analogous to the primates' mirror neuron system. In existing developmental robotics models, the imitation algorithm implicitly implements an instinct to observe others and use this input to update the models' own imitation system.

For the perception of movement, various motion capture methods and artificial vision systems have been used. Motion capture technologies include the Microsoft Kinect, exoskeleton technology, or digital gloves that measure joint angles positions (e.g., the Sarcos SenSuit system for the simultaneous measurement of 35 DOFs of the human body; Ijspeert, Nakanishi, and Schaal 2002), and tracking with the use of magnetic or visual markers (Aleotti, Caselli, and Maccherozzi 2005). In particular, the easy-access, low-cost Microsoft Kinect apparatus and open software system offer a great opportunity for gestural, teleoperation, and imitation studies in robotics (Tanz 2011). Vision-based motion detection systems are typically based on the automatic detection and tracking of body parts (e.g., Ude and Atkeson 2003). Alternatively, Krüger et al. (2010) use parametric hidden Markov models for the unsupervised discovery of action primitives by analyzing the object state space—that is, inferring human action primitives that induce the same effect on the object, rather than focusing on human body part movements.

In addition to a motion perception system, an imitating robot must have the capability to apply selective attention to the specific actions and objects that are the focus of the interaction. This might depend on a combination of bottom-up processes (e.g., saliency of the object/action) and top-down processes (e.g., expectations and goal-directed aspects). In the imitation experiments by Demiris and Khadhouri (2006) described as follows, we see an example of the contribution of bottom-up and top-down mechanisms, in particular where attention helps reduce the cognitive load of the robot.

Finally, to study imitation the experimenter needs to address the correspondence problem (Nehaniv and Dautenhahn 2003), that is, the knowledge required by the robot to convert an action that has been observed into a sequence of its own motor responses to achieve the same result. Breazeal and Scassellati (2002) identify two main approaches to the correspondence problem and the representation of the perceived movements: (1) motor-based representations and (2) task-based representations. The first method uses representations of the perceived movement in motor-based terms, e.g., though the encoding of the demonstrator's movement trajectory into the imitator's motor coordinates. This is the method used by Billard and Matarić (2001) where the excentric frame of reference (relative to the visual tracking system) used for the coordinates of the demonstrator's joints is projected into an egocentric frame of reference for a simulated humanoid robot. The second method represents perceived movements in the imitator's own task-based system. This is the case of the use of predictive forward models, where movement recognition is directly accomplished by the same process that generates the agent's own movements (e.g., Demiris and Hayes 2002).

A variety of combinations of methods for imitation motivation, action perception, and the correspondence problem have led to the proposal of numerous robot imitation experiments. In this section we will focus on some seminal work on robot imitation that takes direct inspiration from developmental psychology studies on imitation. Demiris and colleagues have proposed a computational architecture that incorporates Meltzoff and Moore's (1997) active intermodal matching (AIM) model of the development of imitation abilities in infants. This architecture is called Hierarchical Attentive Multiple Models (HAMMER) (figure 6.7) and incorporates various aspects of the AIM model, primarily the principle of "understanding others by analogy with the self" (Demiris and Hayes 2002, Demiris and Johnson 2003; Demiris and Khadhouri 2006). It also models the developmental stages of imitation behavior in infants, such as the fact that younger infants first imitate only the surface behavior of the demonstrator, and later understand their underlying intentions and therefore can imitate the goal with different behavioral strategies. The AIM component of the HAMMER cognitive architecture has been used for various robot imitation experiments, such as a robotic head that observes and imitates human head movements (Demiris et al. 1997) and a mobile robot capable of learning to navigate by imitating and following another robot (Demiris and Hayes 1996).

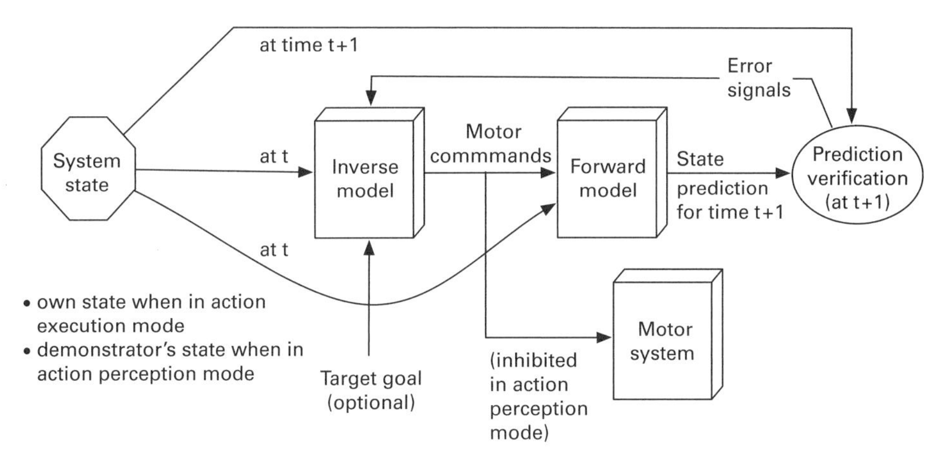

Figure 6.7
Demiris's HAMMER architecture for imitation. Figure courtesy of Yiannis Demiris.

The HAMMER architecture is based on the following principles:

• Basic building blocks consisting of pairs of inverse and forward models, for the dual role of either executing or perceiving an action.
• Parallel and hierarchical organization of multiple pairs of inverse/forward models.
• A top-down mechanism for the control of attention during imitation, to deal with limited access to the sensory and memory capacities of the observer.

Each building block is constituted by an inverse model paired with a forward model. An inverse model takes as inputs the current state of the system and the target goal, and outputs the motor control commands that are needed to achieve the goal. A forward model takes as inputs the current state of the system and a control command to be applied to it and outputs the predicted next state of the controlled system. The forward model works as an internal predictor simulation model. These models can be implemented as artificial neural networks, or other machine learning methods. HAMMER uses multiple pairs of inverse/forward models to observe and execute action. This combination of inverse and forward models has been proposed as a general internal mechanism for motor control, first proposed by Wolpert and Kawato (1998) in the MOSAIC (modular selection and identification for control) model. Wolper and Kawato propose that the brain uses a series of multiple paired modules of forward (predictor) and inverse (controller) models. Demiris's hierarchical model is based on the same idea of paired forward/inverse modules.

When a robot endowed with the HAMMER architecture is asked to imitate a new action via demonstration, the inverse model receives in input the demonstrator's current state as perceived by the imitator. The inverse model generates the associated motor commands necessary to achieve the target goal (the action itself is not executed by the robot during perception/imitation learning). The forward model then uses these motor commands to provide an estimation of the demonstrator's future state at timestep $t + 1$. The comparison between the actual and the predicted states generates an error signal that can be used to adjust the parameters of the action execution and to learn the demonstrator's actions. Learning is achieved through the increase/decrease of the inverse model's confidence value, which indicates how closely the demonstrator's action matches the imitator's action. During imitation learning, many of these pairs of inverse/forward models are active in parallel during a demonstration, with a continuous adjustment of the predictor's model confidence value, strengthening the confidence values of those models that best match the demonstrator's action with its own predictions. The prediction model with the highest confidence value is then chosen for the action execution at the end of the demonstration. If no existing internal models are found that can generate the demonstrated action, the architecture uses its AIM component to learn a new one, using the surface behavior as a starting point.

These pairs of inverse/forward models can be organized hierarchically, where higher-level nodes encode increasingly abstract behavioral aspects, such as goal states (Johnson and Demiris 2004). This has the advantage of simulating the demonstrated action not by following the exact demonstrated movements, but rather their effects on the environment to achieve the goal. This permits the solution of the correspondence problem in imitation, by allowing the robot to choose its own actions to achieve the imitation goal.

To model the effects of selective attention and limited memory capabilities, a top-down attentional mechanism is implemented. Each inverse model only requires a subset of the global state information, that is, some models might specialize for arm movement, others for the torso, and so on. The selection of the states and features of the task to be passed to the inverse model depends on the hypotheses that the observer has on the task being demonstrated. And as there might be multiple parallel hypotheses and state requests, the saliency of each request depends on the confidence value of each inverse model. In addition, this top-down attentional system can be integrated with bottom-up attentional processes, as those depending on the salience properties of the stimulus itself. Here we briefly describe two robot imitation experiments conducted by Demiris and collaborators on the HAMMER architecture. Demiris follows a clear developmental robotics approach by performing comparative analysis between child psychology studies on infants' imitation from a human demonstrator and robotics experiments replicating such phenomena (Demiris and Meltzoff 2008). In particular the child/robot comparison focuses on the initial conditions necessary for imitation (i.e., what is innate in infants, and what functionality must be prewired in the robot) and developmental pathways (i.e., how the performance of infants changes over time, and the corresponding changes in the mechanisms used by the robot to extend imitation skills).

The first experiment (Demiris et al. 1997) explicitly models the face imitation behavior studied by Meltzoff in young infants. The robotic head ESCHeR (Etl Stereo Compact Head for Robot vision) (Kuniyoshi et al. 1995) was used as this is constrained to the human vision system with binocular, foveated wide-angle lenses for a 120-degree field of view, high-resolution fovea of twenty pixels per degree, and velocity and acceleration parameters comparable to human head movements. During the human experimenter's demonstrations, the human head's pan and tilt rotations are estimated first by using an optical flow segmentation algorithm for the vertical and horizontal coordinates, and then a Kalman filtering algorithm to determine the approximate pan and tilt values. To estimate the robot's own head posture, the values of the encoders are recorded, after the head is initialized to its default coordinates of looking straight ahead.

A qualitative imitation approach is used by matching the observed target posture with the current posture. Depending on the difference between the pan and tilt values of the target and of the proprioceptive information on the current head posture, the system activates a series of "move upward" and "move leftward" commands until the target posture has been reached. To extract the representative postures of the sequence of movements necessary to imitate the target behavior, a simple algorithm is used that keeps the posture at time t in memory if either the x or y angle value of the posture represents a local minimum or maximum. During imitation experiments, the robot head successfully imitated the vertical/horizontal movement of the experimenter's head movements, relying only on representations known to exist in primate brains. The model configuration permitted the imitation of a variety of movements, with varying duration and speed. Moreover, the use of the algorithm for extraction of only representative postures allowed the smoother reproduction of demonstrated movements.

A second experiment that tests the HAMMER architecture for the imitation of object manipulation tasks was carried out with the ActivMedia PeopleBot (figure 6.8) (Demiris and Khadhouri 2006). This is a mobile robot platform with an arm and a gripper for object handling. In this experiment, only the on-board camera was used as input sensor, with an image resolution of 160 × 120 pixels and a sampling rate of 30 Hz for a two-second-long demonstration. The human participant demonstrated actions such as "pick X," "move hand towards X," "move hand away from X," and "drop X," where X was either a soft drink can or an orange. The visual properties of these two objects and of the human hand (i.e., hue and saturation histograms) are preprocessed for the objects, to be used later by the inverse models.

To implement the inverse models for these four actions the ARIA library of primitives provided with the ActivMedia PeopleBot was used, in addition to the specification of the preconditions necessary to perform the action (e.g., before picking up an object, the hand needs to move closer). A set of eight inverse models was used, one for each of the

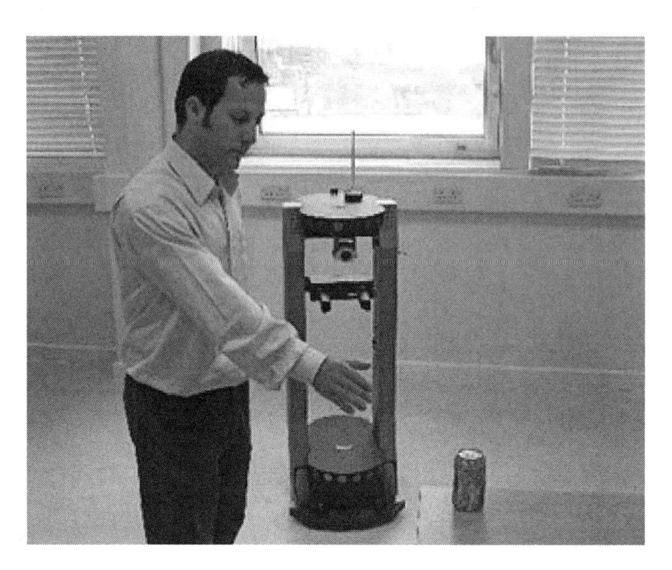

Figure 6.8
A PeopleBot observing an object manipulation act, selectively attending the relevant parts of the demonstration (from Demiris and Khadhouri 2006). Figure courtesy of Yiannis Demiris. Reprinted with permission from Elsevier.

above four actions in combination with each of the two objects. The forward models were implemented using hand-coded kinematic rules to predict the qualitative prediction of the next state of the system expressed as two possible states: "closer" or "away from."

To choose which of the eight inverse models will produce an action (i.e., will win the robot's selective attention), the top-down attentional arbitration mechanism is used, taking into consideration the confidence value of each model (e.g., using the inverse models with higher confidence, or in the alternative "round-robin" equal share method, choosing at each time step one model at a time, in sequential order). Once an inverse model has been selected, this requires a set of object properties related to the objects in the scene: color, motion, and size of the objects (can, orange, or hand). These properties act as biases in the calculation of a combined saliency map, such as with highlighted regions for the positions of the hand and the target object. Using these values, the selected inverse model generates a motor command, which is then passed to the paired forward model to generate the qualitative, binary prediction of the next close/away state. The confidence value of the inverse model is increased/decreased depending on the correct/incorrect binary error value.

To test this implementation of the HAMMER architecture for manipulation behavior, eight video sequences of a demonstrator performing different tasks with the two objects were used with different arbitration methods (round-robin, highest-confidence choice or combinations of both). Experimental results and analyses of the different arbitration methods show that the addition of the attention mechanism produces a significant savings in terms of computational resources. Moreover, the combination of the round-robin and high-priority methods allows an efficient dynamic switch to the highest-confidence inverse model, after some initial model optimization steps.

The testing of the HAMMER model with robot imitation experiments provides a validation of the various developmental and neuroscience phenomena data (Demiris and Hayes 2002), and the AIM theoretical model of imitation proposed by Meltzoff. In particular, the learning of inverse and forward models can be considered part of the infant's developmental processes. The acquisition of forward models is similar to the motor babbling stage of infant development. Robots, like infants, before they fully master reaching behavior go through stages where they generate random movements, and these are associated with their visual, proprioceptive, or environmental consequences. The learning of these associations between actions and consequences can then be inverted to create approximations to the basic inverse models, by observing and imitating others, that is, learning how certain goals correspond to input states (Demiris and Dearden 2005). During later developmental stages, various basic inverse models can be combined in parallel to create multiple inverse models of the interaction with the environment and their hierarchical grouping to handle complex goals and actions.

The HAMMER architecture has also been used to model a range of biological data related to the action perception and imitation systems in primates (Demiris and Hayes 2002; Demiris and Simmons 2006) and extended for a variety of imitation tasks and robotics platforms, such as human-robot collaborative control and imitation for robotic wheelchairs (Carlson and Demiris 2012) and imitation of dancing routines in human-children interaction experiments (Sarabia, Ros, and Demiris, 2011). In addition to this robot imitation architecture, other researchers have proposed developmental-inspired robotics experiments on imitation based on Meltzoff and Moore's AIM model. For example Breazeal et al. (2005) have modeled facial mimicry between robots and adult caregivers, as they see this as a significant milestone in building natural social interactions between robots and humans and a theory of mind. Other researchers have focused on imitation and learning of emotional states. Hashimoto et al. (2006) investigated how a robot can learn to categorize different emotions by analyzing the user's facial expressions and labels for emotional states. Watanabe, Ogino, and Asada (2007) proposed a communication model that enables a robot to associate facial expressions with internal states through intuitive parenting, by imitating human tutors who mimic or exaggerate facial expressions.

6.4 Cooperation and Shared Intention

The development of cooperative and altruistic behavior in children, based on the capability to represent shared intention and to construct shared plans, has been the target of developmental robotics models of social interaction. Specifically, Dominey and collaborators (Dominey and Warneken 2011; Lallée et al. 2010) have modeled the cognitive and sensorimotor skills involved in a cooperative task like the "block launching" task of Warneken et al. (2006) (box 6.1).

In Dominey and Warneken's (2011) experiments, the robotic agent (a 6-DOF Lynx6 robotic arm [www.lynxmotion.com] with a gripper) and a human participant jointly construct a shared plan that allows them to play a game with objects on a table with goals such as "put the dog next to the rose" or "the horse chases the dog." The human and the robot can move four objects (a dog, a horse, a pig, and a duck) and have to place them next to six fixed landmarks (picture of a light, turtle, hammer, rose, lock, and lion). Each moveable object is a wooden puzzle piece with the image of the animal drawn on it, with a vertical bar for the human or robot to grasp. The six landmarks consist of puzzle pieces glued to the table. Their fixed positions allow the robot to more easily determine the location of the objects and its grasping posture.

The cooperating robot's cognitive architecture, and the corresponding experimental setup, are shown in figure 6.9. The central component of this cognitive architecture is the Action Sequence Storage and Retrieval System. The actions of the animal game are organized in a sequential structure (i.e., the shared plan), where each action is associated with either the human or the robot agent. Dominey and Warneken propose that the ability to store a sequence of actions, each tagged with its agent, corresponds to the shared plan. This constitutes part of the core of collaborative cognitive representations and is a unique ability of human primates. The Action Sequence Storage and Retrieval

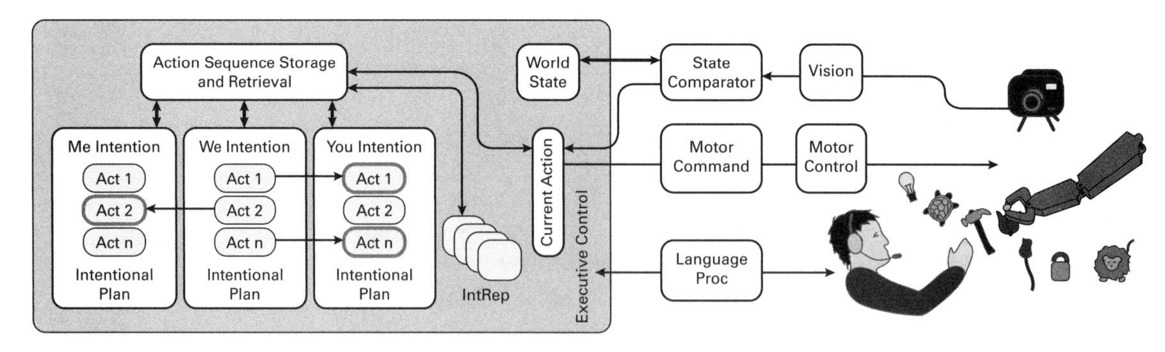

Figure 6.9
Architecture for cooperation system from Dominey and Warneken (2011). Figure courtesy of Peter Dominey. Reprinted with permission from Elsevier.

System design is inspired by neuroscience studies on the involvement of the human brain's cortical area BA46 in real-time storage and retrieval of recognized actions, and its link with sequence processing, language areas, and the mirror neuron system (Rizzolatti and Craighero 2004; Dominey, Hoen, and Inui 2006).

An intentional, goal-directed action representation is used, with the agent specifying each action and with the object and its goal as target location. The goal-directed representation formalism *Move(object, goal location, agent)* can then be used for a variety of tasks such as requesting an action and describing it. Three types of intentional plans are possible: (1) I Intention, (2) We Intention, and (3) You Intention. In the "I" and "You" intention plans, either the robot or the human execute all the actions in a sequence. In the We Intention plan, each action in the sequence is by default attributed either to the robot or the human agent, and the turn taking order is initially fixed. However, following the Carpenter, Tomasello, and Striano (2005) study on role reversal, the robot is equipped with the same capability to reverse the turn-taking order. When the robot is ready to execute a plan, it asks the user if she wants to go first. If the user responds yes, the roles of user and robot remain fixed as in the memorized sequence during the demonstration. Alternatively, the roles are reversed and the robot systematically reassigns the agents to each action.

Another core component of the robot's cognitive architecture is the *World Model*. This encodes the physical locations of the objects in the world, and corresponds to a 3D simulation of the grounded mental model of the agent (Mavridis and Roy 2006). The World Model is continuously updated when the robot detects a change of location of the objects carried out by either the human agent or the robot itself. A vision recognition system allows the tracking of the object locations, and a speech recognition and dialogue management system allows the collaborative interaction between the human and the robot (see box 6.2 for the overview of technical implementation details).

At the beginning of each game, the robot uses the vision system to update the location of each visible object in the world model. To check the correctness of its World Model, the robot lists the object positions by saying "The dog is next to the lock, the horse is next to the lion." Then it asks "Do you want me to act, imitate, play or look again?" If the object description is correct, the human user can ask to act by naming the game (e.g., "Put the dog next to the rose"), and the robot demonstrates it. Alternatively, the human will give a demonstration of a new game involving the four objects. In this second case, the robot memorizes the sequence of actions performed by the human user and then repeats it. During each action demonstration, the human indicates to the agent who has to perform the action by saying "You do this" or "I do this." These roles are assigned in the sequence as the default action roles in the We Intention plan.

A set of seven experiments was carried out to investigate various types of games, interactions, and collaborative strategies (see table 6.3 for a summary overview of the experiment's procedures and main results). Experiments 1 and 2 served to validate the

Box 6.2
Implementation Details of Dominey and Warneken's (2011) Experiments

Robot and Actions

The Lynx6 arm (www.lynxmotion.com) robotic arm with two finger grippers was used in the experiment. Six motors control the degrees of freedom of the arm: motor 1 rotates the shoulder base of the robot arm, motors 2–5 control upper and forearm joints, and motor 6 opens/closes the gripper. The motors are controlled by a parallel controller connected to a PC, through the RS232 serial port. The robot was equipped with a set of action primitives that are combined to (a) position the robot at any of the six locations to grasp the corresponding object (e.g., Get(X)), and (b) move to a new position and release the object (e.g., PlaceAt(Y)).

Vision

To recognize the ten images of the four target objects and six landmarks, the SVN Spikenet Vision System (http://www.spikenet-technology.com) was used. A VGA webcam located at 1.25 m above the robot workspace captures a bird's-eye view of the object arrangement on the table. For each of the ten images, the SVN system was trained offline with three recognition models at different orientations of the objects. During real-time vision processing, once the SVN recognizes all the objects, it returns the reliable (x, y) coordinates of each of the four moveable objects. Subsequently the system calculates the distance of each moveable object to the six fixed landmarks and identifies for each the nearest landmark. Both the human users and the robot are constrained to place the object they move in one of the zones designated next to the six landmarks. This facilitates the robot's capability to grasp that object at the prespecified location (nearest the landmark). During the initial calibration phase, the six target locations are marked next to each of the fixed landmarks. These are arranged on an arc equidistant to the center of rotation of the robot base.

Natural Language Processing (NLP) and Dialogue Management

To communicate with the robot, an automatic speech recognition and dialogue management system is used. This is implemented through the CSLU Rapid Application Development toolkit (http://www.cslu.ogi.edu/toolkit/). This system allows interaction with the robot, via the serial port, and with the vision processing system, via file i/o. To manage the spoken language interaction with the robot, a structured flow of control of the dialogue is predetermined in CSLU (see figure below—courtesy of Peter Dominey). At the start of each interaction, the robot can either choose to Act or Imitate/Play. In the Act state, the human participant utters a request such as "Put the dog next to the rose." Using a grammatical construction template (Dominey and Boucher 2005b) an action is identified through the predicate(argument) formalism as in *Move(object, location)*. To execute the action, the robot has to update the representation of the environment (Update World). In the Imitate state, the robot first has to verify the current state (Update World) and then invite the user to

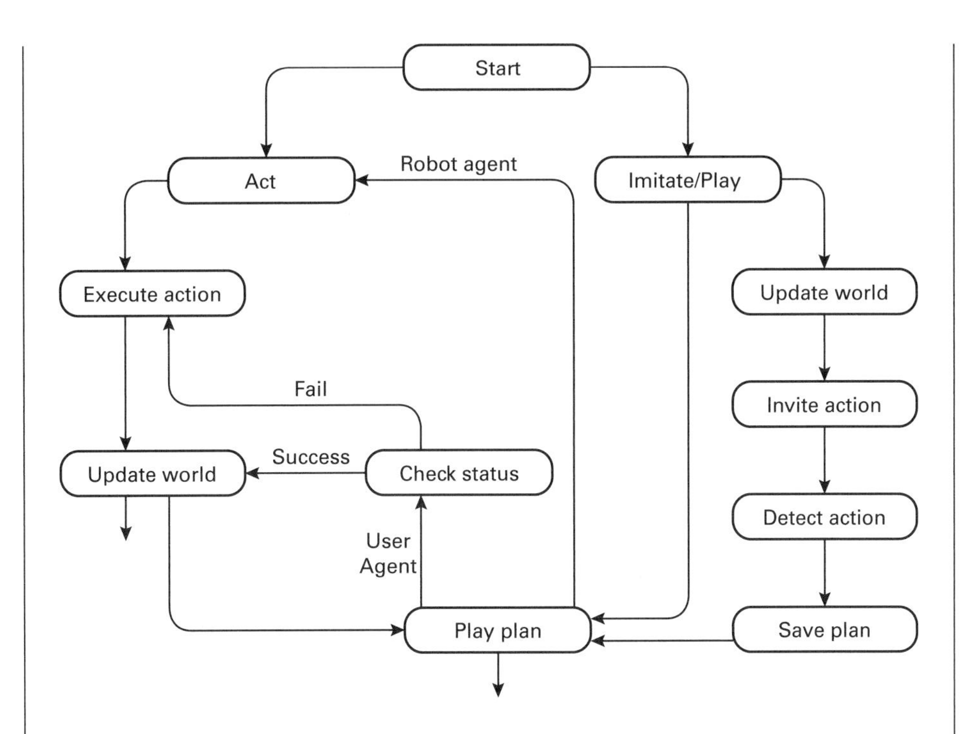

demonstrate an action (Invite Action). The Detect Action procedure is triggered when the robot determines changes in the object location, leading to a saved action (Save Plan) represented by the same "predicate(argument)" formalism. During the demonstration, the user says who the agent should be when the game is to be played (e.g., "You/I do this"). Finally, the save action is executed by the robot (Play Plan).

Example of Experiment 1: Validation of Sensorimotor Control

Act state

- User command "Put the horse next to the hammer."
- The robot requests confirmation and extracts the predicate-argument representation: *Move(Horse, Hammer).*

Execute Action state

- *Move(Horse, Hammer)* is decomposed into the two primitive components of *Get(Horse),* and *Place-At(Hammer).*
- *Get(Horse)* queries the World Model in order to localize Horse with nearest landmarks.
- Robot performs a grasp of Horse at the corresponding landmark target location.
- *Place-At(Hammer)* performs a transport to target location Hammer and releases the object.
- Update World memorizes the new object location.

Table 6.3
Summary of the seven experiments in Dominey and Warneken's (2011) study on collaboration

Experiment	Description	Results
1: Sensorimotor control	User chooses Act and says "Put the horse next to the hammer"; robot extracts *Move(Horse, Hammer)*, and decomposes it into the two primitive *Get(Horse)*, and *Place-At(Hammer)*; robot executes action with World Model checks and update	Ability to transform a spoken sentence into a *Move(X to Y)* command and its constituent primitives; ability to perform visual localization of the target object; ability to grasp the object and put it at the specified location
2: Imitation	User chooses the Imitate state; user performs one action; robot builds a *Move(object, location)* action representation by detecting object location changes in vision system; robot verifies action representation with user; performs action	Ability to detect the final "goal" of user-generated actions as defined by visually perceived state changes; imitation defined as achievement of goal; utility of a common representation of action for perception, description, and execution
3: Cooperative game	Same as imitation, but with a sequence of multiple actions; for each action the user specifies "you do this" or "I do this"; "We Intention" plan with different agents for different actions; robot and user alternate in sequence of actions	Ability to learn a simple intentional plan as a stored sequence of multiple actions with roles assigned to the human and the robot; cooperative, turn-taking execution of actions by user/robot
4: Interrupting cooperative game	Same as before, with one imitation of full alternating sequence; at second imitation, the user does not perform one of her assigned actions; the robot says "Let me help you" and executes the action	The robot's stored representation of the action allows it to help the user
5: Complex game	Same as experiment 3, but with complex game "dog chases horse"; user moves the dog, and robot "chases" the dog with the horse until they both return to their starting places	Learning by demonstration of a complex intentional plan in a coordinated and cooperative activity
6: Interrupting complex game	Same as experiment 4, with complex dog chasing sequence User fails to perform final move, dog back to fixed landmark Robot can substitute user to achieve final goal	Robot's generalized ability to help whenever it detects that the human agent has difficulties
7: Role reversal in complex game	Same game from experiment 5; robot asks, prior to playing the game, "do you want to go first?"; user responds "no" (user was first in demonstrated game); robot starts game and systematically reassigns roles	Ability to benefit from the bird's-eye view representation of the shared intentional plan and take either role in the plan

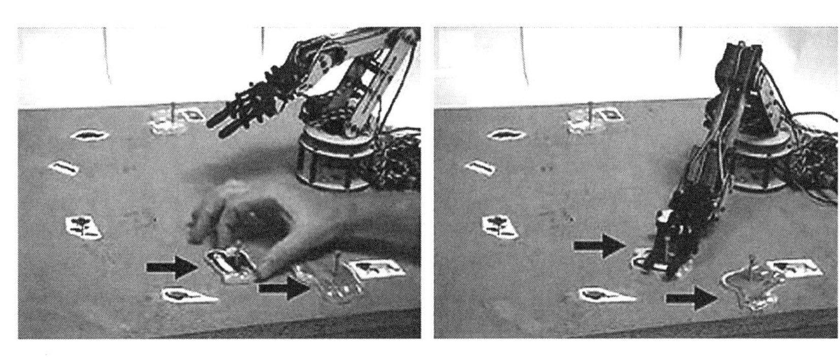

Figure 6.10
Cooperation game between the robotic arm and the human: "the horse chases the dog" in Dominey and Warneken (2011): (a) human participant demonstrating the chasing game, and (b) robot executing the learned game. Figure courtesy of Peter Dominey.

system as a whole and the capability of the robot to play and imitate a single action. Experiments 3 and 4 tested the capability of the robot to build the We Intention plan for a sequence of multiple actions, with the systematic assignment of each action to either "I" (the robot) or "You" (the human user). In particular, in experiment 4, when the user does not perform one of its assigned actions, the robot can use the We Intention plan to substitute the user in the missing action.

In experiments 5 and 6 the sequence learning capability is extended for a more complex game where the target goal of each action is not a fixed landmark, but a dynamically changing target defined by the user. The robot must learn to move the horse to chase the dog moved by the human user (figure 6.10). The robot can also successfully substitute the user in the interrupted chasing game of experiment 6, and return the dog to the fixed landmark. This task is related to the learning-by-demonstration scenario (Zöllner, Asfour, and Dillmann 2004) where agents define a complex intentional plan in a coordinated and cooperative way.

Finally, in experiment 7 Dominey and Warneken (2011) demonstrate the feasibility of the robot's role reversal capability using the "birds' eye view" representation of the shared intentional plan. When the human user fails to start the dog chasing game, as she had done during the previous game demonstration, the robot is able to swap roles with the user and initiate the game by moving the dog. This implements the Carpenter, Tomasello, and Striano (2005) role reversal capability observed in eighteen-month-old children.

This cognitive architecture can be extended to other types of human-robot collaborative tasks. For example, in Lallée et al. (2010), the humanoid robot iCub learns to assemble a table, attaching the four legs to the main board, with the help of the human experimenter (figure 6.11). To achieve this, the robot learns by experience, allowing

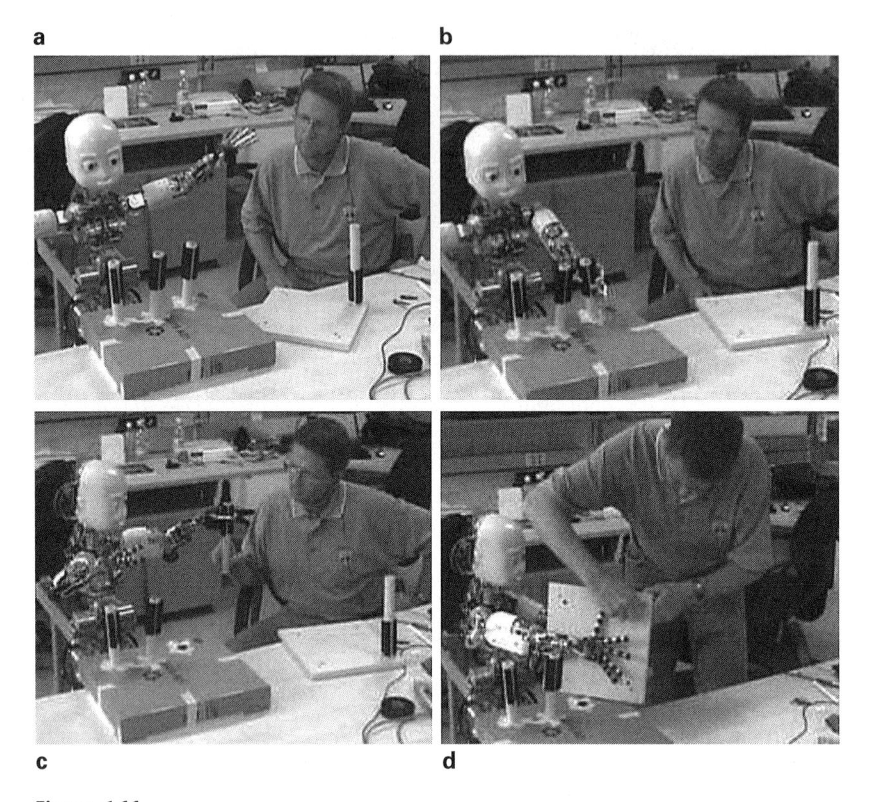

Figure 6.11
The cooperative table construction task using iCub platform in Lallée et al. (2010). Robot and human have to cooperate in order to build a table (a). The robot has to grasp the table legs (b), it passes them to the user (c), and holds the table while the user screws on the legs (d). Figure courtesy of Peter Dominey.

it to anticipate the repetitive activities as the four legs are successively attached. By allowing the iCub the ability to learn shared plans, Lallée et al. (ibid.) provided the robot with the ability to perform the joint task with the human, to demonstrate role reversal, and even to share this acquired knowledge with another iCub remotely over the Internet.

This approach demonstrates that it is possible to build the cognitive architecture of a social robot that is able to observe an action, determine its goal, and dynamically attribute roles to agents for a cooperative task. The design of such a sociocognitive architecture benefits from its direct inspiration from developmental and comparative psychology on the role of altruistic and social skills in early development.

6.5 Theory of Mind (ToM)

The design of theory of mind capabilities in interactive robotic agents is an important extension of social and cognitive capabilities in robots as it allows them to recognize the goals, intentions, desires, and beliefs of others. A robot can use its own theory of mind to improve the interaction with human users, for example, by reading the intention of the others and reacting appropriately to the emotional, attentional, and cognitive states of the other agents, to anticipate their reactions, and to modify its own behavior to satisfy these expectation and needs (Scassellati 2002).

In the previous sections we have looked at the robotic implementation of various capabilities that combined together contribute to the full development of theory of mind in cognitive robots. For example, all the work on gaze behavior modeling (section 6.2) implements a core prerequisite for development of a complete theory of mind. Moreover, the review of the work on face perception in chapter 4 (section 4.2) contributes to the design of key components in the robot's theory of mind. However, notwithstanding the numerous developmental robotic models of individual social skills, there are only a few attempts to explicitly model the integration of many processes for a general robot's theory of mind (Scassellati 2002; Breazeal et al. 2005). Here we will look at Scassellati's (2002) theory of mind architecture for the COG humanoid robot. Scassellati explicitly models theory of mind principles consistent with infants' developmental theories of Leslie and Baron-Cohen (see section 6.1.4). For example, Scassellati focuses on the implementation of Leslie's distinction between the perception of animate and inanimate objects, with subsequent differentiation among mechanical, action, and attitudinal agency. This approach also puts great emphasis on Baron-Cohen's eye detection mechanism.

This robotic theory of mind model is based on the COG robot, an upper-torso humanoid platform with two six-DoF arms, a three-DoF torso, and seven DoFs for the head and neck (Breazeal and Scassellati 2002). The model implements the following behavioral and cognitive mechanisms necessary for theory of mind development:

- Pre-attentive visual routines
- Visual attention
- Face and eye detection (Baron-Cohen's eye direction detector)
- Discrimination of animate and inanimate entities (Leslie's mechanical agency)
- Gaze following
- Deictic gestures

The pre-attentive visual routines are based on saliency map analyses that implement the infant's natural preference for bright and moving objects. The three basic feature detectors implemented in COG are color saliency, motion detection, and skin color detection (Breazeal and Scassellati 2002). Using four color-saturated areas filters (red, green, blue, and yellow), the algorithm generates four opponent-color channels that are subsequently thresholded to produce a smooth-output color saliency map. The robot's visual input is processed at thirty Hertz to extract these three saliency features, because this speed is suitable for handling social interaction with human participants. For motion detection, temporal differencing and region growing is used to generate the bounding boxes of moving objects. For skin detection, the images are filtered using a mask whose color values are the result of the hand classification of skin image regions.

The next stage of visual attention selects the objects in the scene (including human limbs) that require an eye saccade and neck movement. This is achieved by combining the three bottom-up detectors for skin, motion, and colors, with a top-down motivation and habituation mechanism (time-decayed Gaussian representing habituation effects). These visual attentional mechanisms are based on the model of human visual search and attention proposed by Wolfe (1994).

Eye and face detection allow the robot to maintain eye contact with the human experimenter. First a face detection technique is used to identify locations that are likely to contain a face using a combination of the skin and motion detection maps. These regions are processed using the "ratio templates" method (Sinha 1995), which is based on a template of sixteen expected frontal face regions and twenty-three relations among them. The robot then gazes toward the detected face region, which then leads to the identification of the eye subregion (as in Baron-Cohen's EDD module).

The function to discriminate animate and inanimate entities from visual perception of self-generated motion is based on Leslie's mechanical agency mechanism of the Theory of Body module. A two-stage developmental process is modeled. The first stage only uses the spatiotemporal features of object size and motion to track the objects. The second stage uses more complex object features, such as color, texture, and shape. Motion tracking is implemented through the multiple hypothesis tracking algorithm proposed (Cox and Hingorani 1996), where the output of the motion saliency map is processed to produce a labeled trajectory of the coordinates of the centroid of an object over a series of temporal frames. The system is also capable of handling object disappearance from the visual input, for example due to object occlusion, head rotation, or the limited size of the field of view. When a movement is interrupted, a "phantom point" can be created that can later be linked to trajectories of objects that enter, exit, or are occluded within the visual field. A more advanced learning mechanism to detect animate entities has been developed by Gaur and Scassellati (2008).

The gaze following function requires the implementation of three subskills that further process the eye detection operation: (1) extracting the angle of gaze, (2) extrapolating the angle of gaze to a distal target object, and (3) motor routines for alternating between the distal object and the experimenter. This allows the modeling of the incrementally complex infant's gaze-following strategies proposed by Butterworth (1991), starting from a simple sensitivity to the field of gaze toward the fully developed representational capability (see section 6.1.1).

Scassellati (2002) also suggests that a complementary cognitive capability to gaze following is that of understanding, and responding to, deictic gestures. These include imperative pointing and declarative pointing. Imperative pointing involves the pointing gesture toward an object that is out of reach, to implicitly ask the other agent to pick it up and give it to the infant. Developmentally the imperative pointing can be considered a natural extension of the infant's own reaching behavior. Declarative pointing uses a gesture of an extended arm and index finger, to draw attention to a distal object without an implicit request to have it. The implementation of such gesture understanding is core in social robotics, for example to operationalize requests and declarations consistent with the robot's own, and the agent's, beliefs. Although pointing gestures have not been implemented in the cognitive theory of mind model, other researchers have proposed methods to interpret point gestures, as in Hafner and Kaplan 2005.

In addition to the mechanisms proposed by Scassellati in 2002, a robot with a complete theory of mind mechanism also requires additional capabilities, such as self-recognition. This has been recently explored by Gold and Scassellati (2009), with Nico, an upper-torso humanoid robot with the arm and head kinematics of a one-year-old. This baby robot is trained to recognize its own body parts, as well as their reflections in a mirror, using a Bayesian reasoning algorithm. This self-recognition ability allows the robot to recognize its image in a mirror, and to distinguish the experimenter as an "animate other" and static objects as "inanimate." These recent developments on theory of mind capabilities, as well as general progress on developmental robotics models of social skills, are important to support the design of robots capable of understanding the intentions of other agents, such as humans and robots, and integrate them with their own control system for effective human-robot interaction.

6.6 Conclusions

In this chapter we have focused on both developmental psychology and developmental robotics studies on the acquisition of social skills. The literature on child psychology demonstrates that infants are endowed with a strong instinct to interact with other people, as in the case of their capability to imitate facial gestures within the first days of life. This social instinct is strongly supported, and reinforced, by the adults' caregiving behavior of continuous interaction with and stimulation of the infant. This strictly coupled adult-child interaction allows the child to gradually acquire ever more complex joint attention skills (shared eye gaze and pointing gestures), increasingly articulated imitation skills (from body babbling and body movement imitation to repeating the actions on objects and inferring the demonstrator's intentions), altruistic and cooperative interactions with the others (spontaneous help and role reversal behavior), up to developing a full theory of mind (attributing beliefs and goals to other people).

Developmental robotics models have taken direct inspiration from child psychology literature (Gergely 2003), sometimes even through close roboticist-psychologist collaborations (e.g., Demiris and Meltzoff 2008; Dominey and Warneken 2011), to design social skills in robots. Table 6.4 provides an overview of the various skills involved in designing social robots, and how these have been addressed in the robotics studies analyzed in this chapter.

This overview shows that their shared/joint attention, a fundamental capability for social agents, is one of the key phenomena more extensively modeled in robotics experiments. In some studies, shared attention is primarily operationalized as mutual gaze between the caregiver adult (human experimenter) and the infant (robot) as in Kaplan and Hafner (2006b), Imai, Ono, and Ishiguro (2003), and Kozima and Yano (2001). In Triesch et al. (2006), the manipulation of normal and impaired gaze shift patterns is also used to model socio-attentional syndrome such as autism and Williams syndrome. In other experiments, shared attention is achieved by gaze following, as in the Nagai et al. (2003) model. In particular this study provides a nice demonstration of the developmental emergence of different gaze following strategies, as in the study on human infants (Butterworth 1991). The analysis of the robot's gaze following at different stages of the developmental learning process show the robot face first uses an ecological gaze strategy (i.e., looks at an interesting object regardless of the experimenter's gaze direction), then this becomes a geometric strategy (the robot face and human participant achieve joint attention only when the object is in the infant's field of view), and finally leads to the adoption of representational gaze strategies (the infant robot can find a salient object outside own field of view). In addition, joint attention is modeled through the use of pointing behavior, as in Hafner and Kaplan 2005.

Another area of social competence that has attracted much attention in developmental robotics is imitation learning. Imitation has also been one of the main research issues in the wider field of cognitive robotics and human-robot interaction (Breazeal and Scassellati 2002; Nehaniv and Dautenhahn 2007; Schaal 1999). Among the studies with a clear focus on the developmental stages of imitation, the models proposed by Demiris (Demiris and Meltzoff 2008; Demiris et al. 1997) and Breazeal et al. (2005) have explicitly implemented the active intermodal matching model of infant development proposed by Meltzoff and Moore (1997). In Demiris and Hayes (2002), the AIM model is incorporated into the HAMMER architecture, an ensemble of pairs of inverse and forward models, as a means to imitate and learn new inverse models. This robotic architecture also includes a top-down attentional mechanism for the control of attention during imitation, which allows the robot to deal with limited access to the sensory and memory capacities. This allows the robot to integrate top-down attentional biases and bottom-up visual attention cues.

Table 6.4

Mapping of developmental robotics model of social skills and individual social capabilities (two ++ signs indicate main focus of the study, and one + sign a partial explicit inclusion of the cognitive/social skill)

Social and cognitive skills	Kaplan and Hafner 2006b	Nagai et al. 2003	Imai, Ono, and Ishiguro 2003	Kozima and Yano 2001	Triesch et al. 2006	Demiris et al. 1997	Demiris and Khadhouri 2006	Breazeal et al. 2005	Watanabe, Ogino, and Asada 2007	Dominey and Warneken 2011	Lallee et al. 2012	Scassellati 2002	Gold and Scassellati 2009
(robot)	AIBO	Robot head	Robovie	Infanoid	Simu-lation	Robot head	PeopleBot	Leo	Virtual face	Robot arm	iCub	COG	Nico
Mutual gaze	+		++	++	++								
Gaze following		++		+	++							+	
Pointing gesture	++		++	++									
Attention (visual)		++					++					+	
Attention (top-down)		+					++						
Shared/koint attention	+	++	+	+	++							+	
Imitation: face expressions						++		++	+				
Imitation: face emotions									++				
Face detection								+	+			+	
Imitation: body movements							++						++
Cooperation										++	++		
Role reversal										++	++		
ToM								+				++	+
Self-recognition (mirror)													++

Another areas of social development thoroughly investigated through robotic experiments is collaborative and altruistic behavior. Dominey and Warneken (2011) propose a cognitive architecture for human-robot collaboration based on developmental theories and replicate the tasks in the seven infant and chimpanzee experiments carried out by Warneken, Chen, and Tomasello (2006). In the subsequent studies of Lallée et al. (2012), they extend this model for collaborative tasks with the iCub humanoid robot. These experiments also provide a demonstration that the robot is capable of role reversal and bird's-eye view of the collaboration setting (Carpenter, Tomasello, and Striano 2005), in other words, dynamically taking on the demonstrator's role to help her achieve the task.

Finally, some models explicitly address the issue of developing theory of mind capabilities in the robots. In Scassellati 2002, this is achieved by endowing the robot with multiple sociocognitive skills, such as pre-attentive visual routines and visual attention, face and eye detection, discrimination between animate and inanimate entities, gaze following, and deictic gestures. This architecture explicitly tests some of the components in the psychology literature on theory of mind development, as with the operationalization of Baron-Cohen's eye direction detector in the robot's face and eye detection capabilities, and the sensitivity to discriminate between animate and inanimate agency, as in Leslie's mechanical agency concept. The agency animate/inanimate distinction is also used to model robot's self-recognition in Gold and Scassellati's (2009) study.

Overall, this chapter demonstrates that social learning and interaction is one of the most articulated fields of developmental robotics, because the design of social capabilities is an essential precondition for human-robot interaction with robot companions. These experiments on social learning establishes the foundation for the investigation of more advanced capabilities in robots, such as the theory of mind capability to read the intention of the others, and to predict their needs and behavior, thus facilitating effective human-robot collaboration.

Additional Reading

Tomasello, M. *Why We Cooperate.* Cambridge, MA: MIT Press, 2009.

This volume presents Tomasello's theory of human cooperation. It reviews a series of studies on young children and primates aiming at the identification of the specific mechanisms supporting human infants' natural tendency to help others and collaborate. This unique feature, not observed as a spontaneous behavior in our closest evolutionary ancestor, provide the basis for our unique form of cultural organization based on cooperation, trust, group membership, and social institutions.

Nehaniv, C., and K. Dautenhahn, eds. *Imitation and Social Learning in Robots, Humans and Animals*. Cambridge: Cambridge University Press, 2007.

This highly interdisciplinary volume includes chapters on theoretical, experimental, and computational/robot modeling approaches to imitation in both natural (animals, humans) and artificial (agents, robots) systems. This is the successor to the volume edited by Dautenhahn and Nehaniv, *Imitation in Animals and Artifacts* (MIT Press, 2002), and the follow up in the international series of Imitation in Animals and Artifacts workshops organized by the two editors. The 2007 volume includes contributions from developmental psychologists (e.g., M. Carpenter, A. Meltzoff, J. Nadel), animal psychologists (e.g., I. Pepperberg, J. Call), and numerous robot imitation researchers (e.g., Y. Demiris, A. Billard, G. Cheng, K. Dautenhahn and C. Nehaniv).

7 First Words

Language, as the capacity to communicate with others through speech, signs, and text, is one of the defining features of human cognition (Barrett 1999; Tomasello 2003, 2008). As such the study of language learning and language use has attracted the interested of scientists from different disciplines ranging from psychology (for psycholinguistics and child language acquisition) and neuroscience (for the neural bases of language processing) to linguistics (the formal aspects of human languages). Therefore it is not surprising that even in developmental robotics, and in cognitive modeling in general, there has been a great deal of research on the design of language learning capabilities in cognitive agents and robots.

An important issue in language research is the "nature" versus "nurture" debate, that is the opposition between those (nativists) who believe that we are born with knowledge of universal linguistic principles, and those (empiricists) who propose that we acquire all linguistic knowledge through interaction with a language-speaking community. Within the nativist position, some influential researchers have proposed that there are universal syntactic rules or generative grammar principles, and that these are innate in the human brain (Pinker 1994; Chomsky 1957, 1965). For example Chomsky proposed that we are born with a language "brain organ" called the language acquisition device. In Chomsky's Principles and Parameters theory, our linguistic knowledge consists of a series of innate, universal principles (e.g., we always use a fixed word order) with learnable parameters associated with them (e.g., in some languages, the verb always precedes the object, in others the verb follows after the object). The language acquisition device has a set of predefined parameter switches that are set during language development. If a baby develops in an English-speaking community, the word order parameters (switches) will be set to SVO (subject-verb-object) order, while in Japanese-speaking children, this parameter will be switched to SOV (subject-object-verb). Another important aspect of the nativist view is the poverty of stimulus argument. This argument states that the grammar of a language is unlearnable if we consider the relatively limited data available to children during development. For example, a child is never, or rarely, exposed to incorrect grammatical sentences, but she is able to distinguish grammatically correct sentences from those that are ungrammatical. Therefore, the nativist's explanation is that the input during development must be complemented by innate knowledge of syntax (i.e., the principles and parameters of the language acquisition device).

According to the nurture stance, the essence of the linguistic knowledge emerges from language use during development, without any need to assume the existence of innate language-specific knowledge. For example, grammatical competence is seen not as universal, prewired knowledge, as in nativist generative grammar. On the contrary, as Michael Tomasello, one of the primary supporters of the nurture stance, says: "the grammatical dimension of language is the product of a set of historical and ontogenetic processes referred to collectively as *grammaticalization*" (Tomasello 2003, 5). Children's language development depends on the child's own ontogenetic and maturational mechanisms, such as the critical period, and on cultural and historical phenomena that affect the dynamics on a continuously changing shared language. This view of language development (construction) does not exclude that there are innate maturational and sociocognitive factors affecting language acquisition. In fact, genetic predispositions such as the critical period and the biases in categorization and world learning (see section 7.1) do affect language acquisition. However, these are general learning biases, and not innate language- (i.e., syntax-) specific competences. For example, the critical period of language acquisition is one of the crucial phenomena in language development. Typically children only develop full command of a native language if they are exposed to it in during the first few years of life. This phenomenon, and in general the effects of age of acquisition in linguistic competence, has been extensively studied for second language acquisition (e.g., Flege 1987). As for the Poverty of Stimulus argument, in the literature there is evidence that children do get negative examples of impossible, grammatically incorrect sentences, and that parents do correct their children when they make grammar mistakes. In addition, in computational models it has been demonstrated that a reduced, impoverished input can act as a bottleneck that even helps the child discover syntactic regularities in the language (Kirby 2001).

This empiricist view of language learning is normally known as the constructivist, usage-based theory of language development (Tomasello 2003; MacWhinney 1998). This is because the child is seen as an active constructor of his own language system through implicit observation and learning of statistical regularities and logical relationships between the meaning of words and the words used. Within linguistics, this has helped promote cognitive linguistic theories (Goldberg 2006; Langacker 1987). This closely associates semantics with grammar, and demonstrates that syntactic categories and roles emerge out of usage-based regularities in the semantic systems. For example, the general grammatical category of verbs emerges from incremental and hierarchical similarities between verbs sharing common features (e.g., Tomasello's 1992 verb island hypothesis—see section 7.3 for more details).

The constructivist view of language is highly consistent with the embodied developmental robotics approach to the modeling of language learning (Cangelosi et al. 2010). Most of the principles of developmental robotics, as discussed in chapter 1, reflect the phenomena of gradual discovery and acquisition of language observed in studies of children's acquisition of linguistics capabilities, and the role of embodied and situated interaction with the environment in cognitive development. A fundamental concept in robotics and embodied models of language learning is that of the *symbol grounding* (Harnad 1990; Cangelosi 2010). This refers to the capability of natural and artificial cognitive agents to acquire an intrinsic (autonomous) link between internal symbolic representations and referents in the external word or in internal states. By default, linguistic developmental robotics models are based on the grounded learning of associations between words (which are not always encoded as symbols, but may be subsymbolic dynamic representations) and external and internal entities (objects, actions, internal states). As such, these models do not suffer from what Harnad (1990) calls the "Symbol Grounding Problem."

This chapter will explore the close link between embodied, constructivist theories and the developmental robotics model of language learning. In the next two sections we will first briefly review the main phenomena and milestones of linguistic development and the principles involved in conceptual and lexical acquisition (section 7.1). We next discuss in detail one seminal child psychology experiment on the role of embodiment in early word learning (section 7.2). These phenomena and principles will then be mapped to the various developmental robotics studies of language acquisition, respectively for the models of the development of phonetics competence through babbling (7.3), the robot experiments on early word learning (7.4), and the models of grammar learning (7.5).

7.1 Children's First Words and Sentences

7.1.1 Timescale and Milestones

The most significant events of language development are concentrated during the first three to four years. This is not to say that children stop progressing in their linguistic capabilities at the school age. On the contrary, there are important developmental stages associated with primary school age, and beyond, and they mostly regard meta-cognitive and meta-linguistics achievements (i.e., awareness of one's own language system) that gradually lead to adultlike linguistic competence. However, developmental psychology and developmental robotics both focus on the core early stages of cognitive development. These early milestones of language acquisition follow the parallel and intertwined development of incremental phonetics processing capabilities, increasing lexical and grammatical repertoires, and refined communicative and pragmatic faculties.

Table 7.1
Typical timescale and major milestones of language development (adapted from Hoff 2009)

Age (months)	Competence
0–6 months	Marginal babbling
6–9 months	Canonical babbling
10–12 months	Intentional communication, gestures
12 months	Single words, holophrases Word-gesture combinations
18 months	Reorganization of phonological representations 50+ word lexicon size, vocabulary spurt Two-word combinations
24 months	Increasingly longer multiple-word sentences Verb islands
36+ months	Adultlike grammatical constructions Narrative skills

Table 7.1 provides an overview of the main milestones during language development (Hoff 2009). In the first year, the most evident sign of linguistic development is vocal exploration, meaning, vocal babbling. Initially babbling consists of vocal play with sounds such as cooing, squeals, and growls (also known as "marginal babbling"). At around six to nine months of age, children go through the stage of "canonical babbling" (also known as "reduplicated babbling") (Oller 2000). Canonical babbling consists of the repetition of language-like syllabic sounds such as "dada" or "bababa," merging into variegated babbling This is hypothesized to play a role in the development of refined feedback loops between sound perception and production, rather than in communicative purposes, and the transition from the marginal to canonical babbling stage is a fundamental step in phonetic development. Also toward the end of the first year, children start to produce communicative gestures (e.g., pointing) and iconic gestures (e.g., a throwing motion to indicate a ball, or raising the fist to the ear, to mean telephone). These gestures are clear signs of the child's prelinguistic intentional communication and cooperation skills, and first signs of a theory of mind (Tomasello, Carpenter, and Liszkowski 2007; P. Bloom 2000).

Toward the end of the first year, reduplicated babbling, subsequently followed by the richer phonetic combinations of a range of consonant and vowel sounds (variegated babbling), and the restructuring of the language-specific phonetic representations and repertoire, lead to the production of the first single words. The first words are typically used to request an object (e.g., say "banana" to request the fruit), indicate its presence ("banana" to indicate the presence of the fruit), name familiar people, indicate actions ("kick," "draw") and dynamic events ("up," "down"), and ask questions "Whats-that"

(Tomasello and Brooks 1999). These words are normally referred to as "holophrases," as a single linguistic symbol is used to communicate a whole event. In some cases holophrases can correspond to an apparent word combination, as in "Whats-that," though at this stage the child has not yet developed a full independent use of the individual words and of their flexible combination in multiple combination.

During the first part of the second year, children slowly increase their single-word repertoire. This includes a variety of names of social agents (dad, mum), labeling of food, objects, and body parts, words for requests ("more"), and so on. The growth of the child's lexicon is characterized by a nonlinear increase in the number of words, and is called the "vocabulary spurt." The vocabulary spurt (a.k.a. naming explosion), typically observed between months 18 and 24, refers to a steep increase in rate of growth of the vocabulary that happens after the child has learned approximately fifty words (Fenson et al. 1994; L. Bloom 1973). This is typically hypothesized to depend on a restructuring of lexical-semantic representation, leading to qualitative changes in word learning strategies. As the child increases her lexicon repertoire, she is also able to produce two-word utterances. However, during the beginning of the second year, and before the capability to produce two-word combinations is fully developed, children go through a hybrid word/gesture stage when they combine one gesture with a word to express combinations of meanings. For example, a child can say the word "eat" and point to a sweet, to communicate "eat the sweet." Even at these early stages, gestures, and their combination with words, become predictors of future lexical and grammatical competence and general cognitive capabilities (Iverson and Goldin-Meadow 2005).

The first two-word combinations that appear at around eighteen months of age typically follow the structure of pivot-style constructions (Braine 1976). Children produce two-word combination based on a constant element (pivot), such as "more," "look," and a variable slot, such as "more milk," "more bread," "look dog," "look ball."

During the third year the child starts to develop ("construct") more complex grammatical competences. One seminal example of constructivist grammatical development is the verb island hypothesis (Tomasello 1992). Although children at this age are able to use a variety of verbs, these appear to exist as independent syntactic elements, called "verb islands." For example, for some verbs (e.g., "cut") the child is only able to use very simple syntactic combinations of the same verb with different nouns of objects ("cut bread," "cut paper"). Other verbs, instead, can have a richer syntactic use. For example, the verb "draw" is uttered with richer combinations such as "I draw," "draw picture," "draw picture for your," "draw picture with pencil." These differences in the level of complexity and maturity of different verb islands are due to usage-based experience. In the case of the well syntactically developed verb island "draw" the child is exposed to richer combinations of the verb with multiple participant types and multiple pragmatics roles and functions. At this stage, however, the child has not developed the general syntactic and semantic categories of agent, patient, instrument, as in mature adultlike verb categories. Instead, she acquires verb-island specific roles such as "drawer," "drawn object," "draw something for person," and "draw something with." These intermediate syntactic constructions also allow the child to develop more refined morphological and syntactic skills, since for some verb islands there is a richer use of prepositions links to verb as with "on," "by," and so on (Tomasello and Brooks 1999).

From four to six years of age, corresponding to the preschool period in most countries, the child gradually develops mature adultlike syntactic constructions such as simple transitives (agent-verb-patient as in "John likes sweets"), locatives (agent-verb-patient-locative-location as in "John puts sweets on table"), and datives (agent-verb-patient-dative-receiver as in "John gives sweets to Mary") (Tomasello and Brooks 1999). This gradually leads to the development of ever more complex syntactic-morphologic constructions, more abstract and generalized grammatical categories, up to the formation of formal linguistic categories such as word classes. These syntactic skills are accompanied by extended pragmatics and communicative skills, thus leading to refined narrative and discursive capabilities.

7.1.2 Principles of Conceptual and Lexical Development

To be able to achieve the milestones we have outlined during language acquisition, and given that language is closely intertwined with the parallel development of other sensorimotor and social skills, it is important to know which other factors and competences support these developments. As discussed in the initial sections on the nature versus nurture debate, all theories of language development assume that the child relies on some (innate or previously developed) competence to learn the lexicon and syntactical categories. And although there is strong disagreement on the pre-existence of innate language-specific (i.e., syntax-specific) capabilities between the generativist and the constructivist view of language, all developmentalists agree that the child's acquisition of first words and grammar is supported by a set of prelinguistics capabilities. Some of these might be innate, species-specific behavior, while others might just be social and conceptual skills gradually acquired during the early stages of development.

These general cognitive capabilities, often referred to as "biases" or "principles" of conceptual and lexical development (Golinkoff, Mervis, and Hirshpasek 1994; Clark 1993), depend on a combination of perceptual and categorization skills (e.g., the capacity to discriminate and identify objects and entities, and to group them within a category) and social skills (e.g., the instinct to imitate and cooperate). These principles have the main function of "simplifying" the task of word learning by reducing the amount of information that children must consider when learning a new word.

Table 7.2 provides an overview of the main principles that have been demonstrated to contribute to lexical development. This list extends the original set of six principles, as proposed in Golinkoff, Mervis, and Hirshpasek 1994, by including more recent findings from developmental psychology.

Table 7.2
Language acquisition principles (biases) in word learning

Principles (Biases)	Definition	Reference
Reference	Child has some awareness that words are used to map entities in the real world	Golinkoff, Mervis, and Hirshpasek 1994 Mervis 1987
Similarity	Once a label is associated with one instance of an object, this is extended to functionally or perceptually similar exemplars	Clark 1993
Conventionality	Speakers of a common language tend to use the same words to express certain meanings	Clark 1993
Whole-object (Object scope)	Children assume that a novel label is likely to refer to the whole object and not to its parts, substance, or other properties	Markman and Wachtel 1988; Gleitman 1990
Whole-part juxtaposition	Children are able to interpret a novel label as referring to a part, when the novel part label is juxtaposed with a familiar whole-object label	Saylor, Sabbagh, and Baldwin 2002
Segmentation	Infants exploit highly familiar words to segment and recognize adjoining, previously unfamiliar words	Bortfeld et al. 2005
Taxonomic (categorical scope)	Words refer to things that are of the same kind	Markman and Hutchinson 1984
Mutual exclusivity (novel-name nameless category; contrast)	Children assume that nouns pick out mutually exclusive object categories, and so each object category should have only one label	Markman and Wachtel 1988 Golinkoff, Mervis, and Hirshpasek 1994 Clark 1993
Embodiment	Children use their body relationship with the objects (e.g. spatial location, shape of the object) to learn new object-word associations	Smith 2005; Samuelson and Smith 2010
Social cognition	Shared attention, imitation learning, cooperation	Baldwin and Meyer 2008 Carpenter, Nagel, and Tomasello 1998 Tomasello 2008

The *reference principle* is at the basis of word learning and reflects the fact that the child must develop some awareness that words are used to map objects and entities in the real word. Mervis (1987) first observed that at around twelve months of age children learn to name objects just for the sheer pleasure of it. With the additional *Similarity Principle*, once a label is associated with one instance of an object, the same word can be extended to exemplars other than those initially seen, and that share with the original object some functional or perceptual similarities (Clark 1993).

The *whole-object principle* (also known as the object scope principle) starts from the observation that children assume that a novel label that they hear for the first time is likely to refer to some object present in the scene. In particular, the label is assumed to refer to the whole object, rather than to its parts, its substance, or other properties (Markman and Wachtel 1988; Gleitman 1990). The *whole-part juxtaposition principle* is based on the observation that children are able to interpret a novel label as referring to a part, when the novel part label is juxtaposed with a familiar whole-object label (Saylor, Sabbagh, and Baldwin 2002).

The *segmentation principle* (Bortfeld et al. 2005) states that infants, as young as six-month-olds, can exploit highly familiar words, such as their own names or other people's names, to segment and recognize adjoining, previously unfamiliar words from continuous, fluent speech.

The *taxonomic principle* (also the categorical scope principle) states that children assume that a word refers to things that are of the same kind, and that this extends to the members of the basic category to which the original entity belongs (Markman and Hutchinson 1984; Golinkoff, Mervis, and Hirshpasek 1994).

The *mutual exclusivity principle* (also the contrast principle, or the novel-name nameless category principle) starts from the assumption that nouns pick out mutually exclusive object categories, so that each object category should have only one label associated with it (Markman and Wachtel 1988; Clark 1993). Therefore, when a child hears a novel label and sees a new, nameless object, she attaches the new word to the novel object. This is true even in the case when there are two objects, with one already labeled by the child and one with no word associated to it.

The *conventionality principle* refers to the fact that a child assumes that all the speakers of a common language tend to use the same words to express certain meanings, so a child must always use the same label to refer to the same object (Clark 1993).

The *embodiment principle* is based on the observation that children use their body relationship with the external world to learn new object-word associations. For example, the child can use the relations between body posture and spatial location of the objects, even in the temporary absence of the object, to learn a new object-label association (Smith 2005). This principle will be discussed in detail in the next section, with the example of the experimental setup in box 7.1.

Box 7.1
The Modi Experiment (Smith and Samuelson 2010)

Procedure

The parent sits in front of a table, holding the child in her lap. The experimenter sits opposite and during the training she shows, one at a time, two novel, nameless objects (figures below). Two distinct locations on the table are used (left and right). During the test phase, the two objects are shown together in a new location of the table (center). All the participants are children between eighteen and twenty-four months of age, around the typical developmental stage of fast word learning and the vocabulary spurt. We briefly describe here four experiments using this paradigm (see table below for a diagram of the four experiments).

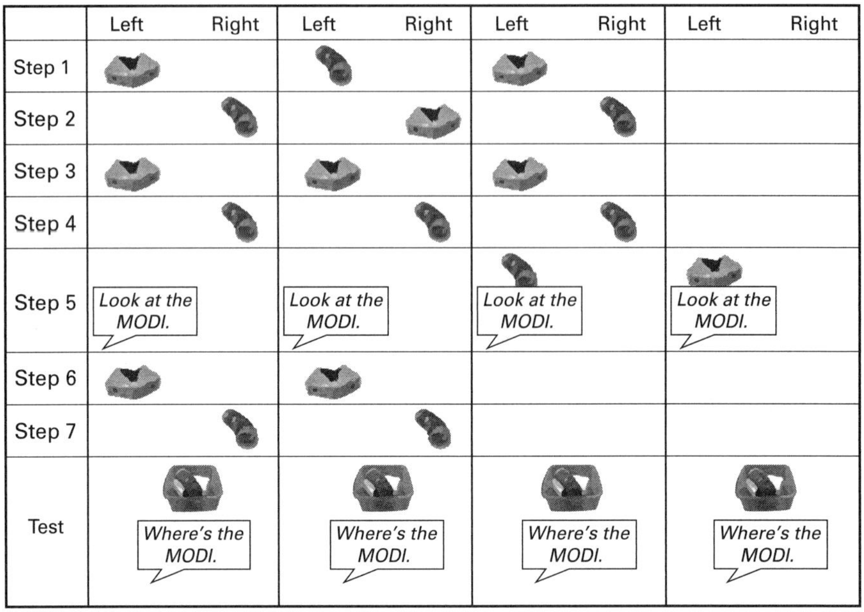

	Left	Right	Left	Right	Left	Right	Left	Right
Step 1								
Step 2								
Step 3								
Step 4								
Step 5	Look at the MODI.		Look at the MODI.		Look at the MODI.		Look at the MODI.	
Step 6								
Step 7								
Test	Where's the MODI.		Where's the MODI.		Where's the MODI.		Where's the MODI.	

Experiments 1 and 2: Object Named in Its Absence

In these experiments, the experimenter shows the two novel objects, one at a time. The name of the object "modi" is only said in the absence of the object. In this first experiment (No-Switch condition) each object is always shown in the same location. The first object is consistently presented on the left of the peripersonal space in front of the child, and the other object is consistently presented on the right side. Two presentations of each individual object are given (steps 1–4). Subsequently (step 5), the child's attention is drawn to one of the now empty locations (left) and simultaneously the linguistic label "modi" is said aloud by the experimenter (e.g., "Look at the modi"). The two objects are then shown again, one at a time (steps 6–7). Subsequently, in the test phase, the child is presented with both objects in a new (central) location and asked: "Can you find me the modi?"

In the second experiment (Switch Condition), the same basic procedure is used, except that the consistency of the left/right location between the two objects is weakened. In the initial presentations, the first object is presented on the right side, with the second object on the left (steps 1–2). The objects' position is then switched in the following two presentations (steps 3–4). In the naming step (step 5) and the final two presentations of the objects (steps 6–7), the initial location of stages 1–2 is used.

Experiments 3 and 4: Object Named while in Sight

In these two experiments, the new label is said at the same time the object is shown to the child. In experiment 3 (Control Condition), a systematic left/right position is used for the two objects throughout the experiment, exactly as in the No-Switch Condition. However, in step 5, the word "modi" is said when the first (yellow) object is shown. This corresponds to the standard object-labeling setup of word naming experiments. In experiment 4 (Spatial Competition Condition), a different group of children is repeatedly presented with the two objects in a systematic left/right spatial location for steps 1–4. At step 5 the second (green) object is now labeled as "modi" while in sight. However, the green object is now located on the left side of the table, that is, the position used by the yellow object.

Results

In experiment 1 (No-Switch Condition), the majority (71 percent) of the children select the spatially correlated object (the one presented in the left side), despite the fact that the name is said in the absence of either object. In experiment 2 (Switch Condition), only 45 percent of the children chose the object shown in the same location highlighted when the word "modi" is said aloud. In the experiment 4 (Control Condition), 80 percent correctly picked the labeled object over the previously unseen object. In experiment 3 (the Spatial Competition condition), a majority of children (60 percent) selected the spatially linked object (yellow) rather than the green object that was actually being attended and labeled at the same time.

Finally, there are a series of observations based on *social cognition principles* that significantly contribute to word learning. These social cognition principles of language learning focus on the child-parent cooperative dyadic relationship, as well as child-peer interactions. For example, Tomasello (2008) and Carpenter (2009) have carried out comparative studies between child and animal (ape) experiments to look at the form and mechanisms of shared attention that exist in human children, but that are absent in primates, and which support language learning in human children. Similar studies have been carried out on the social imitation and cooperation skills (Tomasello 2008). The evidence in support of the core role of joint attention and shared gaze in early word learning is strong. For example, it has been demonstrated that at around eighteen months children pay attention to the speaker's eye gaze direction as a clue to identify the topic of the conversation (Baldwin and Meyer 2008). Also the time infants spend on joint attention is a predictor of the speed of lexical development (Carpenter, Nagell, and Tomasello 1998). Some of these social-cognition biases, and related models in developmental robotics, were discussed in detail in chapter 6.

The word-learning biases discussed in this chapter can also be seen as a manifestation of intrinsic motivation, in this case specifically for the motivation that drives the infant to discover and learn language and communicative behavior. This has been explicitly modeled in developmental robotics, with Oudeyer and Kaplan's (2006) experiments with the AIBO. Their model supports the hypothesis that children discover (vocal) communication because of their general instinct to engage in situations that result in learning novel situations, rather than for intentional communicative purposes. That is, while exploring and playing within their environment, the robots' general motivation to learn makes them select novel tasks not previously learned, as when they are allowed to vocally interact with other agents.

7.1.3 A Case Study: The Modi Experiment

To conclude this concise overview of language development in children, and show how child psychology findings can inform the design of models of language acquisition in developmental robotics, we provide a detailed description of one of the seminal experimental procedures used in child language studies. A developmental robotics experiment, with the replication of the child psychology experimental results, will then be presented in section 7.4. This comparison will demonstrate how a close mapping between empirical and computational studies can benefit both our scientific understanding of language development mechanisms in children and robots, and the technological implication of developing plastic, language-learning robots. In this specific case, the use of a robotic model permits the explicit testing of the role of embodiment and sensorimotor knowledge in early conceptual and word learning.

The experiment is typically referred to as the Binding Experiment, or the "Modi" Experiment (box 7.1). This is an experimental procedure related to Piaget's (1952) well-known A-not-B error paradigm, but more recently used by Baldwin (1993) and Smith and Samuelson (2010) specifically for language development studies. In particular, Smith and Samuelson chose this procedure to demonstrate the role of the Embodiment Principle in early word learning and to challenge the default hypothesis that names are linked to the object being attended to at the time the name is encountered.

The four experiments described in box 7.1 show the systematic manipulation of factors affecting word learning, in other words, the location where the object appears with the label being given in the absence of the object (experiments 1 and 2) and the competition between spatial and temporal associations when the label is presented while the object is in sight (experiments 3 and 4). Results show that children can associate a label to an object even in its absence. In addition, when the spatial/temporal conditions are in competition, as in experiment 4, the embodiment bias based on the child posture is stronger than the presentation of the label at the same time as the object.

Another important observation is that in each of the experiments, changes in the parent's posture from sitting to standing (producing a different schema of the child's spatial-body perspective) can disrupt the children's ability to link the absent object to the name through space, while other visual or auditory distracters do not. This further reinforces the mediation of embodiment factors in language learning. Overall, this study provides clear evidence challenging the simple hypothesis that names are associated to the things being attended to at the time the name is heard. In fact, the experiments provide strong evidence that the body's momentary disposition in space helps binding objects to names through the expected location of that object (Smith and Samuelson 2010; Morse et al. in preparation).

7.2 Robots Babbling

Numerous speech recognition and synthesis applications have been developed in the last twenty years, and these are typically used as natural language interfaces for computers, cars, and mobile phones. Most of the latest speech recognition systems rely on statistical methods, such as hidden Markov models (HMMs), which require offline training of the system with thousands of sample sounds for each word. However, developmental robotics models of speech learning propose an alternative approach to the emergence and development of speech processing capabilities. These developmental models rely on the online learning and imitation in teacher-learner interactions, as during child development, rather than on the offline training with a huge corpora. The design of developmental-inspired speech systems, and their scaling up to deal with large corpora, aims to overcome the bottleneck issues of current speech recognition applications and their limitations and unstable recognition performance in dynamic, noisy environments.

Most of the current robotics models of language learning rely on this type of pre-trained speech recognition system, especially those that focus on the acquisition of words and syntax (cf. sections 7.3 and 7.4). However, a few robotics and cognitive-agent approaches to language learning have specifically focused on the emergence of a phonetic system through interaction between learners and teachers and via the developmental stages of vocal motor babbling. Most of these studies have been based on the use of simulated cognitive agents with physical models of the vocal tract and auditory apparatus, with a more recent focus on the use of developmental robotics approaches.

Oudeyer (2006) and de Boer (2001) have proposed some of the pioneering studies on the emergence of language-like phonetic systems through agent-agent interaction (see also the related work of Berrah et al. 1996; Browman and Goldstein 2000; Laurent et al. 2011). Oudeyer (2006) investigated the evolutionary origins of speech, and in particular the role of self-organization in the formation of shared repertoires of combinatorial speech sounds. This work is based on a brain-inspired computational model of motor and perceptual representations, and how they change through experience, in a population of babbling robots. The agents have an artificial ear capable of transforming an acoustic signal into neural impulses that are mapped into a perceptual neural map (figure 7.1). They are also endowed with a motor neural map that controls the articulatory movements of a vocal tract model. These two Kohonen-like maps are also mutually interconnected. Initially, internal parameters of all neurons and of their connections are random. To produce a vocalization, a robot randomly activates several

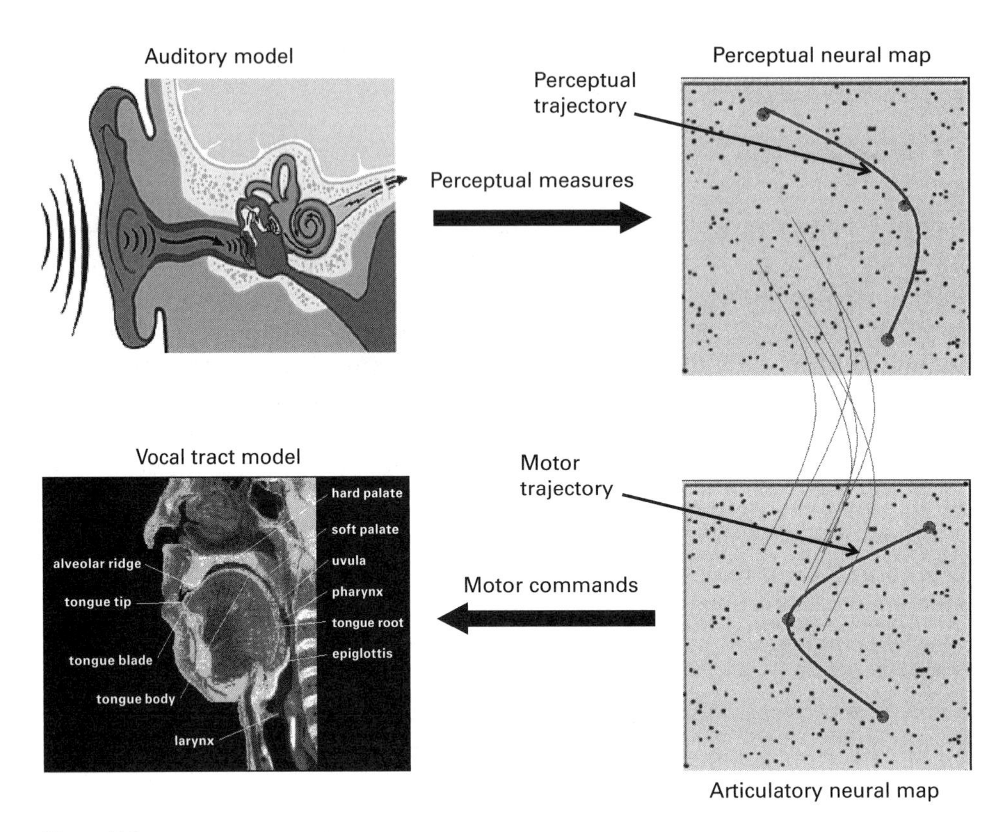

Figure 7.1
Architecture of the model of the self-organization of speech with the ear and articulation models and the perceptual and motor maps. Figure courtesy of Pierre-Yves Oudeyer.

motor neurons, in which internal parameters encode articulatory configurations to be reached in sequence. This produces an acoustic signal through the vocal tract model that can be perceived by the ear model. This is the basis of babbling. These neural networks are characterized by several forms of plasticity: (1) intermodal connections evolve in such a way that each agent learns the auditory-motor mapping when it is babbling; (2) neuron parameters in each map change to model the distribution of sounds heard by the agent; and (3) the distribution of sounds encoded in the motor map follows roughly the distribution of sounds encoded in the perceptual map. Thus, agents have the tendency to produce the same distribution of sounds as that heard in the group of agents. This architecture constitutes a basic neural kit for holistic vocal imitation.

Initially, the agents' random vocalizations are unorganized and spread through the whole speech continuum. This initial equilibrium is unstable and over time symmetry breaks: the population of agents spontaneously generates a shared combinatorial system of vocalizations that map some of statistical regularities and diversity in vocalization systems observed in human languages. The formation of such a discrete speech system also reproduces aspects of the transition from marginal to canonical babbling observed in infants (Oller 2000). These configurations are shown to depend on the interaction between innate morphological and physiological constraints (e.g., nonlinearity of the mapping of articulatory configurations to acoustic waves and auditory perceptions) and the self-organization and imitation mechanisms. In particular, this provides a unified framework to understand the formation of the most frequent vowel systems, which are approximately the same in the robot populations and in human languages.

A similar methodology was used by de Boer (2001) to model specifically the self-organization of the vowel system. More recently, these computational models of speech self-organization have been refined, for example through the use of more realistic models of the larynx position in the speech articulation system of male and female speakers (ibid.) and for babbling experiments (Hornstein and Santos-Victor 2007).

These models mostly focus on the evolutionary and self-organization mechanisms in the emergence of shared language sound systems. More recently, there has been a specific focus on modeling of the developmental mechanisms in speech learning through developmental robotics experiments. In particular, these models have specifically investigated the early stages of babbling and speech imitation in the acquisition of early word lexicons. To provide an overview of the main approaches in the field, we first review a developmental robotic model of the transition from babbling to word forms (Lyon, Nehaniv, and Saunders 2012) with the iCub robot, and then analyze a model of speech segmentation with the ASIMO robot (Brandl et al. 2008).

The study by Lyon, Nehaniv, and Saunders (2010, 2012; Rothwell et al. 2011) explicitly addresses the developmental hypothesis on the continuum between babbling and early word learning (Vihman 1996). Specifically, they carry out experiments to model the milestone when canonical syllables emerge and how this supports early word acquisition. This behavior corresponds to the period of phonetic development in children about six and fourteen months old. These studies model the learning of phonetic word forms without meaning, which can be integrated with the parallel development of referential capability.

The experiments were initially carried out on a simulated developmental cognitive architecture (LESA [linguistically enabled synthetic agent]), and then tested in human-robot interaction studies with the childlike iCub robotic platform. Contingent, real-time interaction is essential for language acquisition. The model starts with the assumption that agents have a hardwired motivation to listen to what is said, and to babble frequently in response. This intrinsic babbling motivation is similar to that shown to emerge developmentally in Oudeyer and Kaplan's (2006) model of the discovery of vocal communication instinct.

Table 7.3
Target words and phonetic transcription (adapted from Rothwell et al. 2011)

Words	Phonetic transcriptions
Circle	s-er-k-ah-l
Box	b-aa-ks, b-ao-ks
Heart	hh-ah-t, hh-aa-rt
Moon	m-uw-n
Round	r-ae-nd, r-ae-ah-nd, r-ae-uh-nd
Sun	s-ah-n
Shape	sh-ey-p
Square	skw-eh-r
Star	st-aa-r, st-ah-r

At the initial stages of the experiment, the robot produces random syllabic babble (Oller 2000). As the teacher-learner dialogue progresses, the robot's utterance is increasingly biased toward the production of the sounds used by the teacher. The teacher's speech is perceived by the robot as a stream of phonemes, not segmented into syllables or words, using an adaptation of Microsoft SAPI. Teachers are asked to teach the robot the names of shapes and colors based on the names of six pictures printed on the sides of boxes (see table 7.3), using their own spontaneous words. It happens that most of the shape and color names are one-syllable words (red, black, white, green, star, box, cross, square, etc.). The teacher is asked to make an approving comment if he/she hears the robot utter one of these one syllable words, and then the reinforced term is added to the robot's lexicon. This process models the known phenomenon that infants are sensitive to the statistical distribution of sounds (Saffran, Newport, and Aslin 1996). The frequency count of each syllable heard by the robot is updated in an internal phonetic frequency table. The robot will continue to produce quasi-random syllabic babble, or novel combinations of syllables, with a bias toward the most frequently heard sounds. This will gradually result in the robot producing words and syllables that match the teacher's target lexicon.

The phonetic encoding is based on four types of syllables: V, CV, CVC and VC (where V stands for vowel, C for consonant or a cluster of consonants), which correspond to the phonetic structure developed early by the infant. It differs in that an infant's phonetic inventory is much more limited by articulatory constraints. Infants can recognize phonemes before they can produce them. The consonant clusters are constrained by the possible combinations in the English language phonetic system. Some combinations only occur at the start of a syllable, such as *skw* as in "square," others only at the end, such as *ks* as in "box," and others in either position, such as *st* as in "star" or "last." Phonemes are represented by the CMU set, using fifteen vowels and

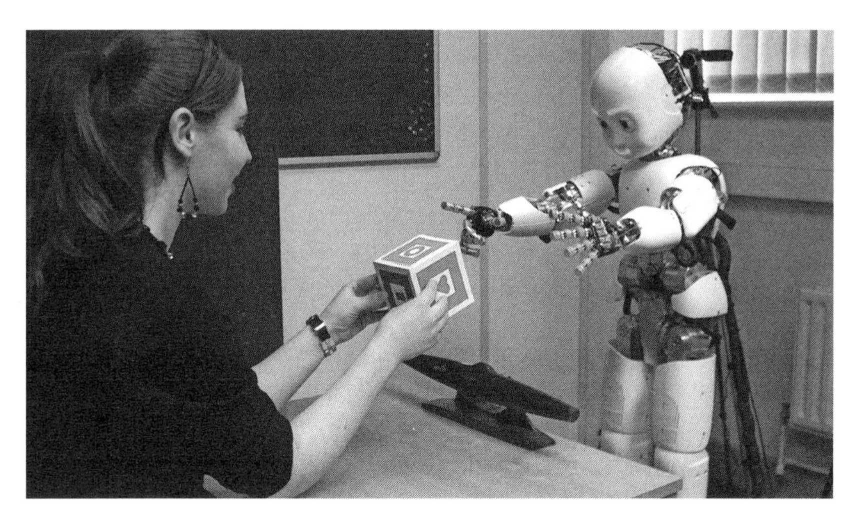

Figure 7.2
A participant interacting with the iCub during phonetic babbling experiments in Lyon, Nehaniv, and Saunders's 2012 study. Figure courtesy of Caroline Lyon and Joe Saunders.

twenty-three consonant sounds (CMU 2008). Table 7.3 shows a few target words and their phonetic transcription. Some words can have more than one encoding, reflecting variability in speech production.

In Lyon, Nehaniv, and Saunders (2012), experiments with thirty-four naïve (not familiar with robots) participants are reported. Participants were randomly assigned to one of five sets. The process was the same for each set except that participants were given slightly different guidelines, for instance whether or not they should take notice of the iCub's expression. During these experiments, the expression of the iCub's mouth (LED lights) was set to "talking" when producing syllables, and reverted to a "smile" expression when listening to the teacher (figure 7.2). The iCub babbled at three-second intervals, through the eSpeak synthesizer (espeak.sourceforge.net).

At the end of the experiments, each lasting two sessions of four minutes each, the robot was, in almost all cases, able to learn some of the word forms. Although the overall learning of the names of shapes and colors was not very high (a contributory factor was the low recognition rate of the phoneme recognizer), some interesting results were observed as a result of the teacher's interaction style. For instance, in experiments reported by Rothwell et al. (2011), a naïve student participant used single-word utterances, such as "moon," repeating these in order to get the robot to learn them. The second participant, who was used to teaching children, embedded the salient words (names of the shapes and colors) within a communicative utterance, such as "do you remember the smile shape it's like your mouth." The repetitive style of the first participant allowed the robot to learn most effectively, with the shortest time and the largest number of words achieved. In the later experiments (Lyon, Nehaniv, and Saunders 2012) this correlation was not significant. Some of the best results came from more verbose teachers, who seemed to produce the salient words in a pronounced manner.

A significant result of the experiments was that teachers often failed to notice when the robot produced a proper word, and so did not reinforce it. There is an issue over perception in human-robot interaction (HRI): in this case it seemed hard to pick out word forms from quasi-random babble; audio perception may be related to intelligibility (Peelle, Gross, and Davis 2013).

Other developmental models of speech learning have specifically focused on the roles of the caregiver's imitation in guiding language acquisition. For example, Ishihara et al. (2009) propose a simulated developmental agent model of vowel acquisition through mutual imitation. This considers two possible roles of the caregiver's imitation: (1) informing of vowel correspondence ("sensorimotor magnet bias") and (2) guiding infant's vowels to clearer ones ("automirroring bias"). The learned agent has an immature imitation mechanism that changes over time due to learning. The caregiver has a mature imitation mechanism, and these depend on one of the two biases. Computer simulation results of caregiver–infant interactions show the sensorimotor magnet strategy helps form small clusters, while the automirroring bias shapes these clusters to become clearer vowels in association with the sensorimotor magnets. In a related model by Yoshikawa et al. (2003), a constructivist human-robot interaction approach to phonetic development in a robotic articulation system is proposed. This study investigates the hypothesis that the caregiver reinforces the infant's spontaneous cooing by producing repetitive adult phoneme vocalizations, which then lead the child to refine her vocalizations toward the adultlike sound repertoire. The robotic agent consists of a mechanical articulation system with five degrees of freedom (DOFs) to control and deform a silicon vocal tract connected to an artificial larynx. The learning mechanisms uses two interconnected Kohonen Self-Organizing Maps respectively for auditory and articulatory representations. The weights of these two maps are trained using associative Hebbian learning. This learning architecture based on Kohonen maps and Hebbian learning has been extensively used in developmental robotics (see also box 7.2). Experiments on the sound imitation interactions between a human caregiver and the robotic agent demonstrate that by simply relying only on the repetitive vocal feedback by the caregiver (with no previous "innate" phonetic knowledge) the robot acquires a gradually refined human-like phonetic repertoire. To resolve the arbitrariness ins selecting the articulation that best matches the perceived human's vocalization sound, the Hebbian learning has to be modified to minimize the toil involved in the articulation (i.e., the torque to deform the vocal tract and the resulting larynx deformation). This toil parameter decreases the arbitrariness between perceived and produced sound, and improves the correspondence between the human's and the robot's phonemes. The

Box 7.2
Neurorobotics Model of the Modi Experiment

This box provides some technical details on the implementation of Morse, Belpeame, et al. (2010) model, to facilitate replication of the neurocognitive experiments on the Modi setup.

Network Topology, Activation Functions, and Learning Algorithm

The neural model can be implemented easily as two separate network types, first the self-organizing map (SOM) and field, which use standard equations, and second the spreading activation model. One SOM receives three inputs corresponding to the average R, G, and B values from the center of the input image, and the posture map receives six inputs corresponding to the pan-tilt of the eyes, head, and torso of the iCub robot. For each SOM, the best matching unit (BMU) has the shortest Euclidean distance between its weights (initially random) and the input pattern at that time (see equation 7.1). The weights of all units within the neighborhood of the BMU are then updated pulling them closer to the input pattern (see equation 7.2).

$$BMU = Max_i\left(1-\sqrt{\Sigma(\acute{a}_j-\grave{u}_{ij})}\right) \tag{7.1}$$

$$\Delta\grave{u}_{ij} = \acute{a}exp\left(-\frac{dist^2}{2size}\right)(a_j\grave{u}_{ij}) \tag{7.2}$$

Speech input is processed using the commercial speech to text software dragon dictate. Each word (as text) is compared to a dictionary of known words (initially empty) and if novel a new entry is made with a corresponding new node in a field. As each word is heard the unique corresponding unit in the word field is activated.

The spreading activation model then allows activity to spreads bidirectionally between the body posture SOM and the other maps (following standard IAC spreading activation, see equation 7.3) via connections (initially at zero) modified online with a Hebb-like learning rule (see equation 7.4), and within each map via constant inhibitory connections.

$$net_i = \Sigma\grave{u}_{ij}a_j + \hat{a}BMU_i \tag{7.3}$$

$$\text{If } net_i > 0,\ \Delta a_i = (\max - a_i)net_i - decay(a_i - rest)$$

$$\text{Else } \Delta a_i = (a_i - \min)net_i - decay(a_i - rest)$$

$$\text{If } a_i a_j > 0,\ \Delta\grave{u}_{ij} = \ddot{e}a_i a_j(1+\grave{u}_{ij}) \tag{7.4}$$

$$\text{Else } \Delta\grave{u}_{ij} = \ddot{e}a_i a_j(1-\grave{u}_{ij})$$

Training Procedure

The SOMs are partially trained using random RGB values, and random joint positions until the neighborhood size is 1, the model then runs with real input from the iCub as it interacts with people in a replication of the experiment described in box 7.1, in other words, we treat the iCub as the child and interact with it in exactly the same way as with the child in the original experiments. The learning rules are continually active and there is no separation between learning and testing phases, the model simply continues to learn from its ongoing experiences. An example of the color SOM, in early pretraining top) and during the experiment (bottom) is shown in the figures below.

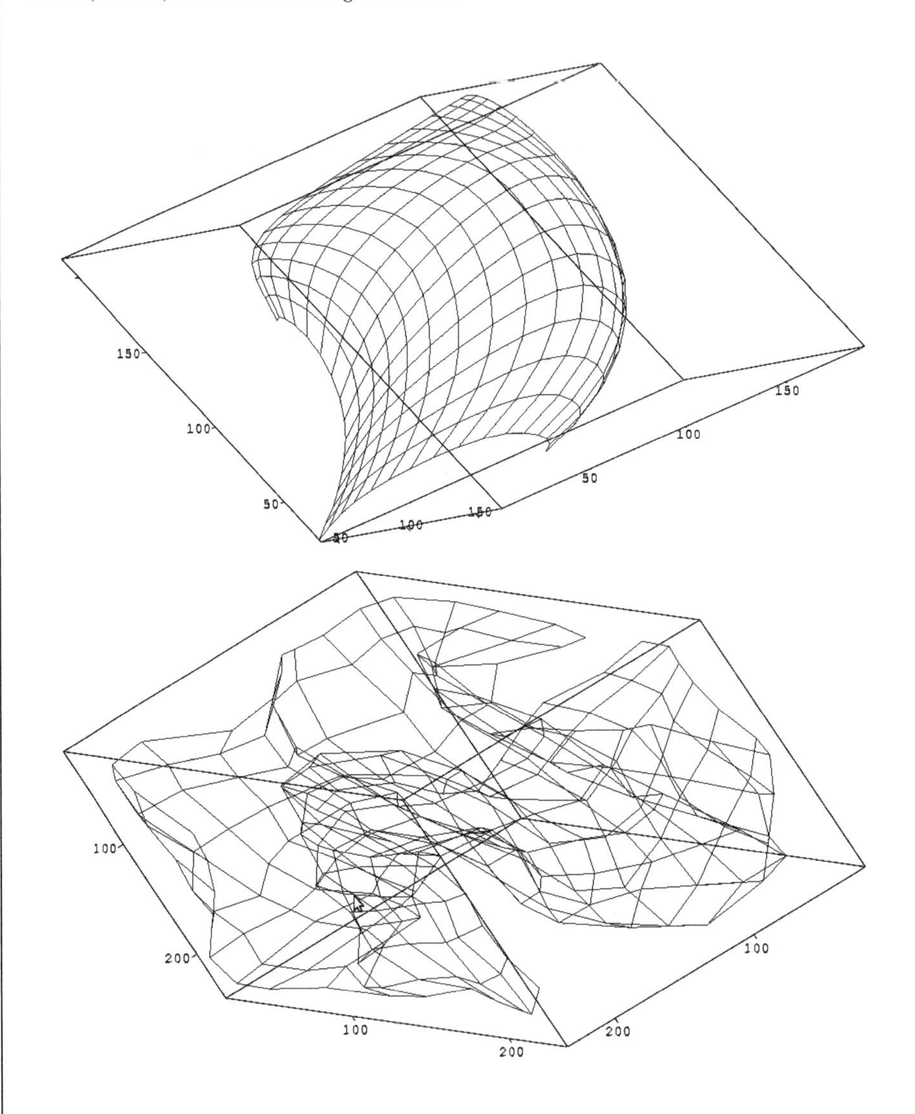

Test and Overview of Results

When the iCub robot is asked to "find the modi" the spread of activation from the "modi" word unit in the word field primes a color map unit via the body posture map. The weights of that externally activated SOM unit correspond to an RGB color and the image is then filtered according to closeness to this target value. If any of the image pixels are a good match to the primed color (above a threshold) then the iCub moves its head and eyes to center those pixels in the image, thus looking at and tracking the appropriate object. In the experiment we simply recorded which object, if any, iCub looked at when asked to find the modi. The experiment was run twenty times in each condition, each time starting with new random connection weights.

work by Yoshikawa et al. (2003) is one of the first developmental robotics models to employ a physical robotic articulation system for the learning of human-like phonetic sounds via direct interaction with human caregivers. For a more recent example of a robot head with a physical model of the vocal tract, see also Hofe and Moore 2008.

Finally, as developmental robotics is based on the principle that cognition is the result of the parallel acquisition and interaction of various sensorimotor and cognitive capabilities, we describe here a model of babbling development that has been extended to include the simultaneous acquisition of phonetic, syllabic, and lexical capabilities. This is the case of the study by Brandl and colleagues (Brandl et al. 2008) on the child-like development of phonetic and word learning in the humanoid robot ASIMO, and other related studies as in the experiments from the ACORN European project (Driesen, ten Bosch, and van Hamme 2009; ten Bosch and Boves 2008). In particular, Brandl and colleagues' work directly addresses the issue of the segmentation of speech into words, which allows infants as young as eight-month-olds to bootstrap new words based on the principle of subtraction (Jusczyk 1999; Bortfeld et al. 2005; see also table 7.2).

Brandl et al.'s (2008) model is based on a three-layered framework for speech acquisition. First the robot develops representations of speech sound primitives (phonemes) and a *phonotactic model* by listening to raw speech data. Second, a *syllable representation model* is learned based on the syllabic constraints implied by the phonotactic model, and by the properties of infant-directed speech. Finally, the robot can acquire a lexicon based on the syllabic knowledge and the word acquisition principles of early-infant language development. This hierarchical system is implemented through a cascade of HMM-based systems, which use statistical information of incomplete speech unit representations at the phonetic, syllabic, and lexical levels.

The raw speech data consists of segments of utterances of different complexity, ranging from isolated monosyllabic words up to complex utterances comprising many multisyllabic words. The model first converts speech segments into sequences of phone symbol sequences based on high frequency concatenations of single state HMMs. These phone representations are the prerequisite to learn the phonotactics of language, that is, the rules that govern the structure of syllables in a particular language. The robot uses the phonotactics model to learn syllable models, as a concatenation of phone HMMs. To bootstrap the early stages of syllable acquisition, the input speech occasionally contains isolated monosyllabic words. The developmental principle of segmentation (Bortfeld et al. 2005) is used to exploit the knowledge of initial syllables to generate new ones. Brandl and colleagues use the example of the segmentation of the "asimo" word into more syllables to explain such a bootstrapping mechanism. Let's assume the robot has already acquired the syllable "si." Given as new speech input the word "asimo," the model starts from the detection of the previously known syllable [si], and then subtracts this spotted syllable generating the sequence [a] [si] [mo]. This allows the robot to learn the two additional syllabic segments [a] and [mo]. Using such an incremental method, the robot acquires increasingly complex and richer syllabic elements. This allows the agent to learn further concatenations of syllables, as for the names for objects.

Brandl et al. (2008) used this to model speech segmentation and syllable use in language grounding experiments with the ASIMO robot, though only monosyllabic words. The ASIMO is able to recognize an object, and detect properties such as object motion, height, planarity, and object location relative to the robot's upper body. It can also use gestures, such as pointing and nodding, to communicate with the human participant (Mikhailova et al. 2008). During the speech segmentation and word learning experiments, the robot first restricts its attention to one of the properties of the objects that should be labeled, such as the object height. To learn the name for this property, each new word is repeated between two and five times. Using the segmentation and novelty detection mechanisms of speech, the robot was able to learn up to twenty words through contingent verbal and gestural interactions alone. Ongoing work with the ASIMO robot at the Honda Research Institute in Germany is now focusing on the extension of such a developmental linguistic mechanism, also using brain-inspired learning architectures (Mikhailova et al. 2008; Gläser and Joublin 2010).

This experiment with ASIMO proposes the explicit exploration of the interaction between phonetics development and word acquisition, based on child-like learning mechanisms. In the next section we will review numerous developmental robotics models of the acquisition of words, with the main focus on the acquisition of referential and symbol grounding capabilities.

7.3 Robots Naming Objects and Actions

Most of the current developmental robotics experiments on lexical development typically model the acquisition of words to name the objects visually presented in the environment and to label their properties (e.g., color, shape, weight). Only a few models investigate the learning of the names of actions performed by the robot or by the human demonstrator (e.g., Cangelosi and Riga 2006; Cangelosi et al. 2010; Mangin and Oudeyer 2012).

In this section we will first review some seminal robotics models of early word acquisition for object labeling. Subsequently, to demonstrate the fruitful close interaction between experimental and modeling data, a detailed description of a study on the direct developmental robotics replication and extension of the Modi Experiment will be given (as described in section 7.1, box 7.1). This child psychology modeling study explicitly addresses a series of core issues within the developmental robotics approach, such as the modeling of the role of embodiment in cognition, the direct replication of empirical data from child psychology literature, and the implementation of an open, dynamic human-robot interaction system. Finally, we will review related work on the robotics modeling of lexical development for the naming of actions, with some directly addressing developmental research issues, while others are more generally related to language learning.

7.3.1 Learning to Name Objects

A rich series of computational cognitive models, typically based on connectionist simulations of category and word learning, has preceded the design of robotics models of lexical development. These include, for example, the neural network models of the categorization and naming of random dot configurations of Plunkett et al. (1997), a modular connectionist model of the learning of spatial terms by Regier (1996), a model of the role of categorical perception in symbol grounding (Cangelosi, Greco, and Harnad 2000) and more recent neural network models for the learning of image-object associations using realistic speech and image data (Yu 2005).

Another important development in computational modeling of language learning, which precedes and has also directly influenced the subsequent work on developmental robotics, is the series of multiagent and robotics models on the evolution of language (Cangelosi and Parisi 2002; Steels 2003). These models investigate the emergence of shared lexicons through cultural and genetic evolution and use a situated and embodied approach. Rather than providing a communication lexicon already fully developed, the population of agents evolves a culturally shared set of labels to name the entities in their environment. These evolutionary models, as in the self-organization models of speech imitation and babbling in Oudeyer (2006) and de Boer (2001), have provided important insights on the biological and cultural evolution of language origins.

Steels and Kaplan's (2002) study with the AIBO robot is one of the first robotics studies on the acquisition of first words that builds directly on previous evolutionary models and employs a developmental setup modeling the process of ontogenetic language acquisition. This study includes a series of language games based on human-robot interaction. A language game is a model of interaction between two language learners (e.g., robot-robot or human-robot) involving a shared situation in the world (Steels 2003). This is based on a routinized protocol of interaction between the speaker and the learner. For example, in the "guessing" language game the listener must guess which object, out of many, is being referred to by the speaker. In the "classification" language game used in this study, the listener only sees one object and must learn to associate the internal representation of the object with a label. In this study the AIBO robot has to learn the labels for a red ball, a yellow smiley puppet and a small model of the AIBO called Poo-Chi. The experimenter uses words such as "ball," "smiley" and "poochi" to respectively teach the robot the names of these objects, while the robot interacts or sees them. The AIBO can recognize other predefined interactional words used to switch between different stages of the interaction in the language game ("stand," "look," "yes," "no," "good," "listen," and "what-it-is"). For example, "look" initiates a classification language game and "what-it-is" is used to test the robot's lexical knowledge. An instance-based method of classification of the objects is used, using multiple instances (views) of the objects, which are classified using the nearest neighbor algorithm. Each image is encoded as a 16×16 2D histogram of the red-green-blue (RGB) values (no segmentation is applied to the image of the object). An off-the-shelf speech recognition system was used to recognize the three object labels and the interactional words. The word-object association and learning is implemented through an associative memory matrix mapping objects' views with their words. During learning, the "yes," "no" words are used respectively to increase/decrease the grade of association between the object in view and the word heard.

Three different conditions are considered for the language learning experiments (figure 7.3): (1) social interaction learning; (2) observational learning with supervision; and (3) unsupervised learning.

In the first condition (social interaction learning), a model of full situated and embodied interaction, the robot navigates in the environment, where the experimenter is also present, and uses an attentional system to interact with the experimenter,

Figure 7.3
Three experimental conditions in Steels and Kaplan's (2002) study: (a) social interaction learning; (b) observational learning with supervision; (3) unsupervised learning. Figure reprinted with permission from John Benjamins.

the three objects, and the surrounding room layout. Human-robot language games, dependent on both the experimenter's and the robot's motivational system, are played to teach the robot the names of the three objects. In the second condition (observational learning with supervision), the robot is left wandering the environment following its own motivational system, with the human experimenter only acting as language teacher when the robot, by chance, encounters any of the three target objects. This generates a set of pairings of a view of an object and of a verbal utterance of the object's name, used for the supervised training. Finally, in the third condition (unsupervised learning) the robot is static and is given a series of images of the objects, which are used by an unsupervised algorithm to classify what is in the image.

This study specifically examines the hypothesis that social learning, rather than individual learning, helps the child bootstrap her early lexicon, and that initial meanings are situated and context dependent. This is highly consistent with the defining principles of developmental robotics, as discussed in chapter 1. The experimental data clearly support the hypothesis. In the social interaction learning experimental condition, after 150 repeated language games, AIBO is able to learn the names of the three objects correctly in 82 percent of the cases. The red ball naming reaches a higher success rate of 92 percent due to the simplicity of the recognition of a uniform red object, with respect to the articulated puppet and the small robot replica. In the observational learning condition, when the experimenter plays a passive role of naming the objects that the robot encounters by chance, a lower average naming rate of 59 percent is achieved. Finally, in the unsupervised condition, the lowest result of 45 percent correct classification and naming is achieved.

Steels and Kaplan's (2002) explanation of the best performance in the social learning condition is based on the analysis of the dynamics of the classification clustering algorithm. Full interaction with the human user allows the robot to gather a better sampling of the objects' views and of label-image instances. In the observational learning condition, the reduced, passive role of the experimenter reduces the quality of the object's view data collected during unguided interactions. Social interaction is therefore seen as a scaffolding mechanism that guides and supports the exploration of the learning space and constraints. As the authors state, "the social interaction helps the learner to zoom in on what needs to be learned" (ibid., 24).

Various other robotics models of language learning, though not necessarily addressing developmental research issues, have been proposed. For example, Billard and Dautenhahn (1999) and Vogt (2000) both use mobile robots for experiments on the grounding of the lexicon. Roy and collaborators (Roy, Hsiao, and Mavridis 2004) teach the RIPLEY robot (a peculiar arm robot with cameras in the wrist) to name objects and their properties relying on an analogical mental model of the state of the physical word in which they interact.

More recently, robotics models of lexicon acquisition have started to directly address developmental research issues and hypotheses. One of the first models in this line of research is Lopes and Chauhan's (2007) work on a dynamic classification system for the incremental acquisition of words (OCLL: One Class Learning System). This study directly addresses the early stages of lexical development for the naming of concrete objects, referring directly to the embodiment language learning bias of the shape properties of early named objects (Samuelson and Smith 2005). This developmental robot model, based on an arm manipulation platform, incrementally develops a lexicon of between six and twelve names through the dynamic adjustment of category boundaries. The robot successfully learns to recognize the names of objects as pens, boxes, balls, and cups relying on the shape categorization bias, and to respond by picking up the named target element out of a set of objects. The use of the open-ended OCLL algorithm is particularly suited, as language development implies a dynamic, open-ended increase of the lexicon size. However, the performance of this learning algorithm diminishes when the lexicon goes above a size of ten or more words. Moreover, this study generically refers to principles of child development, not specifically addressing the current hypotheses and debates in the literature.

A more recent developmental robotics model of early lexicon development has directly targeted the replication and extension of child psychology studies, using the experimental paradigm of the Modi Experiment (Smith and Samuelson 2010; Baldwin 1993), as discussed in section 7.1. This model will be extensively discussed in the following section, to provide a detailed example of a developmental robotics model of language acquisition.

7.3.2 The iCub Modi Experiment

The Modi Experiment described in box 7.1 is directly based on the Embodiment Principle of language acquisition, and strongly supports the hypothesis that the body posture is central to the linking of linguistic and visual information (Smith 2005; Morse, Belpaeme, et al. 2010). For example, Smith and Samuelson (2010) reported that large changes in posture, such as from sitting to standing, disrupt the word-object association effect and reduce the performance in the first experiment to chance levels. In the developmental robotics model of the Modi Experiment (Morse, Belpaeme, et al. 2010) this embodiment principle is taken quite literally, using information on the body posture of the robot as a "hub" connecting information from other sensory streams in ongoing experience. Connecting information via this embodiment hub allows for the spreading of activation and the priming of information across modalities.

In the developmental robotics Modi Experiment, the humanoid robotic platform iCub was used. Although some initial iCub experiments were carried out in simulation through the open source iCub simulator (Tikhanoff et al. 2008, 2011) to calibrate the neural learning system, here the setup and data on the experiments with the physical robot are reported (Morse, Belpaeme, et al. 2010).

The robot's cognitive architecture is based on a hybrid neural/subsumption architecture (figure 7.4), and is based on the Epigenetic Robotics Architecture proposed by Morse and colleagues (Morse, de Greeff, et al. 2010). A set of interconnected, 2D self-organizing maps (SOMs) is used to learn associations between objects and words via the body posture hub. In particular, a visual SOM map is trained to classify objects according to their color, taking as input the data from one iCub's eye camera (the average RGB color of the fovea area). The SOM for the body-posture "hub" similarly uses as input the joint angles of the robot and is trained to produce a topological representation of body postures. Two DOFs from the head (up/down and left/right), and two DOFs from the eyes (up/down and left/right) were used. The auditory input map is abstracted as a collection of explicitly represented word nodes, each active only while hearing that word. These word units are artificially activated though a speech recognition system trained on a subset of ten words.

The neural model forms the upper tier of a two-layer subsumption architecture (Brooks 1986) where the lower tier continuously scans whole images for connected regions of change between temporally contiguous images (figure 7.4). The robot is directed to orient with fast eye saccades and slower head turns to position the largest region of change (above a threshold) in the center of the image. This motion saliency mechanism operates independently from the neural model, generating a motion saliency image driving the motor system. This motion saliency image can be replaced with a color-filtered image to provoke orientation to regions of the image best matching the color primed by the neural model.

The SOM maps are connected through Hebbian associative connections. The weights of the Hebbian connections are trained online during the experiment. Inhibitory competition between any simultaneously active nodes in the same map provides arbitration between multiple associated nodes resulting in dynamics similar to those expressed in Interactive Activation and Competition (IAC) model (McClelland and Rumelhart 1981). As the maps are linked together in real time based on the experiences of the robot, strong connections build up between objects typically encountered in particular spatial locations, and hence in similar body postures. Similarly, when the word "modi" is heard, it is also associated with the active body posture node at that time. The relative infrequency of activity in the word nodes compared with continuous activity in the color map is not a problem given competition is between nodes within each map and not between the maps themselves. Finally at the end of the experiment, when the robot is asked, "Where is the modi," activity in the "modi" word node spreads to the associated posture and on to the color map node(s) associated with that posture. The result is to prime particular nodes in the color map. The primed color is then used to

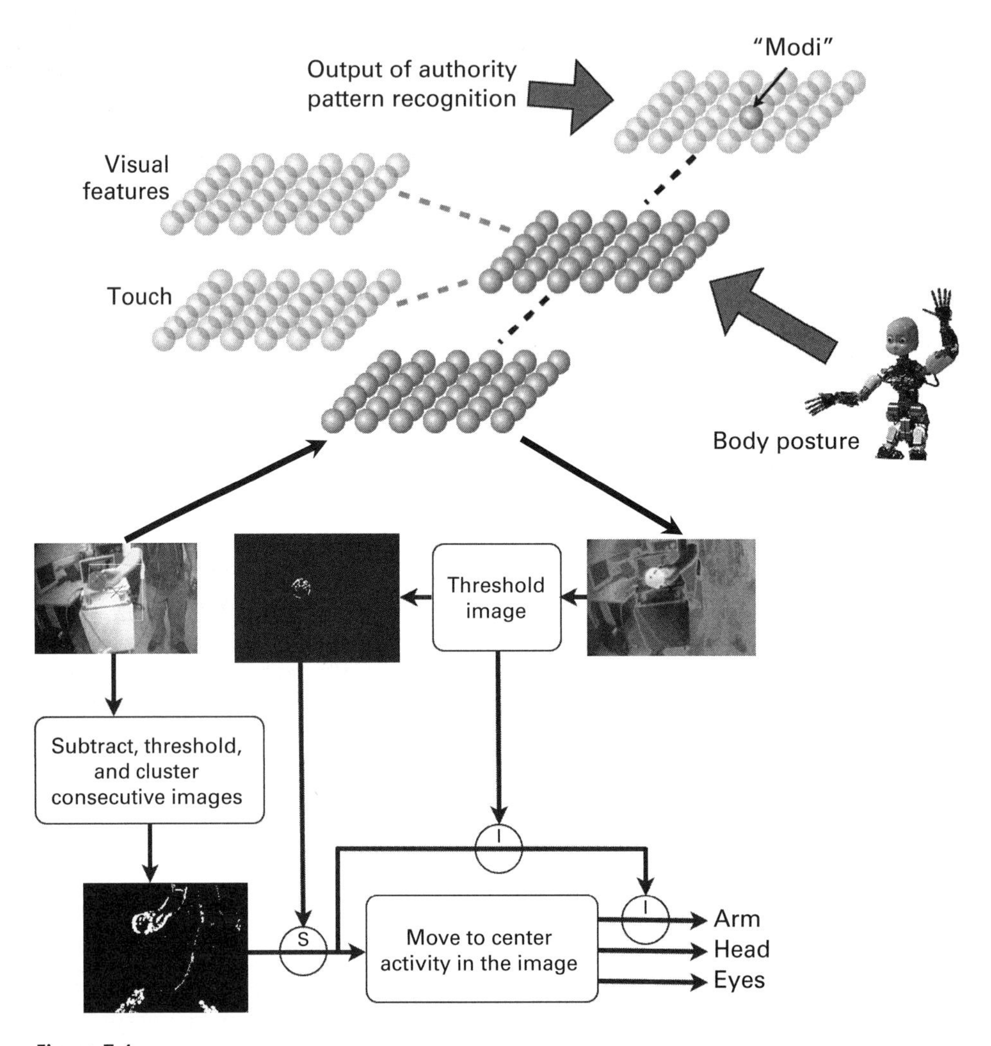

Figure 7.4
Architecture of the robot's cognitive system. The top part visualizes the neural network controller for the learning of word-object associations. The bottom part represents the subsumption architecture to switch behavior between the different stages of the experiment.

filter the whole input image and the robot adjusts its posture to center its vision on the region of the image most closely matching this color. This is achieved using the same mechanism that detects and moves to look at regions of change in the image, replacing the motion saliency image with a color-filtered image. The robot moves to look at the brightest region of the color-filtered image.

Given that the number of associations constructed will grow over time in the absence of negative Hebbian learning and in a changing environment, large changes in body posture are used to trigger a weakening of these associative connections consistent with the eradication of spatial biases in the psychology experiment following changes from sitting to standing. Additionally, external confirmation that the correct object has been selected leads to more permanent connections being constructed either directly between word and color maps or via a second pattern-recognition-based "hub." The Hebbian and the competitive SOM learning implements the Principle of Mutual Exclusivity, as discussed in table 7.2.

The model as described is then used to replicate each condition of Smith and Samuelson's (2010) four child psychology experiments described in section 7.1 (box 7.1). Figure 7.5 shows screenshots of the main stages of the default "No Switch" condition of the first Modi Experiment.

In each condition of each experiment, the results recorded which object, if any, was centered in the robot's view following the final step where the robot was asked to "find the modi." In the No-Switch condition of experiment 1, 83 percent of the trials resulted in the robot selecting the spatially linked object, while the remaining trials resulted in the robot selecting the nonspatially linked object. This is comparable to the reported result that 71 percent of children selected the spatially linked object in the human experiment in the same condition (Smith and Samuelson 2010).

Reducing the consistency of the object-location correlation in the switch condition resulted in a significant reduction in the spatial priming effect with a close to chance performance of 55 percent of the trials finishing with the spatially correlated object being centered in the view of the robot. The remaining trials resulted in the

Figure 7.5
Sample images of the sequential steps of the Modi Experiment with the iCub robot (from Morse, Belpeame, et al. 2010).

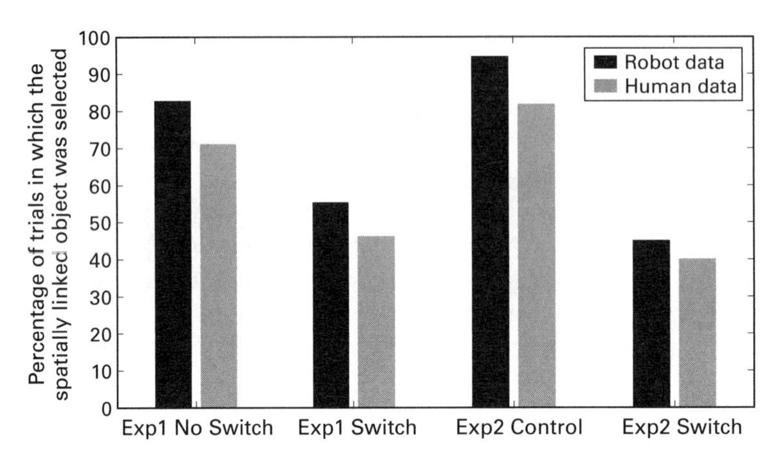

Figure 7.6
Results of the direct comparison between the robot's experiment and the benchmark empirical study of Smith and Samuelson 2010 (from Morse, Belpeame, et al. 2010).

other object being selected. In experiments 3 and 4, when the object was labeled while being attended, the control group resulted in 95 percent of the trials selecting the labeled object, while in the switch condition only 45 percent of the trials resulted in the labeled object being selected. The remaining trials all selected the other object. These results are compared to the reported human child data in figure 7.6.

The close match between the results from the robot experiments and the human child results reported by Smith and Samuelson (2010) supports the hypothesis that body posture is central to early linking of names and objects, and can account for the spatial biases exposed by these experiments. What is of relevance here is that the pattern of results, rather than the absolute percentages, in the various conditions of the experiments are consistent between the human and robot data. As can be seen from figure 7.6 the robot data consistently produced a slightly stronger bias toward the spatially linked objects than the human data, probably due to the fact that the robotic model is only trained on this task. Such a correspondence supports the validity of the cognitive architecture implemented in the robot to achieve the integration of body posture and word learning, proposing potential cognitive mechanisms on embodied language learning.

Moreover, this model has produced novel predictions on the relationship between the space embodiment bias and lexical acquisition. Variations of the posture configuration of the iCub during the different stages of the experiment (corresponding to the sit-down and upright position discussed by Smith and Samuelson) can be used by infants as a strategy to organize their learning tasks. For example, if a second interference task is added to that of the modi learning task, and a different sitting/upright position is separately used for each of the two tasks, then the two postures allows the robot (and the child) to separate the two cognitive tasks and avoid interference. This prediction, suggested by robot experiments, has been verified in new experiments with infants at the Indiana University BabyLab (Morse et al. in preparation).

7.3.3 Learning to Name Actions

The latest advances in developmental robotics models of language acquisition go beyond the naming of static objects and entities. This is because situated and embodiment learning is one of the basic principles of developmental robotics, and as such it must imply the active participation of the robot in the learning scenario, including the naming of actions and the manipulation of objects in response to linguistic interactions. The strict interaction between the language and sensorimotor systems has been extensively demonstrated in cognitive and neural sciences (e.g., Pulvermüller 2003; Pecher and Zwaan 2005) and informs ongoing developments in developmental robotics (Cangelosi et al. 2010; Cangelosi 2010).

A language learning model based on the simulation model of the iCub robot has focused on the development of a lexicon based on names of objects and of actions, and their basic combinations to understand simple commands such as "pick_up blue_ball" (Tikhanoff, Cangelosi and Metta 2011). This study will be described in detail to provide an example of the integration of various perceptual, motor, and language learning modules involved in developmental robotics.

Tikhanoff, Cangelosi, and Metta (2011) use a modular cognitive architecture, based on neural networks and vision/speech recognition systems, which controls the various cognitive and sensorimotor capabilities of the iCub and integrates them to learn the names of objects and actions (figure 7.7). The iCub's motor repertoire is based on two neural network controllers, for reaching toward objects in the peripersonal space of the robot and for grasping objects with one hand. The reaching module uses a feedforward neural network trained with the back propagation algorithm. The input to the network is a vector of the three spatial coordinates (x, y, and z) of the robot's hand, normalized from 0 to 1. These coordinates were determined by the vision system, by means of the template matching method, and depth estimation. The output of the network is a vector of angular positions of five joints that are located on the arm of the robot. For the training, the iCub generates 5,000 random sequences, while performing motor babbling within each joint's spatial configuration/limits. When the sequence is finished, the robot determines the coordinates of its hand and the joint configuration that was used to reach this position. Figure 7.8 (left) shows 150 positions of the endpoints of the robot hands, by representing them as green squares.

The grasping module consists of a Jordan recurrent neural network that simulates the grasping of diverse objects. The input layer is a vector of the states of the hand touch sensors (figure 7.8, right), and the output is a vector of normalized angular

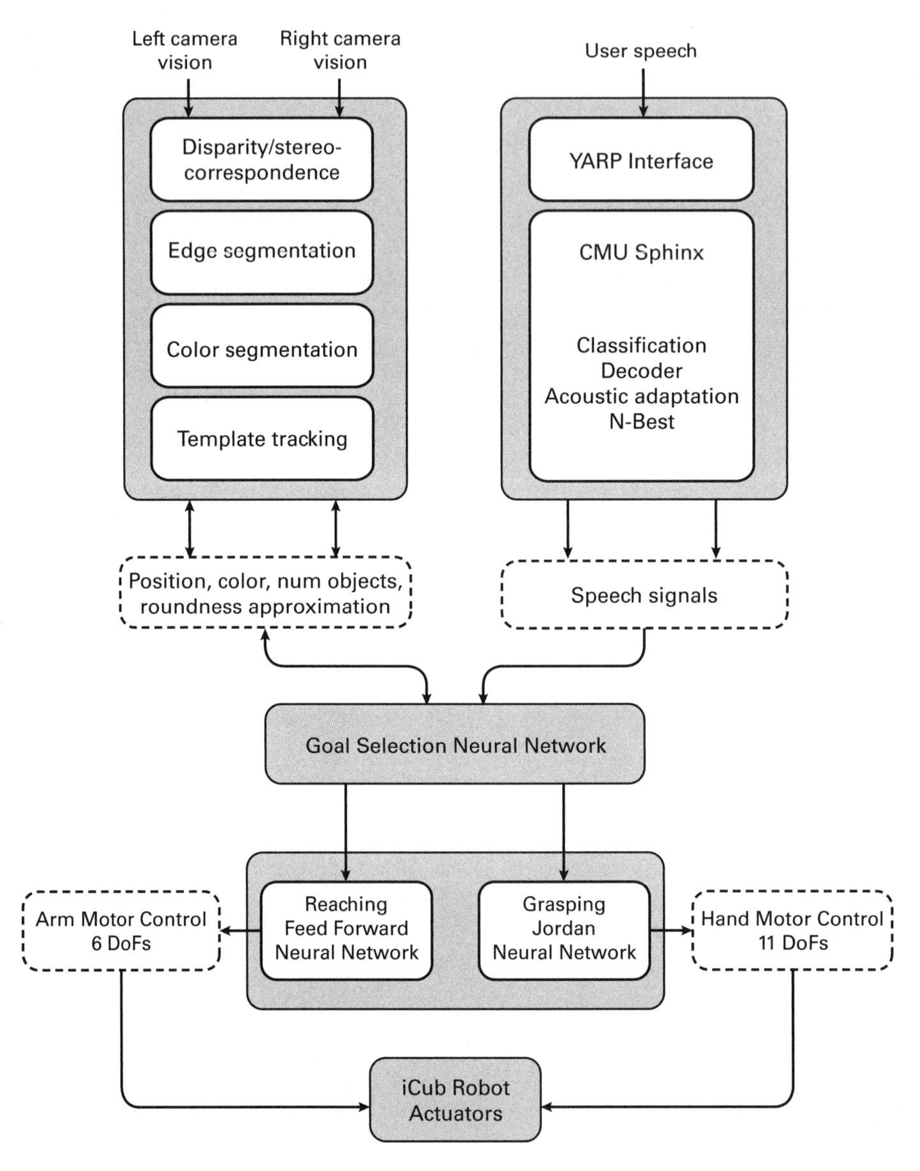

Figure 7.7
Cognitive architecture for the language learning experiment in Tikhanoff, Cangelosi, and Metta (2011). Figure courtesy of Vadim Tikhanoff. Reprinted with permission from IEEE.

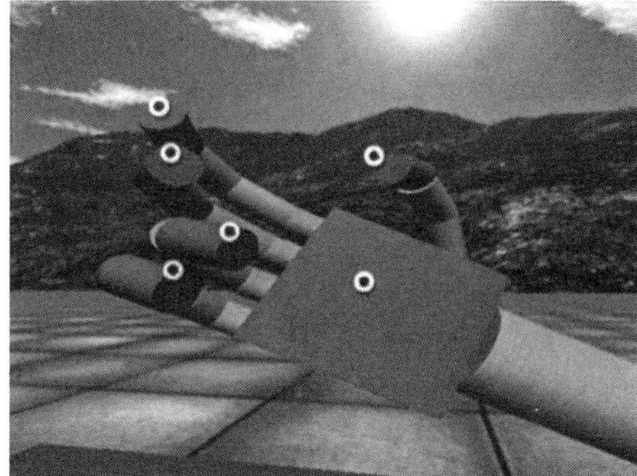

Figure 7.8
Images of the iCub simulator for the motor training. Left: example of 150 end positions of the robot arms during the motor babbling training. Right: location of the six touch sensors on the iCub's hand. Figure courtesy of Vadim Tikhanoff. Reprinted with permission from IEEE.

positions of the eight finger joints. The activation values of the output units are fed back to the input layer, through the Jordan context units. The training of the grasping neural network is achieved online, without the need for predefined, supervision training data. The Associative Reward Penalty algorithm was implemented in the network to adjust the connection weights to maximize the finger positions around the object. During training, a static object is placed under the hand of the iCub simulator and the network at first randomly initiates joint activations. When the finger motions have been achieved, or stopped by a sensor activation trigger, grasping is tested by allowing gravity to affect the behavior of the object. The longer the object stays in the hand (maximum 250 timesteps) the higher the reward becomes. If the object falls off the hand, then the grasping attempt was not achieved and therefore a negative reward is given to the network. Both the reaching and the grasping networks are successfully pretrained, before language is introduced, and are used to execute the reaching and manipulation instructions.

The visual input processing for the object centered in the robot's eye fovea is based on standard vision algorithms for object segmentation based on color filtering and for shape classification (roundness value). The language input is processed through the SPHINX speech recognition system, integrated with the iCub simulator.

The core module that integrates the various processing capabilities to respond to linguistic instructions is the Goal Selection Neural Network. This is an architecture similar to the feedforward networks used in connectionist object-naming experiments (e.g., Plunkett et al. 1992). The input to the network consists of seven visual features encoding object size and location, and the active language input units activated by the speech recognition system. The output layer has four units respectively selecting the four actions: idle, reach, grasp, and drop. The most active output action unit then activates the reaching and grasping modules, trained separately. During the training phase of the Goal Selection Network, the robot is shown an object along with a speech signal. Examples of sentences for handling a blue object are: "Blue_ball," "Reach blue_ball," "Grasp blue_ball," "Drop blue_ball into basket." This network is successfully trained though 5,000 cycles of sentence-object pairs. The final testing demonstrated that the iCub is able to correctly respond to all the combinations of the four action names and object names.

The Tikhanoff, Cangelosi and Metta (2011) model of language understanding in the humanoid robot iCub provides the validation of an integrated cognitive architecture for developmental experiments on vision, action, and language integration. The model also deals with the learning and use of simple sentences based on word combinations. However, these word combinations do not correspond to the full potential of syntactic languages, as the robots would need to be able to generalize behavior to novel combinations of words and the robot's neural controller is explicitly trained on all possible word combinations. Multiword sentences in these studies are more similar to the holophrases of a one-year-old child, than to proper two-word and multiword combinations that appear in later stages of child development, typically in the second year of life. In the next section we will look at developmental robotics models that specifically investigate the acquisition of syntactic capabilities.

Related models have proposed similar cognitive architectures for language grounding of action words. For example, Mangin and Oudeyer (2012) propose a model that specifically studies how the combinatorial structure of complex subsymbolic motor behaviors, paired with complex linguistic descriptions, can be learned and reused to recognize and understand novel, previously untrained combinations. In Cangelosi and Riga (2006), an epigenetic robotic model is proposed for the acquisition via linguistic imitation of complex actions (see section 8.3, this volume). Finally, Marocco et al. (2010) specifically aim at the modeling of the grounding of action words in sensorimotor representation for a developmental exploration of grammatical categories of verbs as names of dynamic actions. They use the simulated version of the iCub to teach the robot the meaning of action words representing dynamical events that happen in time (e.g., push a cube on a table, or hit a rolling ball). Tests on the generation of actions with novel objects with different shape and color properties show that the meaning of the action words depends on sensorimotor dynamics, and not on the visual features of the objects.

7.4 Robots Learning Grammar

The acquisition of grammar in children requires the development of basic phonetic and word learning skills. As shown in table 7.1, simple two-word combinations appear around eighteen months of age, near the period of vocabulary spurts that significantly increases the child lexicon. The developmental robotics models of lexical acquisition discussed previously have not fully reached the stage of reproducing the vocabulary spurt phenomenon, at least for the robotics studies that model the acquisition of the lexicon directly grounded in the robots' own sensorimotor experience. However, a few robotics models have been proposed that start to address the phenomenon of grammar development, though with smaller lexicons. Some of these grammar-learning models focus on the emergence of semantic compositionality that, in turn, supports syntactic compositionality for multiword combinations (Sugita and Tani 2005; Tuci et al. 2010; Cangelosi and Riga 2006). Others have directly targeted the modeling of complex syntactic mechanisms, as with the robotics experiments on Fluid Construction Grammar (Steels 2011, 2012), which permits the study of syntactic properties such as verb tense and morphology. Both approaches are consistent with cognitive linguistics and embodied approaches to language. Here we describe in detail some of the progress in these two fields of modeling grounded grammar acquisition.

7.4.1 Semantic Compositionality

Sugita and Tani (2005) were among the first to develop a robotic model of semantic compositionality, with cognitive robots capable of using the compositional structure of action repertoires to generalize novel word combinations. Compositionality refers to the isometric mapping between the structure (semantic topology) of meanings and the structure (syntax) of language. This mapping is used to produce combinations of words that reflect the underlying combination of meanings. Let us consider a very simple meaning space with three agents (JOHN, MARY, ROSE) and three actions (LOVE, HATE, LIKE). These meanings are isometrically mapped to three nouns ("John," "Mary," "Rose") and three verbs ("love," "hate," "like"). Through compositionality, a speaker can generate any possible combination of noun-verb-noun combinations, such as "John love(s) Mary," and "Rose hate(s) John," directly mapping sentences to agent-action-agent combinations.

The robotics experiments of Sugita and Tani (2005) specifically investigate the emergence of compositional meanings and lexicons with no a priori knowledge of any lexical or formal syntactic representations. A mobile robot similar to a Khepera was used. This was equipped with two wheels and an arm, a color vision sensor, and three torque sensors on both the wheels and the arm (figure 7.9). The environment consists of three colored objects (red, blue, and green) placed on the floor in three different locations (a red object on the left-hand side of its field of view, a blue object in the middle, and a green object on the right). The robot can respond with nine possible behaviors based on the combination of three actions (POINT, PUSH, HIT) with the three objects (RED, BLUE, GREEN) always in the same locations (LEFT, CENTER, RIGHT). The nine possible behavioral sequence categories resulting from the combination of actions, objects, and

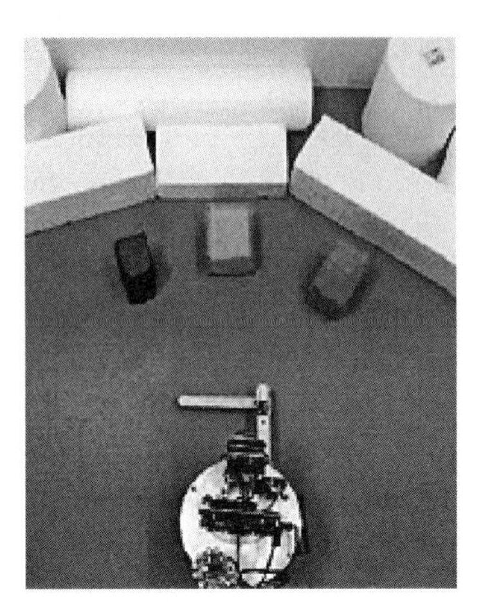

Figure 7.9
Top-down view of the mobile robot with arm and the three objects in front, from the Sugita and Tani 2005 experiment. Figure courtesy of Jun Tani.

locations are shown in table 7.4 (uppercase text). The table also shows the possible linguistic combinations (lowercase).

The robot's learning system is based on two coupled recurrent neural networks (Recurrent Neural Network with Parametric Bias [RNNPB]), one for the linguistic module and the other for the action motor module. RNNPB is a connectionist architecture based on a Jordan recurrent neural network that uses the parametric bias vectors (learned during training) as a compressed representation for the input state activating the behavioral sequence (Tani 2003). In this experiment, the network is taught, through direct supervision with the error backpropagation algorithm, to act on the three objects/location as per table 7.4. The linguistic and motor units respectively encode the sequences of words and actions to be learned. The two RNNPB modules are trained to produce the same parametric bias vectors for each action/word sequence.

The robot experiments are divided in two stages: training and generalization. In the training stage, the robot acquires associations between sentences and corresponding behavioral sequences. The testing stage tests the robot's ability to generate the correct behavior by recognizing (1) the sentences used during training; and (2), most important, the novel combinations of words. A subset of fourteen object/action/location combinations is used during training, with four left for the generalization stage.

After the successful training stage, in the generalization stage the four remaining, novel sentences are given to the robot: "point green," "point right," "push red," and "push left." Behavioral results show that the linguistic module has acquired the underlying compositional syntax correctly. The robot can generate grammatically correct

Table 7.4
Behaviors (capital letters) and linguistic descriptions (lowercase) in the Sugita and Tani (2005) experiment. The elements underlined are not used during the training and are presented to the network for the generalization stage testing.

	Point		Push		Hit	
	behavior	Language	behavior	Language	behavior	Language
RED LEFT	POINT-R	"point red" "point left	PUSH-R	"point red" "point left"	HIT-R	"point red" "point left"
BLUE CENTER	POINT-B	"point blue" "point center"	PUSH-B	"point blue" "point center"	HIT-B	"point blue" "point center"
GREEN RIGHT	POINT- G	"point green" "point right"	PUSH-G	"point green" "point right"	HIT-G	"point green" "point right"

sentences and understand them by giving a behavioral demonstration of the POINT-G and PUSH-R actions. This is achieved by selecting the correct parametric bias vector that produces the matching behavior. Moreover, detailed analyses of the robot's neural representations supporting the verb-noun compositional knowledge are possible by analyzing the space of parametric biases. Figure 7.10 shows the structure of representation for two sample parametric bias nodes for all the eighteen sentences, and separated substructure for the verbs and nouns. In particular, the congruence in the substructures for verbs and nouns indicates that the combinatorial semantic/syntactic structure has been successfully extracted by the robot's neural network. In the diagram of figure 7.10a, the vectors for the four novel sentences are in the correct positions in the configuration of the compositional verb-noun space. For example, the parametric bias vector of "push green" is at the intersection between the substructures for "push" and "green."

These analyses confirm that the sentences are generalized by extracting the possible compositional characteristics from the training sentences (see Sugita and Tani 2005 for further details and analyses). Overall, this robotic model supports the constructivist psychology hypothesis that syntax compositionality can emerge out of the compositional structure of behavioral and cognitive representations. In the next section we will see how an extended syntactic representation capability can support the developmental acquisition of more complex grammatical constructions.

7.4.2 Syntax Learning

An important framework for robot experiments on the development of syntactic knowledge is the Fluid Construction Grammar (FCG) developed by Steels (2011, 2012). The FCG is a linguistic formalism to adapt the cognitive construction grammar for the

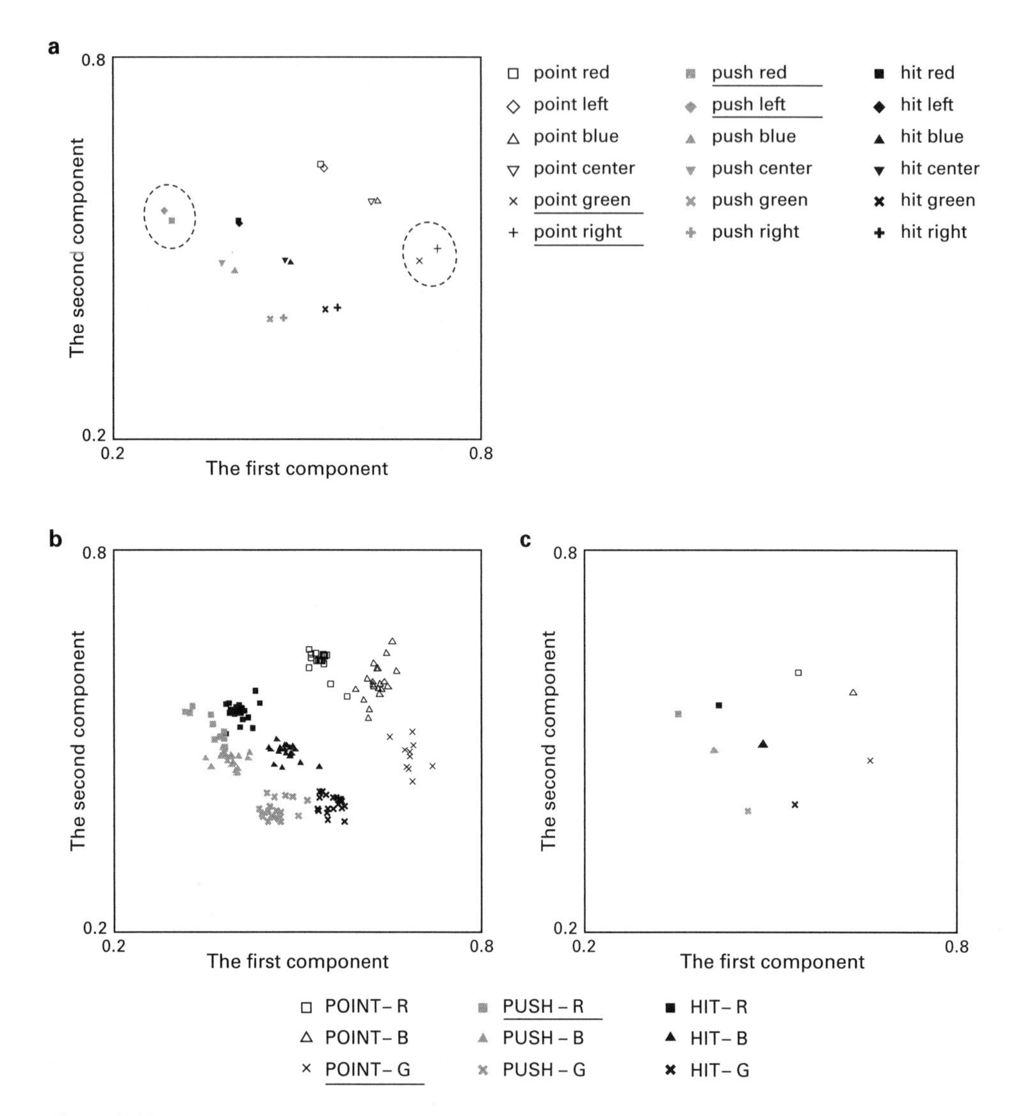

Figure 7.10

Results from the Sugita and Tani experiment. The plots show (a) the PB vectors for recognized sentences in the linguistic module where the vector points within dotted circles correspond to unlearned sentences; (b) the PB vectors for training behavioral sequences in the behavioral module; (c) the averaged PB vector for each behavioral category. Figure courtesy of Jun Tani.

handling of open-ended grounded dialogues, as with robotic experiment on language learning. Although this formalism was originally developed for experiments on the cultural evolution of language (Steels 2012), it is also suitable for language development studies given its focus on the grounding of grammar concepts and relations in situated and embodied robotics interactions. Here we briefly present some properties of the FCG formalism, and a case study on FCG robotic experiments for the acquisition of the lexicon and grammar of German spatial terms (Spranger 2012a, 2012b).

For the overview of the FCG we will primarily follow Steels and de Beule (2006), but refer to Steels (2011, 2012) for an extensive description of the FCG formalism. FCG uses a procedural semantics approach, that is, the meaning of an utterance corresponds to a program that the hearing agent can execute. For example, the phrase "the box" implies that the hearing agents can map the perceptual experience of the box object indicated by the speaker with the corresponding internal categorical (prototypical) representation. This procedural formalism is based on a constraint programming language called IRL (Incremental Recruitment Language) and implements the necessary planning, chunking, and execution mechanisms of constraint networks. The FCG also organizes the information about an utterance in feature structures. For the example "the box," Steels and de Beule (2006) use the following IRL structure:

(equal-to-context ?s) (1)
(filter-set-prototype ?r ?s ?p) (2)
(prototype ?p [**BOX**]) (3)
(select-element ?o ?r ?d) (4)
(determiner ?d [**SINGLE-UNIQUE-THE**]) (5)

These elements are primitive constraints that implement fundamental cognitive operators. In (1), equal-to-context refers to elements in the current context and binds them to ?s. In (2), filter-set-prototype filters this set with a prototype ?p that is bound in (3) to [BOX]. In (4), select-element selects an element ?o from ?r according to the determiner article ?d that is bound to [SINGLE-UNIQUE-THE] in (5), meaning that ?r should be a unique element the speaker is referring to by using the article "the" (rather that the indefinite article "an," which would be used to refer to a generic object).

Figure 7.11 shows the correspondence mapping between the semantic (left) and the syntactic (right) structure of the utterance "the ball." In this feature-based representation system, a unit has a name and a set of features.

In FCG a rule (also called template) expresses the constraints on possible meaning-form mappings. A rule has two parts (poles). The left part refers to the semantic structure formulated as a feature structure with variables. The right part refers to the syntactic structure and is again formulated as a feature structure with variables. For example, among the various types of rules, the *con-rules* correspond to grammatical constructions that associate parts of semantic structure with parts of syntactic structure. These

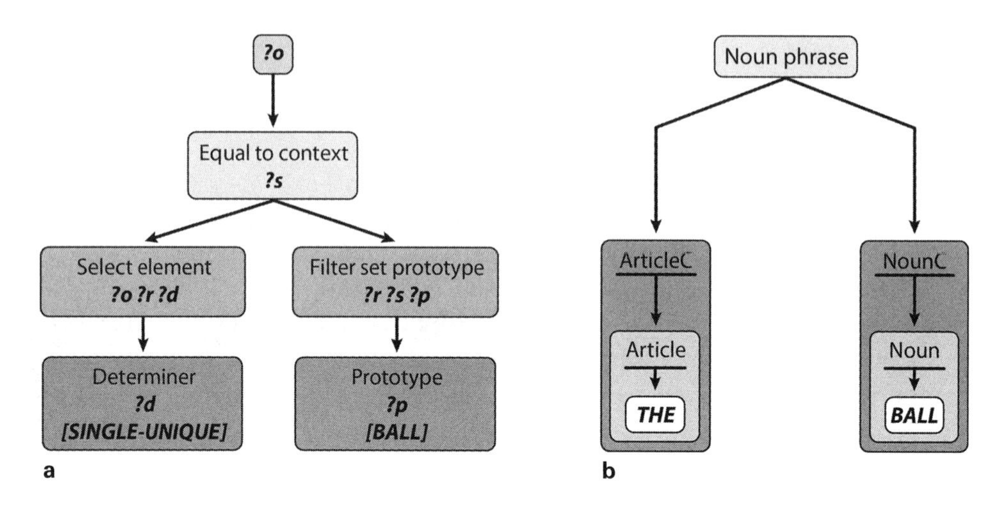

Figure 7.11
Simplified FGC decomposition of the constraint program for "the ball" in the semantic structure (a), and related syntactic structure in the Steels and de Beule (2006) study (b).

rules are used by Unify and Merge operators during linguistic production and parsing. In general, the unification phase is used to check whether a rule is triggered, and the merge phase represents the actual application of the rule. For example, in language production the left pole is unified with the semantic structure under construction, possibly yielding a set of bindings. The right pole is then merged (unified) with the syntactic structure under construction. During understanding and parsing, the right pole is unified with the syntactic structure and parts of the left part are added to the semantic structure. The con-rules are used to build higher-order structure encompassing all the level of complexity of full syntactic representation.

To demonstrate how FCG works in actual robotics language experiments, we will describe the case study on the acquisition of the grammar of German spatial terms (Spranger 2012a). This was carried through language experiments with the humanoid robot QRIO. The robotic agents play a spatial language game, in which they have to identify an object in a shared context. The environment has two robots, a number of blocks of equal size and color, a landmark on the wall, and a box with a visual tag pasted to its front (figure 7.12).

Although FCG was primarily developed for the modeling of the cultural evolution of language, we focus here on the experiments between a speaker and hearer when the meanings and words are predefined by the experimenter (in this case to match the German spatial terms). During each language game, the speaker first draws the attention of the listener to an object using a spatial description, and then the hearer points to the object it thinks the speaker is referring to. The robot's embodiment apparatus (i.e.,

Figure 7.12
Setup for spatial language game with QRIO robots (center) and mental representations for the two robots (sides) from Spranger's (2012a) experiment. Figure courtesy of Michael Spranger. Reprinted with permission from John Benjamins.

perception of distance and angle) determines the main FCG cognitive semantics categories of "proximal-category" (subdivided into near/far subcategories) and "angular-category" (with front/back frontal dimension, left/right lateral dimension, up/down vertical dimension, and north/south absolute dimension). These dimensions are then mapped to the position of the landmark and of the referent object, as any sentence including a spatial term always implies the use of the landmark/referent pair.

All robots are equipped with predefined spatial ontologies and a lexicon for the following basic German spatial terms: *links* ("left"), *rechts* ("right"), *vorne* ("front"), *hinter* ("back"), *nahe* ("near"), *ferne* ("far"), *nördlich* ("north"), *westlich* ("west"), *östlich* ("east"), and *südlich* ("south"). During learning, depending on the success of each language game, the parameters of the FCG elements are refined, such as the width of a spatial category with respect to the central angle/distance. Different experiments are carried out for the learning of separate, or simultaneous, types of categories and spatial dimensions. Results show that the robots gradually learn to map (i.e., align) each spatial term to the target spatial perception configurations (Spranger 2012b).

Once the basic lexicon of spatial terms is acquired through language games, the robots undergo a grammar learning stage. The FCG formalism can help represent the constituent structure, word order, and morphology of the German spatial language.

In the study on grammar learning of Spranger (2012a), two groups of robots are compared: one equipped with the full German spatial grammar, and the second given only lexical constructions. The study compares the performance of the two groups with varying complexity of the scene, for example, with or without allocentric landmarks. Results show that the grammar greatly reduces the ambiguity in interpretation, especially when the context does not provide enough information to solve these ambiguities.

7.5 Conclusions

This chapter frames the current approaches and progress in developmental robotics models of language learning within the state of the art of child psychology studies of language acquisition and consistently with the framework of constructivist theories of language development.

We first gave a brief overview of the up-to-date understanding of the milestones and timescale of language development (table 7.1) and of the main principles used for infant's word learning (table 7.2). This overview shows that for most of the first year, the infant progresses on the development of phonetic competence through various babbling strategies. This leads to the acquisition of a language-like combinatorial phonetics system (e.g., canonical babbling for syllable-like sounds) and the acquisition of the very first words by the end of the first year (e.g., the child's own name). The second year of age is mostly dedicated to the acquisition of words in the lexicon, leading to the vocabulary spurt toward the end of year 2. In parallel, in both the second part of year 2, and in year 3 and subsequent years, the child develops grammatical skills, going from simple two-word combinations and verb islands (around the second birthday) to adultlike syntactic constructions.

The review of the developmental robotics models in sections 7.2–7.4 showed that most of the progress has been achieved for the models of vocal babbling and of early word learning, with some progress in grammar development modeling. Table 7.5 provides a syncretic mapping of child language development stages with the main robotics studies discussed in this chapter. As it can be seen, most robotics models assume that the robot starts with an intrinsic motivation to communicate (but see Oudeyer and Kaplan 2006, and Kaplan and Oudeyer 2007, for an approach specifically focused on intrinsic motivation for communication and imitation). Although robotics language models do not discuss how this intentional communication capability is acquired, some of the works presented in chapter 3 provide operational definitions and experimental testing of the emergence of such types of intrinsic motivations.

The phenomena of babbling and early word acquisition have been extensively covered by various models, though in most studies the focus is on either one or other of the mechanisms. Very few models, as in Brandl et al. (2008), have combined the learning of phonetic representation through babbling with word acquisition skills. As for the emergence of grammatical competence, a few studies have looked at the acquisition and use of two-word combinations, with just one explicitly addressing the learning of adultlike grammar constructs, as in Spranger's (2012a) FCG model of German locatives.

This tendency to produce models that focus on isolated language development phenomena is in part due to the early maturity stages of the developmental robotics discipline, and in part due to the inevitable complexity of an embodied, constructivist

Table 7.5

Mapping of child language development stages and the main robotics studies discussed in this chapter. Two (+) signs indicate the main focus of the work, and one (+) sign the general relevance to the language development stage.

AGE (months)	CAPABILITIES	Oudeyer 2006	Lyon, Nehaniv, and Saunders 2012	Brandl et al. 2008	Steels and Kaplan 2002	Seabra Lopes and Chauhan 2007	Morse, Belpeame, et al. 2010	Tikhanoff et al. 2011	Sugita and Tani 2005; uci et al. 2010	Steels 2012a; pranger 2012b
	(robot)	*Sim. head*	*iCub*	*Asimo*	*Aibo*	*Arm manip.*	*iCub*	*iCub*	*Mobile + arm*	*QRIO*
0–6	Marginal babbling	++								
6–9	Canonical babbling	++	++	++				+		
10–12	Intentional communication	+	+	+	+	+	+	+	+	+
	Gestures			+			+			
12	Single words (objects)			+	++		++	+	+	+
	Single words (actions)			+		++		+	+	
	Holophrases									
	Word-gesture combinations			+						
18	Reorganization of phonetics			++						
	Vocabulary spurt (50+ words)									
	Two-word combinations							++	++	
24	Longer sentences, verb islands									
36+	Adultlike grammar									++
	Narrative skills									

approach to language and the related technological implications. Given the early stages of developmental robotics research, the studies described in this chapter mostly consist of the presentation of a new methodological approach, rather than on the gradual build-up of incrementally complex experiments, and as such will specialize on one cognitive faculty. The exception of the Brandl et al. study, covering both phonetic and lexical development, is explained by the large-scale collaborative project on the ASIMO robot (and also the work by Driesen, ten Bosch, and van Hamme 2009, from the large European project ACORN, www.acorns-project.org). The other issue limiting the breadth of phenomena covered in a single study is the fact that developmental robotics experiments are by their own nature complex, so that the embodied and constructivist view of cognition and language necessitates the parallel implementation of control mechanisms for multiple cognitive capabilities. As this requires complex technological implementations, in most studies researchers tend to start with the assumption of preacquired capabilities, and investigate only one developmental mechanism. For example, in most of the word acquisition studies of section 7.3 the models assume a fixed, preacquired capability to segment speech, thus concentrating only on the modeling of lexico-semantic development (e.g., in Morse, Belpaame, et al. 2010).

A further look at the language development stages that have not currently been modeled by developmental robotics studies (cf. bottom part of table 7.5) shows the areas needing attention in future research. For example, no developmental robotics models exist of the grounded acquisition of large lexicons (fifty-plus words, vocabulary spurt) in robot experiments. Although simulation models have looked at more extensive word repertoires (e.g., 144 words in Ogino, Kikuchi, and Asada 2006), and one long-term human-robot interaction experiment has investigated the learning of 200 object names (Araki, Nakamura, and Nagai 2013), overall the number of words learned in experiments with physical robots and objects has been limited to a few tens of lexical items. The same applies to the learning of incrementally complex grammatical constructions and of narrative skills. However, progress in these areas might benefit from the consideration of significant advancements in other areas of cognitive modeling, especially in the field of connectionist modeling of language (Christiansen and Chater 2001; Elman et al. 1996) and neuroconstructivist approaches (Mareschal et al. 2007). For example, connectionist models exist of construction grammar and the verb island hypothesis (Dominey and Boucher 2005a, 2005b) and of the vocabulary spurt (McMurray 2007; Mayor and Plunkett 2010). Moreover, work on artificial life systems (Cangelosi and Parisi 2002) and virtual agents can inform the development of developmental robotics studies, as in the work on narrative skills in virtual agents (Ho, Dautenhahn, and Nehaniv 2006). The integration of such findings and methodologies in future developmental robotics experiments can help create advances in language learning models.

Additional Reading

Barrett, M. *The Development of Language* (Studies in Developmental Psychology). New York: Psychology Press, 1999.

This edited volume, though not very recent, provides a clear overview of developmental theories and hypotheses on language acquisition. It ranges from a review of phonological acquisition research, to work on early lexical development, constructivist syntax acquisition, conversational skills and bilingualisms, and atypical language development. For a more focused account of usage-based constructivist theory of language development, also read Tomasello 2003.

Cangelosi, A., and D. Parisi, eds. *Simulating the Evolution of Language*. London: Springer, 2002.

This is a collection of review chapters on the simulation and robotics modeling approaches to the evolution of language. The chapters are written by the pioneers in the field, and include a review of the Iterated Learning Model (Kirby and Hurford), early robotic and simulated agent language games (Steels), the mirror system hypothesis for the evolution of the language-ready brain (Arbib), and mathematical modeling of grammar acquisition (Komarova and Nowak). The volume also has a useful introduction on the role of computer simulations for modeling language origins, a tutorial chapter on the main methods for simulating the evolution of language, and a conclusion written by Tomasello on the key facts of primate communication and social learning. A more recent overview of robotics and computer simulation models of language evolution is also available in some chapters in Tallerman and Gibson 2012.

Steels, L., ed. *Design Patterns in Fluid Construction Grammar*. Vol. 11. Amsterdam: John Benjamins, 2011.
Steels, L., ed. *Experiments in Cultural Language Evolution*. Vol. 3. Amsterdam: John Benjamins, 2012.

These two complementary volumes review the theoretical foundations (2011) and the experimental investigations (2012) of the Fluid Construction Grammar framework for the cultural evolution of language. The first volume provides the first extensive presentation of the framework, with the discussion of concrete examples on phrase structure, case grammar, and modality. The second volume includes computational and robotic experiments on the emergence of concepts and words for proper names, color terms, actions, and spatial terms, as well as case studies on the emergence of grammar.

8 Reasoning with Abstract Knowledge

8.1 Children's Development of Abstract Knowledge

Developmental robotics has a strong focus on embodied knowledge, given the utilization of robotic platforms with a rich sensing and actuator apparatus. As such, it is not surprising that most of the achievements in this field have come from studies on basic sensorimotor and motivational skills such as locomotion, manipulation, and intrinsic motivation. However, as we saw in the previous chapter on language, it is possible to start modeling the acquisition of higher-order cognitive skills, such as referential and linguistic skills. Moreover, one of the most influential theories of the development of reasoning and abstract knowledge, proposed by Jean Piaget in the 1950s, roots the origins of reasoning and intelligence in sensorimotor knowledge (Piaget 1952). Piaget's theory is highly consistent with the embodied and situated approach of developmental robotics, and has directly inspired cognitive computational models (e.g., Parisi and Schlesinger 2002; Stojanov 2002).

One of the benefits of using the developmental approach is that we can study how these cognitive skills are integrated, and benefit from, an embodied and situated scenario. This chapter looks at the pioneering studies that try to address such a huge challenge. In particular, these studies use embodied robotics models to understand how a robot can: develop and use abstract representations such as knowledge about quantities and numbers (section 8.2), develop mechanisms supporting the transition from concrete, motor concepts toward more abstract representations (section 8.3), and autonomously develop the capacity to extract and learn abstract representations from the environment to perform decision-making tasks (section 8.4). Moreover, some general cognitive architectures have been developed to specifically model the integration of developmental phenomena ranging from perception to reasoning and decision making (section 8.5). To set the ground for such robotics models, we will first review our current understanding of the child's developmental stages in the ontogenetic origins of reasoning and abstract concepts. Given the limited advances on embodied robotic models of abstract knowledge, the studies presented here only indirectly address the modeling of specific developmental stages. However, most of these studies share the intent to model some of the stages and processes leading to the acquisition of complex skills and abstract representations.

8.1.1 Piaget and the Sensorimotor Origins of Abstract Knowledge

Piaget, a "founding father" of developmental psychology, has proposed one of the most comprehensive theories of the development of reasoning and abstract knowledge rooted in sensorimotor intelligence. He considers three main concepts to explain mental development: (1) schema, (2) assimilation, and (3) accommodation. A *schema* is a unit of knowledge that the child builds and uses to represent relationships between aspects of the world, such as objects, actions, and abstract concepts. Schemas constitute the building blocks of intelligence, and they become more numerous, abstract, and sophisticated during development. *Assimilation* is the process that incorporates new information into previously existing schemas, allowing the child to use the same, though extended, representation to understand new situations in the world. *Accommodation* refers to the process of changing an existing schema into a qualitatively new knowledge representation structure to integrate new information and experiences. For example, a child can initially develop a schema of "dog" that is referred specifically to her own pet dog (e.g., a poodle). Through this simple schema, the child expects that all dogs must be small and fluffy. When the child sees a dog with a very different appearance (e.g., a Dalmatian), she can *assimilate* this new knowledge into a more extended and general schema of dogs. However, when experience of dogs becomes extended, later the child can *accommodate* such an initial schema into separate and more complex categories of dogs, for example, developing two new schemas respectively for the poodle and for the Dalmatian species.

The dynamics of the interaction between the processes of assimilation and accommodation during development leads to increased schema structure and complexity. The interaction between these two processes is called "equilibration" and represents the balance between applying previous knowledge, with assimilation, and changing representations and behavior to account for new knowledge, with accommodation. This then leads to qualitative changes in the knowledge and schemas used by the child. Piaget proposed four stages of cognitive development where the child develops increasingly powerful and sophisticated cognitive and reasoning skills (table 8.1). He uses some very famous experiments to show the qualitative changes in the reasoning strategies used by the children when advancing from one stage to the next. For example, Piaget uses the experiment on the conservation of liquid task to show the difference between pre-operational (2–7 years) and concrete operation (7–12 years) stages. When a five-year old child (pre-operational stage) sees two containers (drinking glasses) with the same level of liquid (milk), she can understand and confirm to the experimenter that the two glasses contain the same quantity of liquid. If then the milk from one

Table 8.1
Stages of mental development in Piaget's theory

Age (years)	Stage	Features
0–2 years	Sensorimotor stage	Recognize distinction between self/world Aware of world entities/features Gradually remember things no longer present No logical inference
2–7 years	Pre-operational stage	Thinking tied to doing Gradually able to think via imagination Egocentric and intuitive thinking, not logical
8–11 years	Concrete operational stage	Mental reverse operation (e.g. understand conservation of liquid task and of transitivity task) Logical reasoning with symbols, though still tied to concrete objects and events Causal reasoning
12+ years	Formal operational stage	Adultlike thinking Full abstract, logical reasoning

glass is poured into a slimmer and taller glass, the child believes that the slimmer glass with a higher level of milk contains more milk. Instead, when the child reaches the concrete operational stage (7 years or older) she will be able to recognize that the quantity of liquid does not change. That is, at the concrete operational stage, the child is able to mentally simulate the reversal of (the pouring) action, thus recognizing that the amount of liquid is still the same.

Notwithstanding some important limitations of Piaget's theory (e.g., he assumes that the child has no innate competences and biases, and also ignores the role of social learning; for example, see Thornton 2008) this remains one of the most influential hypotheses of cognitive development in child psychology. Piaget's theory does not apply to specific cognitive capabilities, but rather is a general and comprehensive view of cognitive development. In the next sections we will look at the developmental timescale of specific capabilities such as numerical cognition and abstract words.

8.1.2 Numbers

Number cognition provides another clear and well-studied example of the gradual development of abstract reasoning skills. Perceiving numbers and quantities is one of the most basic perceptual skills of humans and animals (Dehaene 1997; Campbell 2005; Lakoff and Núñez 2000). This is a nice case study of the development and use of abstract knowledge given the pure and abstract character of the number concept. It provides an exemplar opportunity for the investigation of the role of embodiment and cognitive factors in the transition from perceptual to abstract knowledge (Cordes and Gelman 2005). In line with Piaget's (1952) theory, children start with perceptual cues to group together items of the same perceptual category to count. They subsequently can

Table 8.2
Developmental timescale in the acquisition of number cognition skills

Age (months)	Competence	Reference
4–5 months	Discrimination between sets of different quantities (set size < 4) "Fragile" evidence that infants can compute results of simple arithmetic (1+1, 2–1)	Starkey and Cooper 1980 Wynn 1992; Wakeley, Rivera, and Langer 2000
6 months	Discrimination of sets > 4 if there is a 2:1 numerosity ratio	Xu and Spelke 2000
36–42 months	Development of the one-one and the cardinality principles	Gelman, Meck, and Merkin 1986 Wynn 1990
> 36 months	Development of the order-irrelevance and item-irrelevance principles, after practice and instruction	Cowan et al. 1996
48 months	Good mastering of number directionality (increase in additions, decrease in subtractions)	Bisanz et al. 2005
School age	Gradual development of various arithmetic skills	Campbell 2005

count objects sharing one perceptual feature (e.g., same shape) but differing for other dimensions (e.g., size or color), or count items of the same color but different kind. This results in the child's identification of the concept of entity ("thing"). Gradually, through a series of developmental milestones (table 8.2), the child develops abstract numerical classification ability up to adultlike counting concepts, in other words, when she can count any object of the same numerosity, regardless of perceptual groupings.

In the developmental psychology literature, there is some evidence that infants as young as four months appear to have some (possibly innate?) number cognition abilities. For example, Starkey and Cooper (1980) reported that four-month-old infants show a capability to discriminate between sets of different quantities up to four items. They can discriminate two sets of two and three items each, but not between a set of four and six items. Wynn (1992) used the habituation paradigm to report evidence that five-month-old infants are sensitive to simple arithmetic transformation of the numerosity of a set, such as adding two dolls to form a unique set (1 + 1 = 2), or subtracting one doll from an original set of two dolls (2 – 1 = 1). Xu and Spelke (2000) reported discrimination of sets with more than four items if there is a 2:1 numerosity ratio (e.g., 16 from 8, 32 from 16). However, variability in the replication of these experiments and issues with the methodology appear to suggest that there is only limited evidence that infants can compute these simple arithmetic operations (Wakeley, Rivera, and Langer

2000). Bisanz et al. (2005) suggest that the "fragile" evidence on early infant numerical and arithmetic abilities might be explained by the fact that infants can use non-numerical perceptual and attentional mechanisms (as in Piaget's classification theory) for what appears to be a numerical operation. However, there is general consensus that some kind of preverbal quantification ability appears to be present in preverbal infants, as well as in some animal species (Cordes and Gelman 2005).

Stronger empirical evidence exists on the acquisition of verbal numerical abilities in children that have already developed general linguistic abilities at age three years or older. In particular, to be able to confirm that the child has acquired real numerical knowledge, Cordes and Gelman (2005) propose that these four principles all need to be developed: (1) *one-one*, as each item can only be counted once; (2) *cardinality*, when the last word of the counting sequence represents the numerosity of the set; (3) *order irrelevance*, meaning the items can be counted in any order; and (4) *item-kind irrelevance*, as there are no restrictions on what counts as a countable object. Cordes and Gelman (ibid.) stress that domain-specific pre- and nonverbal counting and arithmetic abilities provide important bases for the subsequent capability to learn the meaning of counting words (numbers).

For the development of the first two principles, Gelman, Meck, and Merkin (1986) and Wynn (1990) reported that three-year-old children start to show mastery of the one-one relationships and the sequentiality of the cardinality principle. The experiments with three-, four- and five-year-olds in the Gelman, Meck, and Merkin (1986) study reveal that children are sensitive to violations of the one-one and the cardinality principles in an experimental task when the child must tell if the counting performance of a puppet is correct or not. In Wynn's (1990) experiment on the "Give-N" task, that is, when the child is asked to give a puppet up to N (six) objects, three-and-a-half-year-old children demonstrate a clear understanding of the cardinality principle.

The development of the order-irrelevance and item-irrelevance principles has been observed in three-year-old children, or older, although the full ability to count objects in various orders is only achieved after practice and repeated instructions from the experimenter (Cowan et al. 1996).

For the full development of arithmetic abilities such as number directionality (whereby addition increases set size and subtraction decreases it), and giving exact answers for additions and subtractions, Bisanz et al's (2005) review article indicates that at four years of age children have developed a stable mastering of such skills. However, due to individual differences and the gradual cognitive development, some evidence of directionality has also been reported in children between two and four years of age. More complex arithmetic abilities will then develop during the school years (Campbell 2005).

In addition to the developmental time perspective, an important issue in numerical cognition is that of embodiment, as various developmental and cognitive psychology experiments have shown the fundamental role of active body involvement in number

Box 8.1

Embodiment Effects in Number Cognition

Number cognition is closely linked to embodiment due to the situated and embodied modes that support the development of counting skills (e.g., pointing at an object while counting, and using fingers to count). Most embodiment effects refer to the interaction between spatial cognition and number representation, resulting from such a developmental experience (Size, Distance, SNARC, Posner-SNARC). Some effects refer to the role of context and function roles in quantity judgment. One effect refers more in general to different attentional and cognitive modalities for quantity recognition, as with the immediate and accurate recognition of small quantities (or subitizing, described later).

Size and Distance Effects

Size and Distance Effects are two of the most common findings from experimental mathematical cognition studies (Schwarz and Stein 1998). They are present in many tasks, and in the context of number comparison they reflect the fact that it is more difficult to compare larger numbers (Size Effect) and numbers that are closer to each other (Distance Effect). Reaction times increase with higher number magnitude and with decreased distance between the numbers compared (see chart below).

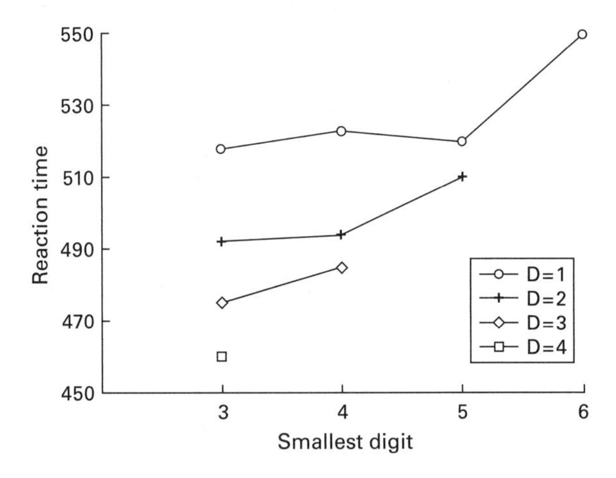

Reaction times of Size and Distance Effects (from Chen and Verguts 2010).

SNARC Effect

The SNARC Effect (Spatial-Numerical Association of Response Codes) is directly related to interactions between number and space. This can be observed in parity judgment and number comparison tasks. Smaller numbers are responded to faster with the left hand than with the right hand, whereas larger numbers are responded to faster with the right hand

(Dehaene, Bossini, and Giraux 1993). The SNARC effect manifests as a negative slope in the reaction time plot.

Posner-SNARC Effect

The Posner-SNARC Effect is based on the attention cueing paradigm (Fischer et al. 2003). A small or large number presented at the fixation acts as a cue and directs attention of the participant toward the left or right side of space, affecting the time needed to detect an object appearing in the visual field after a brief delay. The effect results in faster detection of the target on the left when a small number is presented as a cue, and on the right for large numbers, even though throughout the experiment numbers are not predictive of target locations.

Context and Function Effects

The representation of quantities in tasks on the comprehension and production of vague quantifiers is affected by a range of factors that go beyond the actual number of objects present, and that refer to the context of the objects being quantified and the function they play. These factors include the relative size of the objects involved in the scene, their expected frequency based on prior experience, the function played by the objects present in the scene, and the need to control the pattern of inference of those involved in the communication (Moxey and Sanford 1993; Coventry et al. 2005).

Subitizing Effect

The *subitizing effect* refers to the rapid, accurate, and confident judgments for small numbers of items, typically between 1 and 3, without explicit counting (Kaufman et al. 1949; Starkey and Cooper 1995). This is hypothesized to depend on the existence of a preverbal numerical system that uses global, analog magnitude representation to rapidly recognize small numbers of objects (and their corresponding word number). Above the subitizing effect threshold, the number estimation becomes less precise, or requires explicit counting strategies.

Gesture and Finger Counting Effects

Embodiment effects have been reported connected with the role of gestures in helping children to learn to count (Graham 1999; Alibali and Di Russo 1999) and for the contribution of hand motor circuit and finger counting strategies in numerical cognition (Andres, Seron and Olivier 2007).

representations. Gelman and Tucker (1975) showed that if the child counts slowly and can touch the items, this improves her counting performance (see also studies by Alibali and DiRusso 1999, discussion to follow). In adult cognition, various number embodiment phenomena have been reported. These include the size, distance, and SNARC effects (Spatial-Number Association of Response Code; Dehaene, Bossini, and

Giraux 1993), as well as contextual effects in quantification judgments (see box 8.1 for full explanation). These phenomena have been modeled in connectionist simulations (e.g., Chen and Verguts 2010; Rajapakse et al. 2005; Rodriguez, Wiles, and Elman 1999). More recently, these embodiment effects have been investigated through embodied developmental robotics experiments, showing the interaction between space and number representations following the acquisition of interrelated sensorimotor and counting skills (Rucinski, Cangelosi, and Belpaeme 2011—see section 8.2).

8.1.3 Abstract Words and Symbols

Here we provide a brief overview of current research in developmental psychology on abstract concepts and words, which will later inform developmental robotics experiments on abstract word learning. We will first look at the dichotomy of abstract vs. concrete concepts, and then focus on one special case of abstract concepts: negation, and its role in language development.

In child psychology it is normally accepted that children follow a concrete to abstract developmental pathway. The classical view supports the hypothesis that the acquisition of concrete words precedes the mastering of abstract words (e.g., Schwanenflugel 1991; Caramelli, Setti, and Maurizzi 2004) as children learn through sensorimotor interaction with the world. Children start by learning about concrete objects and events, such as dog, teddy, and water, and gradually move toward the learning of more abstract concepts such as peace and happy. Moreover, children's earlier understanding of the world is based on concrete observations of spatiotemporal interaction between physical objects (Gelman 1990). This concrete-to-abstract pathway does not necessarily apply to all domains, as observations have been reported of the opposite abstract-to-concrete developmental transitions, for example, in the Simons and Keil (1995) study on natural/artifact reasoning.

Studies on the concrete/abstract words are based on various methodologies. For example, the mode of acquisition (MoA) (Wauters et al. 2003; Della Rosa et al. 2010) method has been used to characterize the way in which children learn the meaning of words: (1) through perception; (2) through linguistic information; or (3) via a combination of perceptual and linguistic means. For example, in a study with elementary school children, Wauters et al. report that with school age progression there is a shift from reading texts with a higher proportion of perceptually acquired words toward a higher proportion of meanings that are acquired purely though the linguistic MoA. Della Rosa et al. (2010) have simultaneously looked at a series of other constructs in addition to MoA, such as age of acquisition and familiarity, to understand the concrete/abstract dichotomy. Moreover, various researchers support the view that the distinction between concrete and abstract words is more of a continuum, rather than a dichotomy. For example, there are words such as "physician" or "mechanic" that are of intermediate concrete/abstract value with respect to strongly concrete nouns (dog, house) and highly abstract concepts (e.g., odd number, democracy) (Wiemer-Hastings, Krug, and Xu 2001; Keil 1989). Moreover, apparently concrete words such as "push" and "give" significantly differ in the level of concreteness and motor modality: "push" is more univocal and linked with the action of pushing with a hand, while "give" implies multiple motor instances of the process of passing an object to another person with one hand, two hands, the mouth, and so on. Similarly, the word "use," as in "use a pencil" or "use a comb" is more abstract and general with respect to more specific action descriptions such as "draw with a pencil" and "brush with a comb." This concrete/abstract continuum is highly consistent with theoretical and empirical evidence from embodied cognition studies of abstract concepts (Borghi and Cimatti 2010; Borghi et al. 2011). For example, Borghi et al. (ibid.) have demonstrated that concrete words tend to have a more stable meaning, as they are strictly coupled to sensorimotor experience (e.g., perception, or motor response, linked to an object), while the meaning of abstract words such as "love" can vary more easily due to cultural contexts. This concrete/abstract continuum will be examined though developmental robotics models of the direct and indirect grounding of words on sensorimotor knowledge (see Stramandinoli, Marocco, and Cangelosi 2012; section 8.3).

A special case of the study of abstract words is that of function words, that is, words from a language that play a syntactic role and do not carry meaning in isolation. This is the case of prepositions and short words such as "to," "in," "by," "no," "is." We will take as an example here the learning of the negation function words "no" and "not." One of the fundamental properties of language is its propositional nature, meaning, the fact that, through language, we can assert facts that are either true or false through speech acts (Harnad 1990; Austin 1975). Therefore, in addition to the capacity to understand and describe true events, it is essential to have the complementary awareness of negating nonexisting facts, those of negation. Many types of negation events observed during early child development provide a clear indication that linguistic abilities are grounded in volitional or affective states (Förster 2013; Förster, Nehaniv, and Saunders 2011; Pea 1978, 1980). Moreover, Nehaniv, Lyon, and Cangelosi (2007) have hypothesized an important role of negation in the evolution of human language.

Although the developmental linguistic literature on negation is rather limited, there have been a few key studies that have systematically investigated the early emergence of negation phenomena with both gestures and speech. In particular, Pea (1980) was one of the first to provide a systematic analysis and classification of negation acts, and to stress the fact that negation provides an example of expression of "motor-affective sensorimotor intelligence." Other taxonomies have been discussed by Choi (1988) and Bloom (1970). Table 8.3 provides an overview of the various types of negation that broadly follows these taxonomies, in particular Förster, Nehaniv, and Saunders's (2011) analysis and adaptation to developmental robotics research.

Table 8.3

Different types of frequent negations and their expression and main characteristics (adapted from Pea [1980] and from Förster, Nehaniv, and Saunders [2011])

Type of negation	Linguistic/ behavioral expression	Characteristics
Rejection	"No!" Also with/as gesture	Action-based negations to reject objects (or persons, activities, or events) in the immediate environment First type of negation to emerge developmentally (with gestural and nongestural predecessors of the word "no") Strongly affective/motivational type aversion
Self-prohibition	"Don't" Approach and hesitation	Used about objects or actions previously forbidden by the caretaker It requires internal representation of the preceding external prohibition Motivational/affective component
Disappearance (Nonexistence)	"Gone," "All gone" Empty hand gesture	It signals the disappearance of something in the immediate past It requires an internal representation of the disappeared object on shorter timescale
Unfulfilled expectation	"Gone," "All gone"	It signals that objects are absent from their expected or habitual location. Also used when an activity is unsuccessful in contrast to previous success (e.g. broken toys). It requires representations on longer timescales
Truth-functional denial (Inferential negation)	"It's not true"	Response to a proposition not held to be true by the child In the case of inferential negation, the child assumes that the conversation partner holds the statement to be true (without having actually heard the statement) Most abstract and the last to emerge developmentally It requires logical reasoning and truth-conditional semantics It is independent of the child's attitude towards it (not affective/motivational). It requires representations of facts in the present, past, or future

This taxonomy shows that the first type of negation to emerge developmentally is the Rejection, and that typically rejection gestures precede the use of the word "no." On the other end, the advanced Truth-functional denial emerges in the later stages of cognitive development, when the child has acquired a complex understanding of compositional semantics. Another distinction is that the first three forms of negation are affective- and motivational-based phenomena, while the last one is more abstract and independent from the child's affective state. This taxonomy has informed recent developmental robotics studies of the role of affective behavior in the acquisition of negation (see Förster 2013; section 8.3.2).

8.2 Robots That Count

Developmental and cognitive robotics permit the modeling of the link between embodiment and (higher-order) cognition such as learning to count and use abstract numerical symbols. Rucinski Cangelosi, and Belpaeme (2011) have developed a developmental robotics model of number cognition based on this approach. This constitutes the first attempt to model the development of number cognition using the embodied approaches of cognitive robotics (see box 8.2).

In Rucinski's study, a simulation model of the iCub is first trained to develop a body schema of the upper body through motor babbling of its arms. The iCub is subsequently trained to learn to recognize numbers by associating quantities of objects with their alphanumerical representation "1," "2," and so on. Finally, the robot has to perform a psychological-like experiment and press a left or right button to make judgments on number comparison and parity judgment (figure 8.1). This setup allows the investigation of embodiment effects on number/space interaction, as discussed in box 8.1.

The robot's cognitive architecture is based on a modular neural network controller (figure 8.2). This extends previous connectionist models of number cognition, specifically the neural network model by Chen and Verguts (2010) and the cognitive robotics model of Caligiore et al. (2010). Following Caligiore's TRoPICALS architecture, the processing of information is split into two neural pathways: (1) "ventral," responsible for processing of the identity of objects as well as task-dependent decision making and language processing; and (2) "dorsal," involved in processing of spatial information about locations and shapes of objects and sensorimotor transformations that provide online support for visually guided motor actions.

The "ventral" pathway is modeled in a very similar way to the components of the Chen and Verguts (2010) neural network model. It consists of the following components: (1) a symbolic input (designated in figure 8.2 as INP) that codes for the (alphanumerical) number symbols, using place coding and fifteen neurons; (2) a mental number line (designated as ID) that codes for number identity (the meaning of the symbol)

Box 8.2

Neurorobotics Model of Number Embodiment

This box provides some technical details on the implementation of Rucinski, Cangelosi, and Belpaeme's (2011) model, to facilitate replication of the neurocognitive experiments on number learning and embodiment.

Network Topology, Activation Functions and Learning Algorithms

The artificial neural network shown in figure 8.2 was implemented using the firing rate model, in which the activity of each unit corresponds to an average firing rate over a group of neurons. In figure 8.2 dark gray areas represent all-to-all connections between layers. Neural activations propagate from bottom to top. All neurons use a linear activation function. Thus, the activity of neurons in the ID layer for example, can be described by the following differential equation:

$$I\dot{D}_l = -ID_i + \sum_{j=1}^{15} w_{i,j}^{INP,ID} INP_j \text{ for } i = 1,2,\ldots,15$$

where *IDi* and *INPi* designate activations of the *i*-th neuron in the semantic and input layers and $W^{INP,ID}$ is the matrix of connections between these layers. Equations for the other layers are constructed in an analogous way.

Training of Embodiment Network

The training of the dorsal pathway consisted of two phases. First, the Kohonen maps were constructed using the standard unsupervised learning algorithm, using the input data as described in the main text, to implement a motor babbling task for the iCub robot (figure below). Subsequently, the activations in the developed maps were used to obtain weights $W^{GAZ,LFT}$ and $W^{GAZ,RGT}$ via Hebbian learning, which were then used throughout the model testing. For simplicity, after training of the Kohonen maps was finished, in subsequent training and testing, the input vectors to those maps were replaced by corresponding map activations.

$$I\dot{D}_l = -ID_j + \sum_{j=1}^{15} w_{i,j}^{INP,ID} INP_j \quad \text{for } i = 1, 2, \ldots, 15$$

Left: simulated iCub robot performing motor babbling. Right: colored circles encode unit locations in SOM topology.

The Kohonen maps used the hexagonal internal topology and Gaussian neighborhood function. LFT and RGT maps were trained for 24,000 epochs with a learning coefficient of 0.001 and the neighborhood spread parameter decreasing linearly throughout training from 6.75 to 0.5. The same training parameters were used for the GAZ map, except for the

number of epochs (4,000) and the learning coefficient (0.006). In the Hebbian learning phase, Kohonen maps used the exponential activation function, with the input normalized with respect to the dimensionality of the input space. The training lasted for 1,000 epochs and the learning rate was 0.01. After the training, the connection strengths were normalized in such a way that the total activation propagated through the weights when exactly one neuron of the GAZ map was fully active and did not exceed 1.

Hebbian learning of the $W^{INP,GAZ}$ weights, representing the number-space association, used fifty rows of objects (placed randomly along the vertical coordinate) and lasted for 500 epochs with the learning rate of 0.01. After training, the weights were normalized so that the total activation propagated through the weights was equal for every number.

Training of Number Network

The goal of the training of the ventral pathway was to establish the connections $W^{ID,DEC}$ that allow the model to perform number comparison and parity judgment. The training strongly resembled the one applied by Verguts, Fias, and Stevens (2005), and entailed employing the Widrow-Hoff delta learning rule according to the activations of the units in the INP and DEC layers after they reach a stable state. The only difference between the training applied by Rucinski et al. was that no error threshold was applied; all other training parameters, including the learning coefficient and the distribution of the training data, were the same as used by Verguts, Fias, and Stevens (ibid.).

Test and Overview of Results

The core element of model testing was the acquisition of response times of the model in number comparison, parity judgment, and visual target detection tasks. These were obtained by integrating the equations describing the model for the particular task and taking as the response time of the robot the moment of time in which the value of one of the nodes in the RES layer exceeds a predetermined threshold (this threshold was equal to 0.8 in visual target detection task and 0.5 in all other tasks). Obtained response times were analyzed in standard ways in order to assess the presence of the size, distance, SNARC, and Posner-SNARC effects (see Chen and Verguts 2010).

with linear scaling and constant variability, with the same number of fifteen neurons; (3) a decision layer (DEC) for the number comparison and parity judgment tasks (four neurons, two for each task); (4) a response layer (RES), with two neurons for left/right hand response selection, integrating information from both pathways and responsible for the final selection of the motor response. For the number comparison tasks, which require more than one number to be processed at the same time, short-term memory was implemented by duplicating necessary layers.

The "dorsal" pathway is composed of a number of neuronal maps that code for spatial locations of objects in the robot's peripersonal working space, using different frames of reference (Wang, Johnson, and Zhang 2001). One map is associated with gaze direction (GAZ in figure 8.2), and two maps respectively for each of the robot's left and right arm (LFT and RGT). These maps are implemented as 49-cell (7×7) 2D Kohonen self-organizing maps (SOMs) with cells arranged in a hexagonal pattern. Input to the

Figure 8.1
Robotic simulation setup for number learning study. Figure courtesy of Marek Rucinski.

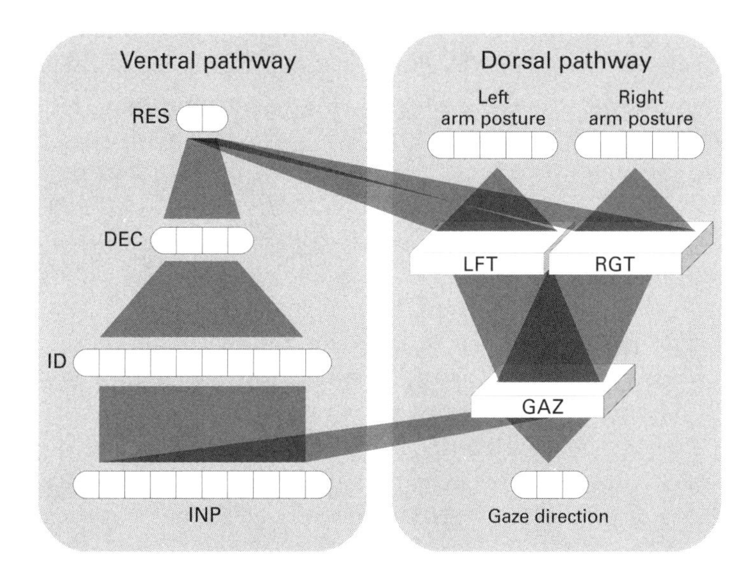

Figure 8.2
Neural network control architecture for number cognition in the iCub. Acronyms: INP—input layer; ID—semantic layer; DEC—decision layer; RES—response layer; GAZ—gaze direction map; LFT—reachable space map for the left arm; RGT—reachable space map for the right arm. See text for explanation. Figure courtesy of Marek Rucinski.

gaze map arrives from the 3D proprioceptive vector representing the robot gaze direction (azimuth, elevation, and vergence). The input to each arm position map consists of a seven-dimensional proprioceptive vector representing the position of the relevant arm joints: shoulder pitch, roll and yaw, elbow angle and wrist pronosupination, pitch and yaw. The gaze map is linked to both arm maps to implement the transformation of spatial coordinates between frames of reference corresponding to these body parts (so that a position in the visual field can be translated into an arm posture corresponding to reaching to this position and vice versa). This is the core component of the model where the embodied properties of the model are directly implemented as the robot's own sensorimotor maps.

To model the developmental learning processes, a number of sequential training phases corresponding to different stages of development of a human child are defined. First, spatial representations for sight and motor affordances have to be built and correspondences between them established. Later, the child can learn number words and their meaning. Usually in late preschool years, children learn to count (table 8.1). More or less at the same stage the child may be taught to perform simple numerical tasks such as number magnitude comparison or parity judgment. All these stages are reflected in the model.

In order to build the gaze and arm space maps, the robot performs a process equivalent to motor babbling (von Hofsten 1982). The child robot refines its internal visual and motor space representations by performing random movements with arms, while observing its hands, to reach for toys in its visual field. This enables the robot to perform tasks like visually guided reaching later in life. This stage of development was implemented in the robot by selecting ninety points uniformly distributed on what has been assumed to be the robot's operational space (a part of a sphere in front of the robot with 0.65 m radius, centered between the robot's shoulder joints, and spanning 30 degrees of elevation and 45 degrees in azimuth). These points served as target locations for directing gaze and moving both arms of the robot using inverse kinematic modules. After a trial in which the robot reaches a random position, the resulting gaze and arm postures were read from proprioceptive inputs and stored. Between each trial, the head and arms of the robot were moved to the rest position in order to eliminate any influence of the sequence in which the points have been presented on the head and arm posture at the end of the motion. These data were used to train the three SOMs using the traditional unsupervised learning algorithm. In order to reflect the asymmetry between reachable space for the left and right arm (some areas reachable by the right arm cannot be reached by the left arm and vice versa), only two-thirds of the extreme points corresponding to an arm were used when building a spatial map for this arm (e.g., leftmost two-thirds of all points for the left arm). Learning parameters were adjusted manually based on the observation of the learning process and analysis of how well resulting networks span target spaces.

Transformations between the visual spatial map for gaze and the maps of reachable left and right spaces were implemented as connections between the maps and were learned using the classical Hebbian rule. In a process similar to motor babbling, gaze and appropriate arm were directed toward the same point and resulting co-activations in already developed spatial maps were used to establish links between them.

The next developmental training stage regards the learning of number words and their meaning. This corresponds to establishing links between number words, modeled as activations in the input ventral layer, and number meaning, being activations in the mental line hidden layer.

Subsequently the robot is taught to count. The goal of this stage is to model the cultural biases that result in an internal association of "small" numbers with the left side of space and "large" numbers with the right side. As an example of these biases we considered a tendency of children to count objects from left to right, which may be associated with the fact that European culture is characterized by left-to-right reading direction (Dehaene 1997; Fischer 2008). In order to model the process of learning to count, the robot was exposed to an appropriate sequence of number words (fed to the ventral input layer of the model network), while at the same time the robot's gaze was directed toward a specific location in space (via the input to the gaze visual map). These spatial locations were generated in such a way that their horizontal coordinates correlated with number magnitude (low number presented on the left, large numbers on the right) with a certain amount of Gaussian noise. Vertical coordinates were chosen to uniformly span the represented space. While the robot was exposed to this process, Hebbian learning established links between number word and stimuli location in the visual field.

Finally, the model is trained to perform a number of reasoning tasks, such as number comparison and parity judgment, which corresponds to establishing appropriate links between the mental line hidden layer and neurons in the decision layer.

The learning architecture was validated through iCub simulation experiments of the three tasks used by Chen and Verguts (2010). These tasks involved measuring response times (RT) of the model, calculated as the amount of activity needed to exceed a response threshold in one of the two response nodes.

The first experiment investigates the size and distance effects observed in number cognition. These are two of the most common findings from experimental mathematical cognition studies (e.g., Schwarz and Stein 1998). They are present in many tasks, but in the context of number comparison they mean that it is more difficult to compare larger numbers (size effect) and numbers that are closer to each other (distance effect). This should be evident in RTs growing with number magnitude and with decreased distance between numbers being compared. RTs obtained from simulating the experiment in our model are reported in figure 8.3.

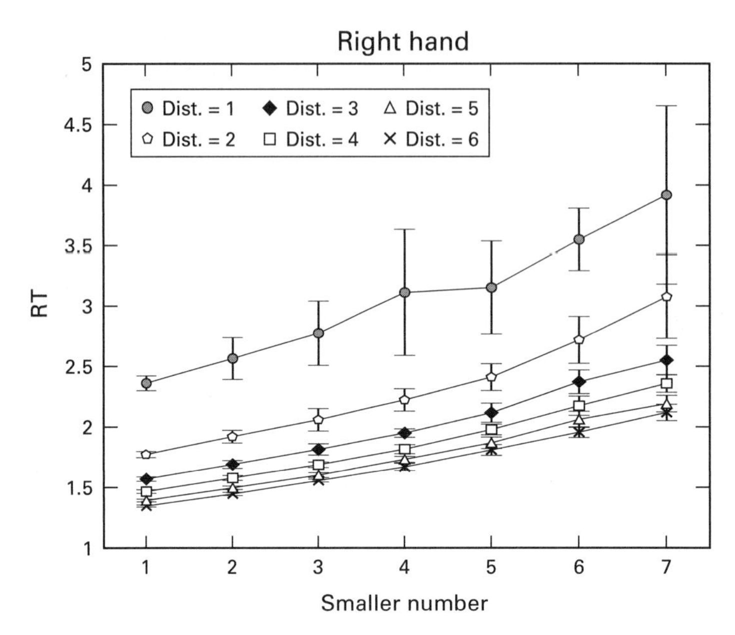

Figure 8.3
Simulation results of the size and distance effects in the number comparison task for the right hand. The robot's reaction time (RT) increases with larger numbers and for smaller distances between numbers. Figure courtesy of Marek Rucinski.

Response times were measured for all pairs of numbers from 1 to 7 for left and right hands. The chart in figure 8.3 shows that both size and distance effects are present in the model. The sources of the size and distance effects depend on the monotonic and compressive patterns of weights between the mental line and decision layers.

The second experiment focuses on the SNARC effect. This is more directly related to interactions between number and space, than size and distance effects. The RTs obtained by the iCub model in parity judgment and number comparison tasks were computed. The difference between right hand and left hand RTs for the same number in both congruent and incongruent condition is reported. The SNARC effect manifests itself in a negative slope as in figure 8.4.

In this model, the presentation of a number word leads to an automatic activation of the relevant parts of the visual space representation, due to the links established during development (learning to count stage), with the left side associated to small numbers, and the right side for large ones. Visual space representations in turn are linked to both motor maps, although not symmetrically. As some parts of the visual space that can be reached by the right arm cannot be reached by the left arm, and vice versa, when transformation from the visual space map to arm maps occurs, both arm-related

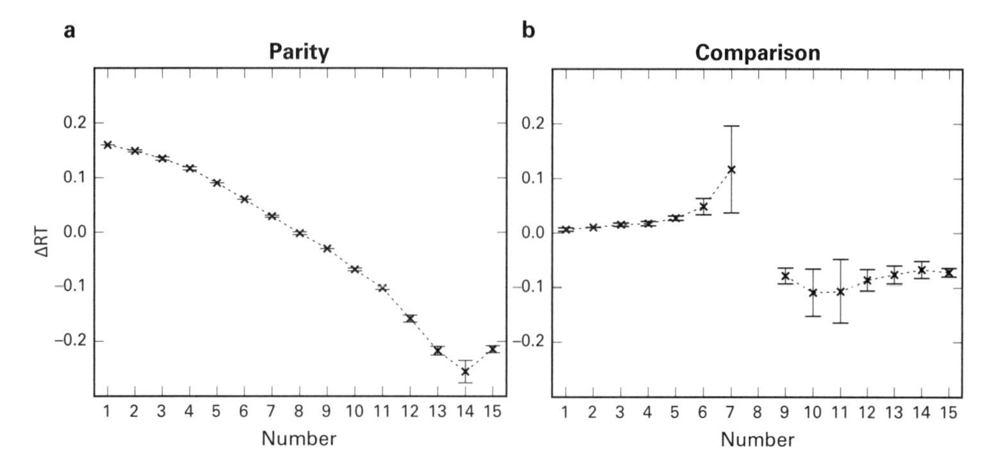

Figure 8.4
Simulation results of the SNARC effect in the parity judgment task (a) and the magnitude comparison task (b). Figure courtesy of Marek Rucinski.

representations will be activated to a similar degree only for the areas in the center of the visual map. For the areas placed to the sides of the visual space, the map associated with one arm will be activated more strongly than the other, as it overrepresents that side of space. This is a natural consequence of the robot morphology.

Because there is a significant overlap between represented areas, the effect is not sudden, but connections between visual and motor maps form a gradient from left to right. The links to the left arm map become weaker, while those to the right become stronger. Thus, for instance when a small number is presented, internal connections lead to stronger automatic activation of the representations linked with the left arm than that of the right arm, thus causing the SNARC effect. The same model is also used to replicate the Posner-SNARC effect (Rucinski, Cangelosi, and Belpaeme 2011).

A subsequent extension of this model has specifically focused on investigating gestures in learning to count (Rucinski, Cangelosi, and Belpaeme 2012). The simulated iCub is trained to count by either looking at objects being counted (visual condition) or pointing at objects while counting them (gesture condition). Gestures are implemented as a proprioceptive input of the gesturing hand position to the robot's neural architecture. The iCub's architecture uses recurrent neural networks as this task requires the learning of incremental number sequences. The comparison of the visual vs. gesture performance demonstrates that the proprioceptive input signal connected with gesture improves its learning performance as this carries information that may be exploited when learning to count. Moreover, the comparison of the behavior of the model with that of human children (reported in Alibali and DiRusso 1999) reveals strong similarities in terms of the effect of gesture and the size of the counted set, though with some differences in the detailed patterns of errors (Rucinski, Cangelosi, and Belpaeme 2012).

8.3 Learning Abstract Words and Concepts

Very few developmental models have started to look at the acquisition of abstract words, as the default main focus of developmental robotics has been on sensorimotor skills and the naming of concrete objects (see chapter 8). Here we will discuss two early models of abstract word learning. The first looks at the grounded acquisition of abstract concepts, within the concrete/abstract continuum, via the mechanism of symbol grounding transfer (Cangelosi and Riga 2006; Stramandinoli, Marocco, and Cangelosi 2012). The second developmental robotics model addresses the acquisition of the word "no" and of the concept of negation in the early stages of language development (Förster, Nehaniv, and Saunders 2011). These studies, by exploring the grounded origins of abstract words and concepts, constitute the bases for future robotics investigations on the development of abstract knowledge for reasoning and decision-making skills.

8.3.1 Toward the Grounding of Abstract Words

In chapter 8 we looked at the robotics model of language development. Most of those studies focused on the acquisition of words for individual objects and actions, and only a few also investigated the emergence of simple semantic and syntactic compositional structures. However, one of the main properties, and benefits, of language use is that of "linguistic generativity." This is the capacity to combine words to express (i.e., invent) new concepts. For example, I can learn, via direct grounding and sensorimotor experience, the word "horse" when I ride a horse, the word "stripes" when I look at striped patterns, and the word "horn" via pictures of bull's and goat's horns. The grounded learning of these symbols allows me to invent and communicate new concepts via linguistic descriptions based on combination of these three words. I can communicate the concepts of zebra to somebody who has never see this animal by saying "zebra = horse + stripes," and even invent a new animal species such as unicorn by saying "unicorn = horse + horn." The process of transferring the meaning of words directly grounded via experience to the meaning of words generated via word combinations is called "symbol grounding transfer" (Cangelosi and Riga 2006), and follows Barsalou's (1999) mental simulation model of concept combination. Generativity via novel word combination can be used to describe practical concepts, such as animals, but also to describe words with abstract components, such as "*use* an object," "*accept* a present," ideally up to less concrete, pure abstract concepts such as "beauty," "happiness," "democracy."

Here we describe an early model of linguistic generativity using the symbol grounding transfer for the learning of new, complex actions (Cangelosi and Riga 2006), and

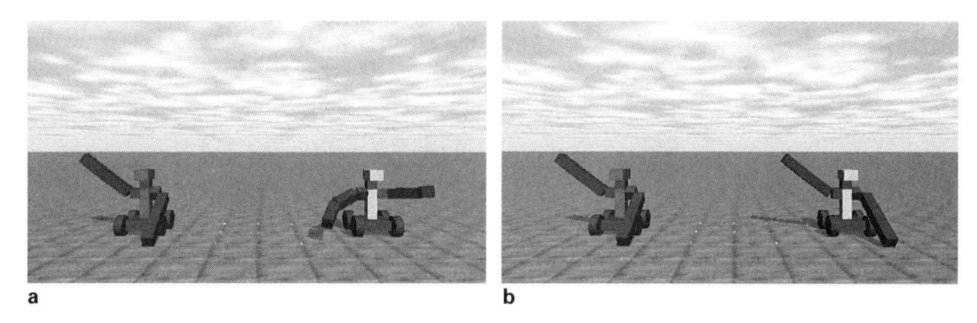

Figure 8.5
Simulated robot setup for the grounding transfer of composite action concepts. The teacher is on the left, and the learner on the right of each image. Scene before (a) and after (b) the action is learned, namely, when the learner robot is capable of imitating the action of the teacher. From Tikhanoff et al. 2007. Reprinted with permission from IEEE.

the model's subsequent adaptation for the gradual learning of more abstract words such as "accept," "refuse" (Stramandinoli, Marocco, and Cangelosi 2012). Cangelosi and Riga (2006) used two agents, a teacher and a learner, implemented as simulated humanoid robots. The teacher has been preprogrammed by the experimenter to demonstrate motor actions and to teach the names of actions to the learner (e.g., figure 8.5). The learner first learns to name basic motor primitives via the motor and linguistic imitations of the teacher. Subsequently, the learner agent uses language instructions to autonomously combine the grounded meaning of basic action names and learn to perform complex actions through the grounding transfer mechanism. The model consists of a computer simulation of two robotic agents embedded in a virtual environment based on the physics simulator Open Dynamics Engine.

The learner agent, controlled by a neural network, learns to imitate the teacher's actions and the names of these actions. The simulation consists of three training stages and a test phase. Training is incremental and is loosely inspired by developmental stages. The three training stages are as follows:

1. Basic grounding (BG): This refers to the imitation learning of the capacity to repeat the basic actions visually demonstrated by the teacher, and to the learning of the names of basic actions. Six basic action primitives are used, such as "close_left_arm," "close_right_arm," "lift_right_upperarm," "move_wheel_forward."
2. Higher-order grounding 1 (HG-1): This regards the learning of complex actions via word combinations, rather than through direct visual demonstration. The teacher produces linguistic utterances such as "grab = close_right_arm" + "close_left_arm," and the agent learns to produce these composite actions though grounding transfer.
3. Higher-order grounding 2 (HG-2): This corresponds to the HR-1 stage, with the difference that a new HG-2 concept consists of the combination of the name of one

BG primitive and one HG-1 word. For example, the agent learns that "carry = move_wheel_forward" + "grab."

The learning agent successfully learns all BG and HG action categories, and in the test stage it is able to autonomously execute all the basic and the higher-order actions, following the input of the corresponding action names. The core mechanism behind this capacity is the grounding transfer mechanism. This is based on a neural network implementation (Cangelosi 2010) of the perceptual symbols system hypothesis by Barsalou (1999). This implements the use of mental simulation to combine sensorimotor concepts into more complex, higher-order mental representations.

The use of this symbol grounding transfer for the generation of complex actions via word combination was further extended to generate larger sets of composite actions, for example, up to a repertoire of 112 complex actions based on the semaphore alphabet gestures (Cangelosi et al. 2007; Tikhanoff et al. 2007).

The same symbol grounding transfer has been used to look at the learning of more abstract words. Using the concrete/abstract continuum, we can go from one extreme of very concrete words, such as "hammer," "nail," "stone," and "give" to very abstracts concepts such as "democracy," "beauty," and "love." Between the two extremes, we can consider different levels of abstractness. For example, the word "accept" (an in "*accept* a present") is an extension of the concrete action of picking an object but in a friendly social context. And the word "use" (as in "*use* a hammer," "*use* a pencil") is a more abstract version of the concepts of "hammer with a sledgehammer" or "drawing with a pencil" (though still linked to action such as hammering and writing). Stramandinoli, Marocco, and Cangelosi (2012) developed a model for the grounding of these intermediate abstract concepts such as "accept," using the same symbol grounding mechanism described earlier.

This model was based on simulations of the iCub robot and broadly follows Cangelosi and Riga's (2006) symbol-grounding transfer method. The iCub first has to learn, through direct grounding, words for concrete motor primitives such as "push." It then has to communicate, via a linguistic generative method and higher-order grounding, more abstract concepts such as "accept." Specifically, during the BG training stage the robot learns the names associated with the eight action primitives through direct sensorimotor experience. The names of action primitives, given in input to the robot's neural network, are "push," "pull," "grasp," "release," "stop," "smile," "frown," and "neutral."

The two different HG stages were implemented, to allow different levels of combining basic and complex actions. In the first HG-1 stage, the robot learns three new higher-order action words ("give," "receive," "keep") by combining only basic action primitives, such as "keep [is] grasp [and] stop." In order to obtain the transfer of grounding from basic actions to higher-order words, the network calculates separately the output corresponding to the words contained in the description ("grasp," "stop") and stores it. Subsequently, the network receives as input the higher-order word "keep" and as target the outputs previously stored. During the second HG-2 stage, the robot learns three higher-order actions ("accept," "reject," "pick") consisting of the combination of basic action primitives and higher-order action words (e.g., "accept [is] keep [and] smile [and] stop").

The robot's successful demonstration of the higher-order complex actions demonstrates that the symbol grounding mechanism not only is able to transfer the grounding to concrete actions (e.g., "grab," as in Cangelosi and Riga 2006), but also works for gradually less concrete, and more abstract concepts (as for the learning of "accept" in Stramandinoli, Marocco, and Cangelosi 2012. However, although the same mechanism can be used for the grounding of other less concrete concepts, such as "make" and "use," the real challenge remains the grounded modeling of highly abstract words such as "beauty" and "happiness."

8.3.2 Learning to Say No!

This model aims at the grounding of another category of abstract words, such as "no" and the related concept of negation, using a combination of sensorimotor, social, and affective mechanisms. As we saw in section 8.1.2, many types of negation that are observed during early child development are a clear indication that linguistic abilities are grounded in volitional or affective states (Förster, Nehaniv, and Saunders 2011; Pea 1980). This is the case of the examples of negation such as prohibitive speech acts, acts of refusal, and acts of motivation-dependent denial.

A scenario for the investigation of the sensorimotor and affective bases of negation in developmental robots has been proposed by Förster, Nehaniv, and Saunders (2011). They distinguish some key properties differentiating the various types of negation listed in table 8.3. First, negations can be distinguished through their relatedness to affect or volition, or to more abstract knowledge of events. So the modeling of the first type of negation (Rejection, Self-prohibition) requires the presence of a mechanism for the level of emotional and motivational state, and its manifestation through facial or body gestures. The second property differentiating negations is the increasing complexity with regard to the required memory, and the level of internalization that the child has of the presence of an object's functional/location properties. For example, Rejection requires short-term memory and a purely reactive behavior, while Unfulfilled expectations and Denial require longer-term memory perspectives. This distinction suggests the need to model different memory strategies and the internalization of the negated object/event's properties.

This provides an operational modeling framework to understand how the capacity to learn and use abstract concepts such as the words "no," "gone," "don't" is related to their grounding on affective and volitional states, rather than on pure sensorimotor (action) as in the case of concrete words, and on the role of memory and internal representations of the external world.

Förster (2013) used a developmental framework based on the iCub robot to test two alternative hypotheses on the development of negation: (1) the negative intent interpretation of rejection; (2) physical and linguistic prohibition. The negative intent interpretation states that negation developmentally originates in the caregiver's linguistic interpretation of a child's motivational state (intent interpretations). This has been explained by Pea (1980) with the hypothesis that negation is acquired developmentally when the caregiver interprets the child's physical rejection behavior as a manifestation of negation, and the corresponding child-caregiver engaging in linguistic expressions such as "no," "no, don't want it." The utterance of negation words is therefore associated with a negative motivation state and rejection behavior, and often has prosodic saliency (Förster 2013). An alternative developmental theory on negation is that the child's use of negation is rooted in the parental prohibition. This is based on Spitz's (1957) hypothesis that negation is acquired when the caregiver prohibits a child from doing something. To test these two hypotheses and the mechanisms necessary to acquire negation in robots, Foerster designed two sets of experiments (rejection and prohibition) to contrast these hypotheses.

The negation experiments use the cognitive architecture depicted in figure 8.6. This is based in part on the LESA architecture employed in the Lyon, Nehaniv, and Saunders (2012) language learning experiments (see section 7.2). Below we give a brief description of the main modules. For the motivation of this architecture and full details, see Saunders et al. 2011 and Förster 2013.

The Perceptual System module processes the image from the robot's cameras to identify the most salient object (using an object saliency algorithm), the shape of the image in one of the sides of the cube (through the ARToolKit package; www.hitl.washington.edu/artoolkit), and the user's face (faceAPI software, seeingmachines.com). The perceptual system also processes the motor encoders and the detection of external force on the robot's arm (e.g., when the user moves the robot's arm away to stop it from getting closer to the object). This system is also used to identify the two crucial events of "picking up" and "putting down" an object, which trigger different behaviors related to negation interactions. Moreover these behaviors serve to focus attention on one object at a time (the one picked up) when multiple objects are present.

The core component for the negation experiments is the Motivational System. This uses a simplified representation of the motivation state that can take on three discrete numerical values: -1 (negative), 0 (neutral), and 1 (positive). Each experimental session starts with a neutral motivation state, which is then changed when the valence of the picked-up object is either positive or negative. The object-valence association is predetermined by the experimenter. In the prohibition scenario, the valence of the object is further modulated by the robot's perception of external pressure force on its arm.

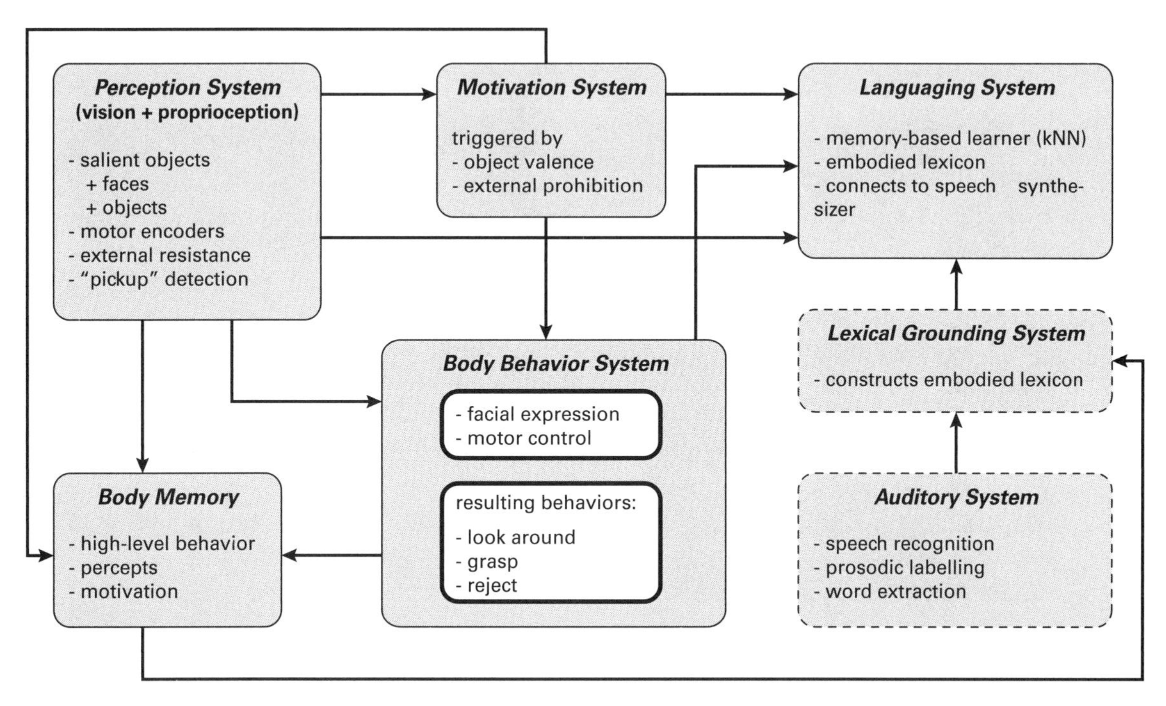

Figure 8.6
Cognitive architecture used in the negation experiments by Förster (2013). See text for full explanation. Figure adapted from Förster 2013.

The Body Behavior System controls the production of five behaviors: Idle, Looking around, Reaching for object, Rejecting, and Watching. These behaviors can be accompanied by three facial expressions (happy, neutral, sad) depending on the negation experiment and the valence of the object. To produce a realistic and believable interaction behavior, a trial-and-error fine-tuning of the time constants (for the duration and synchronization of the five behaviors) and the coordination of subactions was conducted. Within the Prohibition Scenario, the participants are taught to restrain the robot's arm movement as soon as the iCub tries to reach for the object (figure 8.7).

The Languaging System controls the robot's utterance of words, depending on the interaction with the user and the object and on its motivation state. Words learned during the interaction are saved in the Lexicon Grounding subsystem, with the use of the Auditory subsystem, and is in part based on the Lyon, Nehaniv, and Saunders (2012) language learning approach. A SensoriMotor-Motivational (SMM) vector holds information about the current state of the robot's active behavior, attended object, its motivational state, and the visual and motor perception states. The robot speaks when an object is being picked up and put in front of it. Using a memory model that matches

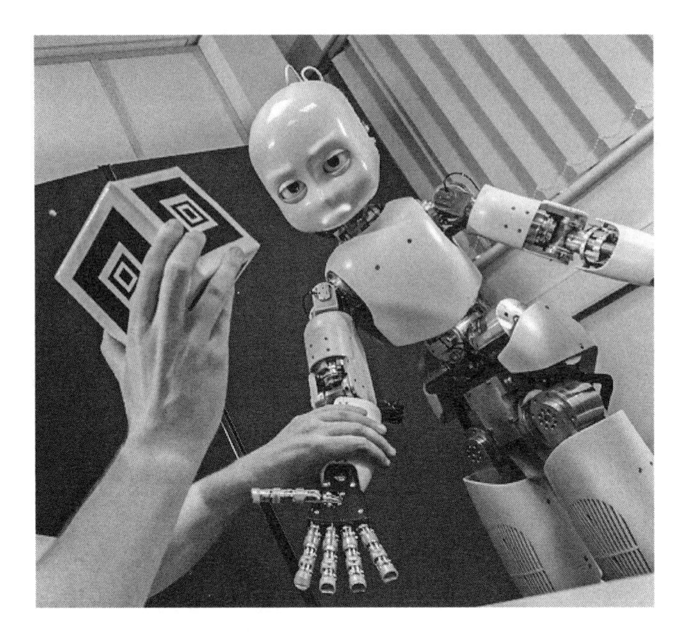

Figure 8.7
Example of the physical restraint of the robot's arm when the iCub tries to reach a forbidden object with negative valence. Figure courtesy of Frank Förster. Photo by Pete Stevens.

the current perceptual and motivation states with the previously learned lexicon, the robot utters the word that best matches the context. It then utters a second word (the next in the memory-matching algorithm) until the situation changes or the object is put down by the participant. The system is based on an efficient implementation of the k-nearest-neighbor algorithm (Tilburg Memory-Based Learner [TiMBL]; Daelemans and van den Bosch 2005) that matches the state of the SMM vector with the words in its grounded lexicon.

Two different experiments were carried out to investigate the acquisition of the earliest developmental types of negation and of the use of the word "no." One experiment used a rejection scenario, and one a prohibition scenario, and these were explicitly planned to test the two alternative hypotheses on the origins of negation. In both experiments, the naïve participants are instructed to teach the iCub robot (named "Deechee") the names of five objects, that is, different black-and-white shapes printed on boxes (star, heart, square, crescent moon, and triangle). No reference to the negation/prohibition aims was mentioned, and participants only knew that Deechee may like, dislike, or feel neutral about specific objects and this would be shown through its facial emotions.

The rejective scenario experiment was designed to test Pea's (1980) hypothesis on negative intent interpretation from rejection. This concerns the utterance of rejective words such as "no," to mean "don't want X" when the child is offered the object X that it dislikes. It also concerns motivation-dependent denial when it says "no" to a question such as "do you want X?" From a developmental timescale, this starts with children's nonverbal behavior such as turning the head away from an aversive object, or rejecting and pushing away this object. The caregiver's intent interpretation of this behavior is rejection. This leads to their production of sentences like "no" or "no, don't want it," which the children learn via imitation. To implement such a rejection scenario, the iCub robot elicits the participant's rejection interpretation with nonverbal behaviors toward a disliked object by frowning, after a brief look at the aversive object, and turning its head away. During five interaction sessions between each participant and the robot, a predefined set of object-valence values for each object were defined, with the dislike/neutral/like (-1, 0, 1) values changing between sessions.

The prohibition scenario experiment was designed to test Spitz's (1957) hypothesis on the acquisition of the negation construct from the combination of physical and linguistic prohibition. This states that the first type of negation originates from the prohibition behavior and utterances initiated by the participant. The prohibition experiment is based on the rejection scenario, with the additional constraint that two of the objects are declared to be forbidden. The participant is therefore instructed to keep the robot from touching these two prohibited objects, including physically restraining the iCub and pushing its arm back to avoid contact with the objects. This event was called the "resistance event" and caused the robot's motivation value to be set to negative. This negative emotional state is communicated by the robot to the participant through an accentuated frowning face (brighter sad mouth and eyebrow LED intensity) and longer gaze at her face.

To compare the results of the two experiments, the final two sessions of the five consecutive rejection interactions were considered the baseline (control) condition. In the five sessions of the prohibition experiment, the first three sessions implemented the rejection-plus-prohibition scenario, while the final two sessions only implemented the rejection scenarios and were used to directly compare the two groups from the rejection and prohibition experiments. These studies were based on between-subject comparisons, as each individual participant was only involved in the rejection or the prohibition scenarios.

A variety of analyses were carried out for the data within each experiment (e.g., comparison between first and last sessions) and for comparisons between the rejection and the prohibition experiments. These concerned three main dependent variables: human participants' mean length utterance (MLU), utterances per minute (u/min), and number of distinct words (#dw). In the rejection/prohibition experiments, although the MLU measures do not show any significant difference, the u/min variable shows a higher number of utterances per minute in the prohibition group as compared with the rejection group. Moreover, although the two groups start from a similar u/min values, the difference becomes significant in the later stages, demonstrating that participants in the prohibition condition talk significantly more in the last session, while there is no change in talking frequency in the rejection condition. For the #dw variable, the prohibition group produces more distinct negative words (no difference for the total #dw). Moreover, in the comparison between the two negation experiments and another related study on the learning of object nouns with the same architecture (Saunders et al. 2011), it is seen that both negation setups elicit a higher number of negative utterances. This can be explained by the use of the emotional display of facial expression and of body behavior linked to motivational states.

To use the two experiments to validate the two alternative hypotheses on the developmental origins of negation, the robot and human utterances were classified along Pea's negation types, after advanced analyses on inter-rater classifications of the robots' and human's utterances (see Förster 2013 for details). In particular, in the analysis of the human participants' utterances, the five most frequently observed negation types were considered: (1) negative intent interpretations, (2) negative motivational questions, (3) truth-functional denials, (4) prohibitions, and (5) disallowances (see Pea's negation type classification in table 8.3). Not surprisingly, the main difference between the first three sessions of the rejection and prohibition conditions is the presence of linguistic prohibition and of disallowance utterances solely in the rejection scenario, while the negative intent interpretations and negative motivational questions were observed more often in the rejection scenario. Moreover, a statistically significant low frequency of negative intent interpretations is reported in the prohibition experiment. Overall these results show that the experimental setup and the cognitive architecture used achieve the experimenter's aim to create two different scenarios for the development of negation construct in robots. The human participants produce linguistic interpretations of the robot's motivational and emotional states and produce questions and descriptions of these motivational negation states as "No, you don't like that." More interestingly, in the comparison of the last two sessions (when both sets of participants use a rejection-only interaction) the analyses show that the participants of the previous rejection sessions produce a statistically higher number of truth-functional denial utterances as compared to those from the prohibition sessions. Motivation-dependent utterances do not differ between the two experiments. Förster explains the lower level of truth-denial utterances (typically produced when the robot gives the wrong label for an object) in the prohibition experiment with the hypothesis that in the prohibition scenario the agents focus more on the motivational interactions than on neutral object-naming verbal exchanges.

The analyses of the robot's utterances suffer from the low inter-rater agreement on the classification of the robot's linguistic production. This was mostly due to the lack of agreement on the coders' classification of the robot's intention to desire, or not, a particular object. Notwithstanding this data coding issue, a systematic pattern of the production of negation utterances can be observed. In the rejection experiment, the iCub produces negation words with seven out of the ten participants, while in the prohibition scenario all interactions involve the robot's use of negation words. An analysis of the "felicity" rate coding of the robot's utterances (i.e., the successful, adequate pragmatic performance of a speech act, regardless of its truth value—see Austin 1975) can further shed light on the difference between the final two sessions of both experimental conditions. The felicity (adequacy) rate of all negative utterances is much lower in the prohibition experiment (30 percent) than in the rejection experiment (67 percent). This is surprising, as it counterintuitively suggests that the prohibition scenario inhibits the robot's production of negation speech acts. This is explained by Förster with the analysis of the temporal relations between the robot's motivational state and the participant's production of prohibition and disallowances utterances. Data show that in most cases of the prohibition scenario the robot was in a positive and neutral state when it was prohibited from touching the forbidden object. Moreover, there is a lack of temporal synchrony between the participant's pushing of the robot's arm and its utterance of negation words. In the rejection scenario, instead, negative intent interpretations and negative motivational questions were often temporally associated with negative motivational states.

Overall this pattern of results indicates that the rejection scenario is the condition that best facilitates the acquisition of negation in this human-robot interaction setup, though both hypotheses and mechanisms appear to contribute to the development of negation concepts and skills. The observed lack of synchrony between motivational states and negative utterances in the prohibition scenario does not allow the model to be used as data in support of the validation of the first hypothesis (intent interpretation) and rejection of the second hypothesis (linguistic prohibition). Moreover, Förster also suggests that this lack of temporal synchronization might also happen in real caregiver-child negation interactions. Therefore an important implication of this study is the necessity to thoroughly analyze numerous aspects of the interaction (motivational states, utterance frequency/type/length) in developmental and psycholinguistic studies of negation in human participants. This would require long video-recorded sessions of caregiver-child negation interactions in everyday settings where parental prohibition naturally occurs.

8.4 Generating Abstract Representations for Decision Making

Decision making, in the form of problem solving, heuristics, and planning, has been extensively used in the classical approach to artificial intelligence. These methods typically rely on the use of state-value vectors, heuristic functions and search strategies or logic approaches. They have been demonstrated to work very effectively, as in human-level chess reasoning and playing (Hsu 2002), although they rely on computationally intensive, and redundant, processes. Moreover, logic and symbolic-based methods and representations (Byrne 1989), and mental models methods (Bucciarelli and Johnson-Laird 1999) have been used to model human reasoning, such as syllogistic reasoning tasks typical of philosophical problems.

In a cognitive robotics approach, especially with a developmental perspective, it is important that the internal representations used for reasoning and decision making are generated by the agent through cognitive processing and generation of abstract representations, rather than using tables and heuristics generated offline by the experimenter. This is similar to the symbol grounding problem of language learning (in chapter 7), where the words need to be grounded in the agent's own interaction with the environment. In the case of decision making for cognitive robots, Gordon, Kawamura, and Wilkes (2010) have proposed a neuro-inspired approach to the grounded generation of abstract internal representations (e.g., task-based grounded dynamic features) for decision making, and have validated this in robotics experiments. Gordon and colleagues (ibid.) suggest that the result of the agent's interaction with the word generates task-based dynamic features that associate situations with task-dependent appraisals. These features can adaptively tune the control parameters for decision making. They identify three key appraisal processes involved in this task: (1) the *relevance* in the current situation, given the current goals; (2) the *utility* that should be attached to response options; and (3) how *urgently* the robot must perform the actions. These respectively correspond to the concepts of abstraction, of learning reward, and of the trading between decision time and task performance.

Gordon, Kawamura, and Wilkes developed a neurally plausible model of cognitive control in cognitive robots to improve task performance. The term "cognitive control" is used to refer to the modeling of top-down executive processes. This is based on attentional mechanisms and working memory, planning and internal rehearsal, and error correction and novelty detection. This cognitive control model also includes emotions for the robot's affective evaluations of the task. Affective states are used to evaluate the current situation and past experience with respect to relevance, urgency, and utility of the specific action that needs to be selected. As such the cognitive control model first processes experience, stored as episodic memory, to identify relational information that can be used to derive situation-based appraisals. Subsequently it represents the identified relations and appraisals for its use in online decision making.

These principles inform the design of the ISAC cognitive architecture for decision making in cognitive robotics (Gordon, Kawamura, and Wilkes 2010; Kawamura et al. 2008; figure 8.8). This architecture has three distinct control loops that provide reactive, routine, and deliberative control. It also has three types of memory systems: short-term, long-term, and working memory. The long-term memory system consists of procedural knowledge (e.g., a task described as a series of sequential steps), episodic (e.g., memory of the temporal and spatial properties of past events), and semantic

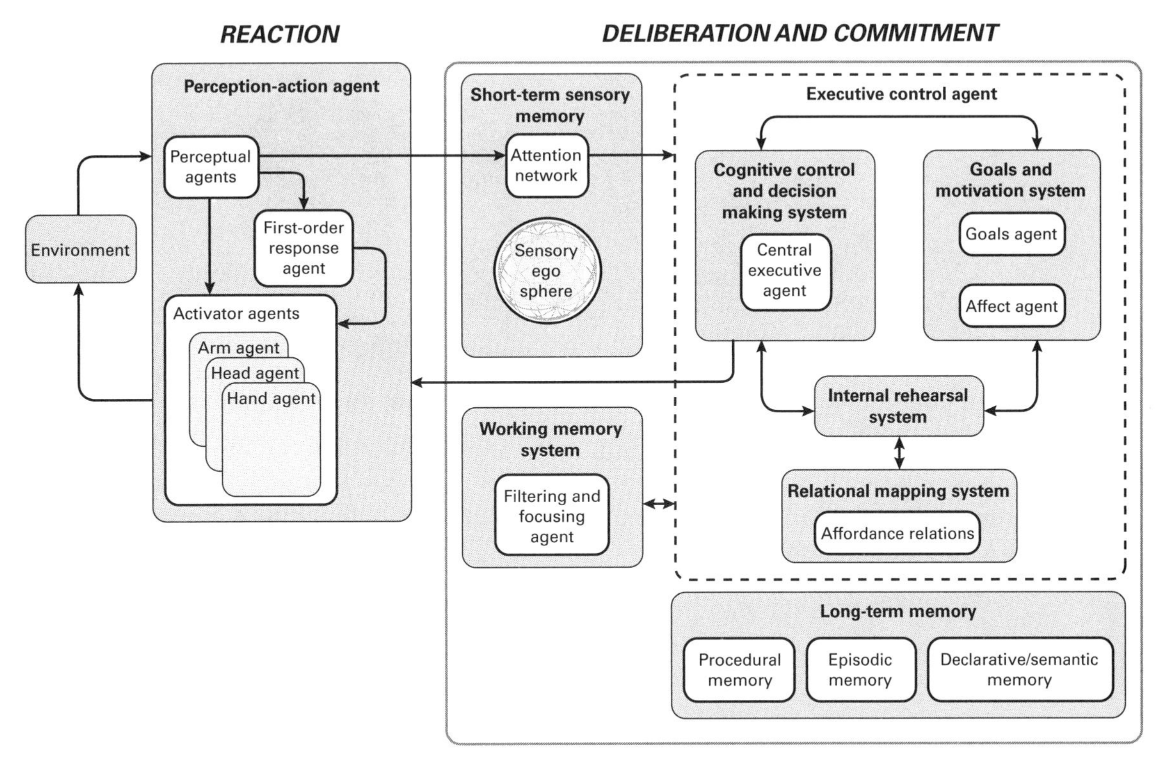

Figure 8.8
The ISAC cognitive architecture in Gordon, Kowamura, and Wilkes 2010. Figure courtesy of Stephen M. Gordon. Reprinted with permission from IEEE.

(e.g., declarative knowledge about entities and their features). The short-term and the working memory systems provide the initial buffers for perceptual and state information. In addition, the creation of goal-relevant dynamic representations (i.e., feature vectors) is fundamentally linked to relevance appraisals and is assigned to working memory. These subsystems are controlled by a complex, higher-order executive control system. This executive control agent, which contains the internal rehearsal system and the goals and motivation system, assigns goals and motivations, generates plans, and selects responses. The utility signals are used to prioritize responses, make predictions, and recursively generate plans.

Many of these components are implemented as spreading activation and neural network modules, given the neuro-inspired approach of the cognitive architecture.

To test this neuro-inspired model of decision making, Gordon, Kawamura, and Wilkes (2010) carried out experiments on the everyday task of bagging groceries with the ISAC humanoid robot (Kawamura et al. 2008). The robot is situated at the end of a

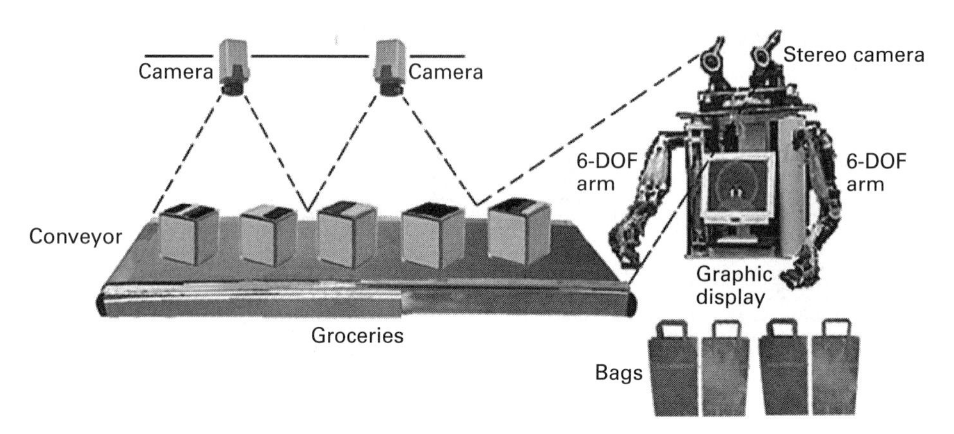

Figure 8.9
Experimental setup for the decision making experiment on grocery bagging, from Gordon, Kowamura, and Wilkes 2010. Figure courtesy of Stephen M. Gordon. Reprinted with permission from IEEE.

conveyor belt and must successfully bag all groceries that appear on the belt (figure 8.9). During one training episode, as groceries are being bagged, additional random ones are presented until a prespecified number is reached. At each step of the planning process, the robot is unaware of how many and what type of groceries may appear next, thus requiring decision making. The grocery items were twenty-five colored cardboard boxes, each with nine attributes (e.g., weight, size, soft/hard material) known by the robot. Two predetermined complex actions were stored in procedural memory, with an independent controller responsible for performing the individual steps constituting the following action:

BagGroceryLeftArm(grocery_x,bag_y) and *BagGroceryRightArm(grocery_x,bag_y)*, with the use of the various grocery items *grocery_x* and individual bags *bag_y*.

To achieve this task, three constraints were imposed and used by a critic observer: (1) prevent lighter items of food from being destroyed/crushed by heavier ones; (2) do not overload a bag with more than twenty pounds of items; and (3) minimize the number of bags used.

Experiments were carried out in simulation and with the physical robot ISAC. In each experiment, the robot goes through a series of training episodes. In each episode, a predefined number of groceries had been selected and bagged, with the number of groceries presented randomly selected from the range of ten to fifteen items. Retraining was performed every ten episodes using all episodes encountered up to that point. The robot is required to learn multiple concepts in parallel and to apply the learned knowledge to its ever-increasing corpus of experience. Thus the agent is evaluated on whether its performance improves, worsens, or stagnates over time.

The results reported in Gordon, Kawamura, and Wilkes (2010) indicate that neurally inspired decision making, based on the parallel learning and evaluation of multiple events, supports the design of effective and generalizable decision-making capabilities. The ISAC humanoid robot endowed with these was able to learn the grocery bagging task by learning its own appraisal of what was relevant for that task, how utility should be associated with different situations, and the urgency that should be attached to each situation. Although this learning decision-making system does not explicitly address the developmental stages of the acquisition of decision-making skills, it provides a testable cognitive control architecture for the future investigation of how the various memory, affective, and control modules develop to reach adultlike reasoning performance.

8.5 Developmental Cognitive Architectures

The preceding sections review specific experiments and modeling approaches to the study of specific knowledge-processing skills in developmental robots. These models typically focus on a single, specialized abstract knowledge capability, such as numbers, abstract words, or decision making. However, an important field of cognitive modeling concerns the development of general cognitive architectures, which can be employed to simulate and integrate a variety of behavioral and cognitive tasks using the same cognitive mechanisms.

The term *cognitive architecture*, originally proposed in the field of computational cognitive modeling and artificial intelligence, refers to a broadly scoped, general-purpose computational model that captures the essential structure and process of the mind. As such, it can be employed for wide-ranging, multiple-level, multiple-domain modeling and simulation of behavior and cognitive phenomena, rather than focusing on separated, individual skills (Sun 2007). Such an integrated modeling framework and the simultaneous use of the same processes to simulate a variety of cognitive capabilities supports the attempt to unify many findings into a single theoretical framework, which can then be subject to further testing and validation (Langley, Laird, and Rogers 2009). Such a type of architecture would typically include a knowledge representation system, different memory stores (long-term and short term), knowledge manipulation processes, and possibly learning methods. Langley and colleagues (ibid.) identify nine functional capabilities for a prototypical cognitive architecture (table 8.4). These cover the whole, broad spectrum of sensorimotor and cognitive capabilities.

Vernon and colleagues (Vernon, Metta, and Sandini 2007; Vernon, von Hofsten, and Fadiga 2010) analyze three paradigms for cognitive modeling, with a related set of associated general cognitive architectures: (1) cognitivist; (2) emergentist; and (3)

Table 8.4
Core functional capabilities of a general cognitive architecture, as proposed by Langley, Laird, and Rogers 2009

Functional capability	Main features
Recognition and categorization	Recognition processes to identify if a situation matches a stored pattern Categorization to assign objects, situations, and events to known concepts
Decision making and choice	Representation of alternative choices or actions Process for selecting among alternatives Process to determine if an action is allowable and matches the precondition/context
Perception and situation assessment	Perception (visual, auditory, tactile) to sense the external world Knowledge about own sensor dynamics and configuration Selective attention system to allocate and direct its limited perceptual resources to relevant information Interpretation of whole situation to construct a large-scale model of the current environment
Prediction and monitoring	Ability to predict future situations and events accurately Needs a model of the environment and the effect actions have on it Monitoring of the situation and changes over time
Problem solving and planning	Internal representation of a plan as an ordered set of actions and their effects Construct a plan from components available in memory Problem solving as the multistep construction of a problem solution via search strategies Mixture of internal planning and external behavior Learning to improve future problem solving Modification of existing plans in response to unanticipated changes
Reasoning and belief maintenance	Reasoning as capability to draw mental conclusions from other existing beliefs or assumptions Representation of (logical or probabilistic) relationships and structure among beliefs Mechanisms to draw inferences using knowledge structures (deductive and inductive inferences) Belief maintenance mechanisms to update representations
Execution and action	Representation and storage of motor skills for action Execution of simple and complex skills and actions in the environment From fully reactive, closed-loop behavior to automatized, open-loop behavior Learning and revision of skills
Interaction and communication	Communication system as natural language (and nonverbal communication) Mechanisms for transforming knowledge into the form and medium of communication Conversational dialog and pragmatic coordination between agents
Remembering, reflection, and learning	Meta-management mechanisms for all the above capabilities Remembering as the capability to encode and store, retrieve and access the results of cognitive processing in memory (episodic memory) Reflection on knowledge as explanation of own inferences, plans, decisions, or actions Reflection as meta-reasoning about the processes used for the generation of inferences and the making of plans Learning for generalization of new skills/representations beyond specific beliefs and events

hybrid. The cognitivist paradigm, also known as the symbolic paradigm, is based on the classical view of artificial intelligence (GOFAI: Good Old-Fashioned Artificial Intelligence) and on the information processing approach in psychology. Such a paradigm states that cognition fundamentally consists of symbolic knowledge representations and of symbol manipulation phenomena. Computationally, this approach is based on formal logic and rule-based representation systems and, more recently, on statistical machine learning and probabilistic Bayesian modeling methods. Prototypical cognitivist architectures are Soar (Laird, Newell, and Rosenbloom 1987), ACT-R (Anderson et al. 2004), ICARUS (Langley and Choi 2006), and GLAIR (Shapiro and Ismail 2003). In this field, a cognitive architecture is also proposed as a unified theory of cognition, as it tries to capture the common fundamental (symbolic) processes underpinning all aspects of behavioral and cognitive capabilities. Given the strong focus on symbol representation and manipulation, these architectures mostly do not address key phenomena linked to developmental robotics as embodiment and learning and development. These architectures propose symbolic (nongrounded) representation of perceptual and action concepts, with minimal or no learning processes, except for the implementation of processes to learn and develop new models in ICARUS.

The emergentist paradigm, on the contrary, views cognition as the resulting, emergent self-organizing phenomena resulting from the interaction of the individual with its physical environment and social world. This historically started as a reaction and alternative to the classical symbolic cognitivist models, replacing the von Neumann information processing view of cognition with parallel, decentralized cognitive processing frameworks (Bates and Elman 1993). The emergentist approach is actually based on a varying family of paradigms, which include connectionist neural network models, dynamical systems, embodied cognition, and evolutionary and adaptive behavior systems. Such paradigms mostly provide general theories and principles of cognitive processing, rather than classical cognitive architectures. For example, the connectionist paradigm proposes the use of artificial neural networks as general distributed and parallel processing systems capable of modeling the wide variety of cognitive skills, from sensorimotor learning, to categorization, language learning, and even implementing rule-based systems. Given its emergentist nature, such paradigms pose a strong focus on learning, on decentralized, subsymbolic, and sensorimotor representation and on adaptation.

Hybrid paradigms try to integrate mechanisms from both cognitivist and emergentist approaches. For example, the CLARION architecture (Sun, Merrill, and Peterson 2001) combines emergentist neural network modules to represent implicit knowledge and symbolic production rules to model explicit symbol manipulation processes.

Cognitive architectures are increasingly being proposed and developed within the field of cognitive systems and robotics, though with only a limited extension to developmental robotics. Vernon and colleagues (Vernon, Metta, and Sandini 2007; Vernon, von Hofsten, and Fadiga 2010) proposed a classification of general cognitive architectures for artificial cognitive systems along the cognitivist/emergentist/hybrid dimension. Table 8.5 extends this original classification with the addition of more recent architectures with a key focus on development, including the iCub architecture by Vernon, von Hofsten, and Fadiga 2010 and other architectures presented in this volume (e.g., ERA, LESA, HAMMER).

The first set of architectures (top row of table 8.5) lists architectures not directly addressing either general robotics or developmental robotics modeling (the cognitivist architectures Soar, ACT-R, ICARUS, the emergentist Self-Directed Anticipative Learning (SDAL) architecture, and the hybrid CLARION). Although these have had a significant impact in numerous areas of cognitive modeling, their impact in robotics has been minimal to date. This is mostly due to the fact that these architectures, especially the cognitivist ones, are specialized for performing sequential search operations, not always suitable to handle concurrent and distributed information processes in robots' sensors and actuators. Moreover, cognitivist architectures put a great emphasis (and computational overhead) on the high-level reasoning and inference mechanisms, and less importance on the low-level sensorimotor capabilities, which are essential in any robotics model.

For the cognitivist architectures that have more specifically been proposed for robotics, but not directly addressing developmental issues, the ADAPT architecture (Adaptive Dynamics and Active Perception for Thought) (Benjamin, Lyons, and Lonsdale 2004) takes a strong embodiment stance and combines reasoning features from Soar and ACT-R with short-term working memory of sensory data to handle task goals and actions. GLAIR (Grounded Layered Architecture with Integrated Reasoning) (Shapiro and Ismail 2003) is an architecture that integrates a high-level symbolic Knowledge Layer with a mid-level Perceptuo-Motor Layer and the lower-level Sensori-Actuator Level. The CoSy Architecture Schema (Hawes and Wyatt 2008) focuses on the organization of information processing modules (subarchitectures) into specific schemas for robot assistive tasks, and is capable of modeling local and global goals and motivation.

Among the emergentist architectures for general robotics studies, one of the most influential approaches has been the Subsumption Architecture (Brooks 1986), based on behavior-based robotics. This approach (also called Autonomous Agent Robotics in Vernon, von Hofsten, and Fadiga 2010) consists of hierarchical layers (each subsuming the lower one) of competing behaviors. The Global Workspace framework (Shanahan 2006) is based on a connectionist architecture that implements internal simulations of agent-environment simulations. The Darwin architecture (Krichmar and Edelman 2005) is strongly modeled on a brain-inspired structure and organization of the brain, where behavior results from the interaction of neural mechanisms with sensorimotor contingencies. This, as the Global Workspace model, is strongly focused on learning. Moreover, the DAC architecture (Distributed Adaptive Control; Verschure, Voegtlin,

Table 8.5
Classification of cognitive architectures used in cognitive modeling, cognitive robotics, and developmental robotics

	Cognitivist	Emergentist	Hybrid
General	Soar (Laird, Newell, and Rosenbloom 1987) ACT-R (Anderson et al. 2004) ICARUS (Langley and Choi 2006)	SDAL (Christensen and Hooker 2000)	CLARION (Sun, Merrill, and Peterson 2001)
Robotics	ADAPT (Benjamin, Lyons, and Lonsdale 2004) GLAIR (Shapiro and Ismail 2003) CoSy (Hawes and Wyatt 2008)	Subsumption (Brooks 1986) Darwin (Krichmar and Edelman 2005) Cognitive-Affective Schema (Morse, Lowe, and Ziemke 2008) Global Workspace (Shanahan 2006) DAC (Verschure, Voegtlin, and Douglas 2003) TRoPICALS (Caligiore et al. 2010)	HUMANOID (Burghart et al. 2005) Cerebus (Horswill 2002) Kismet (Breazeal 2003) LIDA (Franklin et al. 2014) PACO-PLUS (Kraft et al. 2008)
Developmental robotics	ISAC (Gordon, Kawamura, and Wilkes 2010)	iCub (Vernon, von Hofsten, and Fadiga 2010) ERA (Morse, de Greeff, et al. 2010) LESA (Lyon, Nehaniv, and Saunders 2012) Shared Gaze (Nagai et al. 2003) HAMMER (Demiris and Meltzoff 2008) Cooperation (Dominey and Warneken 2011) SASE (Weng 2004) MDB (Bellas et al. 2010)	Cognitive theory of mind (Scassellati 2002)

and Douglas 2003) is based on a neural implementation of learning mechanisms organized around three hierarchical levels: Reactive Layer (for reflexes and autonomic control), Adaptive Layer (stimulus/action association and classical conditioning) and Contextual Layer (planning and operating conditioning). The Cognitive-Affective Schema (Morse, Lowe, and Ziemke 2008) also uses an emergentist approach based on autonomy-preserving self-maintenance homeostasis processes.

A variety of paradigms have been proposed that combine symbolic and emergentist methods for general cognitive robotics, as with the HUMANOID (Burghart et al. 2005), Cerebus (Horswill 2002), LIDA (Franklin et al. 2014), and PACO-PLUS (Kraft et al. 2008) cognitive architectures. Moreover, the cognitive architecture proposed to control the Kismet social robot (Breazeal 2003) puts emphasis on the role of drives, affective appraisal, and emotion for expressive behavior in human-robot interaction.

As for the cognitive architectures specifically proposed to model developmental robotics phenomena, we have seen a few examples in this volume. Apart from the symbolic ISAC architecture used by Gordon, Kowamura, and Wilkes (2010) in the decision-making experiments (see section 8.4), most use an emergentist and hybrid approach. In chapter 6 on social development we saw examples of strong emergentist cognitive modeling architectures: the cognitive control architecture for Shared Gaze in the Nagai et al. (2003) model (see section 6.2), the HAMMER model for imitation learning (Demiris and Meltzoff 2008—section 6.3), the cognitive architecture for Cooperation (Dominey and Warneken 2011—section 6.4), and Scassellatti's (2002) theory of mind architecture for the COG robot (section 6.5). Chapter 7 on language development included the presentation of the LESA architecture for phoneme and word acquisition (Lyon, Nehaniv, and Saunders 2012—section 7.2; also used in the negation experiments discussed earlier) and the ERA architecture (Morse, de Greeff, et al. 2010) for word learning experiments (section 7.3). Additional developmental-inspired cognitive architectures include SASE (Weng 2004) and Multilevel Darwinist Brain (MDB; Bellas et al. 2010).

An additional general cognitive architecture not directly discussed in the previous chapters, but that has had an important influence in the field of developmental robotics, is the one presented by Vernon, von Hofsten, and Fadiga 2010) for the iCub robot. They also offer a research roadmap and a set of guidelines for the evaluation of cognitive architectures in developmental robotics. This architecture is based on the initial work done on the iCub robot, though it has a general structure that can be applied to other baby robots. The cognitive architecture comprises twelve modules for various perceptual, motor, and cognitive capabilities (figure 8.10). The architecture also includes the additional iCub Interface module, which is the software middleware system, based on the YARP protocol, to control the physical robot (see section 2.5.1).

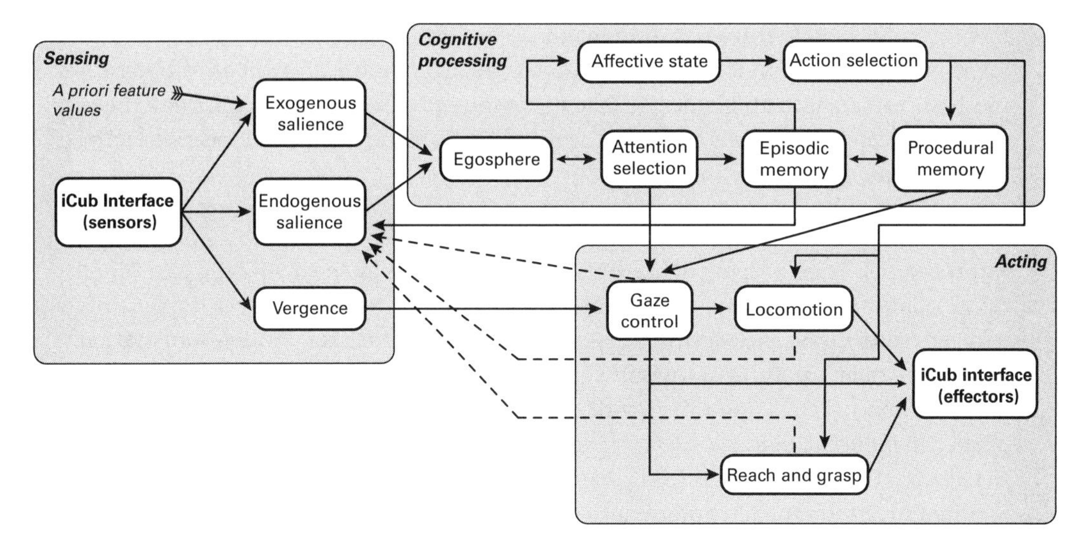

Figure 8.10
The iCub cognitive architecture (adapted from Vernon, von Hofsten, and Fadiga 2010).

The perceptual-cognitive functions are organized around these five modules:

• *Exogenous salience*. The exogenous nature of this perceptual module refers to the role of external visual and auditory stimuli. This model includes a multimodal saliency-based bottom-up visual attention system (Ruesch et al. 2008). An image from the robot's camera is processed to extract features such as intensity, double-opponent color, and local orientation (at several scales), to perform center-surround filtering for local contrast, and produce a normalized feature map. For the auditory salience processing, the origin of a sound source is calculated through inter-aural spectral and time differences.
• *Endogenous salience*: This module refers to the dependence of attention on internal states. The endogenous salience module identifies the location of the most salient point in the egosphere through a simple color segmentation algorithm for the fovea region in log-polar images.
• *Egosphere*: This is the result of the combination of the multimodal visual and acoustic saliency maps into a continuous robot-centered spherical surface. This converts an external 3D map into an egosphere representation centered at the iCub's body. The egosphere also serves the role of short-term memory for salient regions to allow the iCub a more natural shifting of attention from current salient region to previous salient points.
• *Attention selection*: This module is used to identify the final salience map in the combined egosphere representation using a combination of winner-take-all and inhibition

mechanisms. A winner-take-all non-maxima suppression process is implemented to identify the most salient location in the egosphere. In addition, an inhibition-of-return mechanism is used to attenuate the saliency of the egosphere locations that have been attended to previously, thus allowing the robot to explore novel salient locations in its environment.

The motor cognition functions use four modules:

• *Gaze control.* This component controls the head and eye by specifying the gaze direction (azimuth and elevation angles) and the vergence of the two eyes. It implements two modes of head gaze control: (1) saccadic motion, first for fast eye orientation toward the salient point and the subsequent neck rotation and eye counterrotation to keep the image stable, and (2) smooth pursuit motion for continuous slow following of the salient point/object.
• *Vergence.* This module uses binocular vision to measure the horizontal translation required to register the left and right images and keep the central areas aligned. It outputs the disparity of the selected region to the Gaze control module for head and eye movement initiation.
• *Reach and grasp.* The implementation of this module include a task-space reaching controller, to compute the arm and torso configuration of the desired pose using a nonlinear optimizer and human-like trajectory generation, and a grasping component that learns visual descriptions of grasping points from experience.
• *Locomotion.* As the iCub was originally designed for crawling on hands and knees, this module implements a dynamical system approach (see section 5.4) to switch between four stable behaviors: (1) crawling, (2) transition from sitting to crawling, (3) transition from crawling to sitting, and (4) reaching for a target object. This component is currently being extended to allow bipedal walking.

The anticipation and adaptation components are used to accomplish both anticipation and adaptation via the episodic and the procedural memory modules.

• *Episodic memory.* This component uses a visual memory of autobiographical events expressed as a set of saved log-polar images of seen objects (visual landmark).
• *Procedural memory.* This is implemented as a set of associated perception-action-perception triplets. During learning, the robot associates a temporally ordered pair of images and the action performed for the transition between the first and the second image. During a recall session, either of the images will recall the action associated with it and the corresponding second image. The combined episodic and procedural memory systems allow the realization of a simple form of mental simulation of sensorimotor states.

The architecture also includes two modules for motivational states and for autonomous action selection.

- *Affective state.* This module receives inputs from the Episodic memory module, for previous and current active events and images, and from the Action selection component for current selected action. The Affective state is implemented as a competitive network of three motivational states: (1) curiosity, which is governed by exogenous factors, (2) experimentation, dominated by endogenous factors, and (3) social engagement, for a balanced set of exogenous and endogenous factors.
- *Action selection.* This module selects the next state among curiosity, experimentation, and social motivational states discussed previously. When the curiosity state is selected, the robot is in a learning mode, while the experimentation state activates the prediction mode.

The architecture described in Vernon, von Hofsten, and Fadiga 2010 has been significantly extended since 2010, given the growing community of researchers working on the iCub. These for example include new manipulation strategies (see sections 5.2 and 5.3 on iCub reaching and grasping developmental models) and new sound and speech processing functionalities (for example sections 7.2 and 7.3 on word learning experiments).

With respect to the long tradition and wide use of cognitivist architectures, research on emergentist and hybrid robotics architectures is still in its infancy. With the exception of a few examples of cognitive modeling frameworks comprehensively covering the whole spectrum of motivation, affective, sensorimotor, and reasoning capabilities (as in the iCub architecture described earlier), most cognitive architectures specialize in a subset of cognitive capabilities. For example, the ERA architecture integrates visual and motor skills with simple linguistic behavior, but does not include mechanisms to address motivation and affective behavior, or reasoning tasks. Morse, de Greeff, et al. (2010) have proposed a hierarchical implementation of the ERA architecture with the potential to address higher-order cognitive and reasoning skills, though this has not been fully implemented in robotic reasoning and decision-making scenarios. However, the ongoing efforts to test the emergentist and hybrid architectures in a variety of increasingly complex robot experiments can lead to the definition of a comprehensive cognitive modeling paradigm suited to the modeling of online, open-ended learning in developmental robotics.

8.6 Conclusions

The developmental robotics models discussed in this chapter constitute some of the few experimental attempts to model the development of abstract knowledge in autonomous robots. Interestingly, they address quite different aspects of reasoning with abstract knowledge, ranging from the learning of numerical concepts, to the grounding of abstract words, and the generation of abstract representation for decision-making tasks.

These studies provide sound evidence of the benefits of using the embodied and situated approach of cognitive robotics to model higher-order cognition. For example, in Rucinski, Cangelosi, and Belpaeme's 2011 experiments on the acquisition of number concepts, it has been shown how abstract numerical representations directly rely on the spatial body schema previously developed by the robot during motor babbling. This results in the replication of human data on embodiment effects such as the size, distance, and SNARC effects. The model shows that the learned relationship between abstract concepts, such as the relative distance between numbers, is grounded in sensorimotor representation of the distance between positions of the arms. The negation experiments by Förster, Nehaniv, and Saunders (2011) explicitly address the modeling of different types of negation constructs by relying on the robot's own conceptual, affective, and social representations. Again, abstract representations linked to the word "no" are grounded on situated and embodied experience. Finally, the Gordon, Kawamura, and Wilkes (2010) model of decision making in the ISAC humanoid platform shows that the robot can develop autonomous decision-making strategies by generating abstract representations from the practical training experience of grocery packing tasks. This is achieved though the robot's own appraisal of the task relevance, utility, and urgency, within a neuro-inspired learning architecture.

The relatively low number of robotics models of abstract knowledge, and the limited complexity of the reasoning capabilities currently modeled, illustrates some of the open challenges in this field. An example is the challenge to use embodiment approaches to model the robot's acquisition of highly abstract concepts, which has no evident and direct link to sensorimotor representations. In fact, the grounding of abstract concepts, such as *democracy* and *freedom*, has historically been one of the main criticisms of embodied cognition (Barsalou 2008). Another key issue in the embodied modeling of reasoning and abstract knowledge acquisition is the development of various forms of reasoning strategies such as logic, inductive and deductive reasoning, and insightful problem solving. Moreover, although the robotics studies discussed in this chapter take direct inspiration from empirical evidence and theories on the developmental acquisition of numerical and abstract representations, they do not explicitly address key developmental science hypotheses. Both Rucinski's and Förster's studies try to model the sequential and incremental learning of various skills (e.g., motor babbling before number learning), however they do not aim at a faithful modeling of the various stages of number cognition development, as shown in section 8.1.1. This is even more evident in Gordon's study on decision making, with very loose inspiration from developmental work.

These pioneering studies, and a closer look at the developmental psychology literature, do provide some important methodological and scientific tools for future progress on the modeling of abstract reasoning in robots. Specific areas for future developmental robotics models include:

- Modeling of gesture, touch, and finger use in the acquisition of counting and number concepts (Fischer 2008; Andres, Seron, and Olivier 2007; Gelman and Tucker 1975)
- Investigation of the role of emotions and modes of acquisition along the concrete/abstract continuum (Della Rosa et al. 2010; Borghi et al. 2011; Kostas et al. 2011)
- Role of internal simulations for concept combination and the grounding of abstract concepts (Barsalou 2008; Barsalou and Wiemer-Hastings 2005)
- Grounding the acquisition of function words, extending the scenario for the development of negation (Förster, Nehaniv, and Saunders 2011)
- Explicit modeling of Piaget's concrete and formal operational reasoning stages (Piaget 1972; Parisi and Schlesinger 2002)

Increased and focused efforts on these areas of research can lead to a better scientific understanding of how sensorimotor knowledge bootstraps cognition, and to technological advances in the autonomous development of abstract reasoning skills in robots. Moreover, the growing development and use of cognitive architectures within cognitive and developmental robotics contribute to the modeling of the integration of numerous sensorimotor and cognitive skills in robots. This is also in line with the developmental robotics principle of modeling online, cross-modal, continuous, open-ended learning, which can lead to cognitive bootstrapping and the emergence of higher-level, abstract knowledge manipulation skills (see section 1.3.6).

Additional Reading

Lakoff, G., and R. E. Núñez. *Where Mathematics Comes From: How the Embodied Mind Brings Mathematics into Being.* New York: Basic Books, 2000.

This is a book on the link between embodied cognition and abstract mathematical concepts. The key tenet is that conceptual metaphors play a central role in the development and use of mathematical ideas, concepts, and skills (including our embodied understanding of the mathematical concept of infinity). It draws on evidence from developmental psychology, numerical learning in animals, cognitive psychology, and neuroscience to provide a rich philosophical analysis of mathematical cognition.

Langley, P., J. E. Laird, and S. Rogers. "Cognitive Architectures: Research Issues and Challenges." *Cognitive Systems Research* 10 (2) (June 2009): 141–160.

This article provides a systematic analysis of (1) the core functional capabilities in a cognitive architecture, (2) their main properties for knowledge management, (3) the evaluation criteria, and (4) the open research issues. It also includes an appendix with a succinct but comprehensive summary of the main cognitive architectures used in cognitive modeling. Although the article has a general approach bias toward cognitivist cognitive architectures, it addresses many important points relevant to emergentist and developmental architectures.

Vernon, D., C. von Hofsten, and L. Fadiga. *A Roadmap for Cognitive Development in Humanoid Robots*. Cognitive Systems Monographs (COSMOS), vol. 11. Berlin: Springer-Verlag Berlin, 2010.

This book provides a comprehensive review of the iCub cognitive architecture. It starts from a review of developmental psychology and neuroscience findings on sensorimotor and cognitive development, to introduce a research roadmap for developmental robotics. This roadmap and its research guidelines inform the design of the iCub cognitive architecture. The volume also discusses the classification of cognitive architectures into the cognitivist/emergentist/hybrid dimensions, putting a special emphasis on the emergentist paradigms for cognitive systems and robotics, within which the iCub architecture is framed.

9 Conclusions

This volume started with the introduction of the main theoretical principles underpinning developmental robotics. These included the view of development as a self-organizing dynamical system; the integration of phylogenetic, ontogenetic, and maturation phenomena; the strong emphasis on embodied, grounded, and situated development; the focus on intrinsic motivation and social learning; the nonlinearity of qualitative changes in development; and the importance of modeling online, open-ended cumulative learning. All these posit fundamental questions on the nature of development in both natural and artificial cognitive systems.

These principles have inspired past and current developmental robotics models, though with a varying degree of involvement and a primary focus on one or few of these design principles. Only a few studies have attempted to take into consideration all of these principles simultaneously and to embed their features in the developmental computational models and architectures. This is for example the case of the studies based on the iCub emergentist cognitive architecture (Vernon, van Hofsten, and Fadiga 2010) that we reviewed in chapter 8. Such an architecture included principles from self-organization and emergence, intrinsic motivation, social learning, embodied and enactive cognition, and online, cumulative learning. The architecture, however, only indirectly addresses the interaction among phylogenetic, ontogenetic and maturational phenomena.

Most of the existing developmental robotics models only focus on some of these principles, providing a clear operationalization of only one or few of these principles in specific developmental mechanisms. In this final chapter we will first reflect on the progress and achievements reached to date on each of the six principles presented in chapter 1. We will then look at other general achievements related to methodological and technological progress. This analysis will inform the identification and discussion of the open scientific and technological challenges for future research in developmental robotics.

9.1 Main Achievements on the Key Developmental Robotics Principles

9.1.1 Development as a Dynamical System

The dynamical systems approach proposed by Thelen and Smith (1994) sees development as multicausal change within a complex dynamic system. The growing child can generate novel self-organizing behaviors through its interaction with the environment, and these behavioral states vary in their stability within the complex system. Such a principle is based on the mechanism of decentralized control, self-organization and emergence, multicausality, and nested timescales.

One study directly implementing and testing the dynamical systems hypothesis is the model of fetus and neonatal simulations by Kuniyoshi and Sangawa (2006). This model specifically instigates the hypothesis that partially ordered dynamical patterns emerge from the chaotic exploration of body-brain-environment interactions during gestation. This hypothesis is based on the observation of the spontaneous "general movements" reported in the two-month-old (eight to ten weeks) human fetus, and their role in the emergence of meaningful infant motor behaviors such as rolling over and crawling-like movements. The fetus's neural architecture consists of one Central Pattern Generator (CPG) per muscle. These dynamic CPGs are functionally coupled via the body's interaction with the physical environment and the muscle's response, rather than through direct connections. The behaviorally coupled CPGs produce periodic activity patterns for constant input, and chaotic patterns for nonuniform sensory input.

A similar dynamical system approach based on CPGs has been employed in various studies on the development of motor and locomotion skills (section 5.1.4). Righetti and Ijspeert (2006a; 2006b) used CPGs with the iCub to demonstrate the development of crawling strategies that correspond to the gait pattern produced by a human infant. Such a dynamical approach leads to the acquisition of the ability to combine periodic crawling movement with shorter, ballistic actions, such as reach-like hand movements (section 5.4). Li et al. (2011) extended this dynamic CPG approach to model crawling locomotion abilities in the NAO robot, Taga (2006) used similar CPG mechanisms to account for the stages of walking, by freezing and freeing DOFs, and Lungarella and Berthouze (2004) investigated the bouncing behavior. All these studies show the contribution of CPGs in modeling dynamical systems phenomena in development.

9.1.2 Phylogenetic and Ontogenetic Interaction

Progress on the modeling of the interaction of evolutionary (phylogenetic) and developmental (ontogenetic) phenomena in robotics has been limited, as most cognitive robotics models either focus on the evolutionary or the developmental learning timescale. For example, numerous models exist in evolutionary robotics (Nolfi and Floreano 2000), which investigate phylogenetic mechanisms and the evolutionary emergence of sensorimotor and cognitive capabilities. These tend to be separated from the developmental robotics models, which only emphasize the ontogenetic acquisition of behavioral and cognitive skills. However, some studies based on evolutionary robotics methods have been proposed that explore developmental issues. This is the case, for example, of the models of intrinsic motivation of Schembri, Mirolli, and Baldassarre (2007) on the evolution of a reinforcement-learning intrinsic reward mechanism in child and adult robotic agents. A set of evolutionary models have also been proposed that look at cultural-evolution phenomena, as with robotics and multiagent models of language origins (Cangelosi and Parisi 2002; Steels 2012). In this volume we analyzed Oudeyer's cultural evolution model of the emergence of shared phonetics repertoires.

Other pioneering work on phylogeny/ontogeny interaction phenomena, such as the Baldwin effects and heterochronic changes (e.g., Hinton and Nolan 1987; Cangelosi 1999), looks at general learning/evolution interaction, but not at developmental issues.

Some progress, instead, has been achieved in the field of modeling maturation changes in development. One key model that addresses both ontogenetic and maturational changes is the work by Kuniyoshi and colleagues on fetus and neonatal robots. The first model developed by Kuniyoshi and Sangawa (2006) consists of a minimally simple body model of fetal and neonatal development. The subsequent model by Mori and Kuniyoshi (2010) provides a more realistic rendering of the fetus's sensorimotor structure. Both models offer a useful developmental robotics research tool to investigate pre-birth sensorimotor development, as they are based on a realistic representation of the fetus's sensors (1,542 tactile sensors!) and actuators, and the reaction of the body to gravity and to the womb environment. The first model has one main parameter related to the developmental timescale, the gestational age that distinguishes the agent between a thirty-five-week old embryo and a 0-day old newborn.

Other developmental robotics models that implement some kind of maturational mechanism include the Schlesinger, Amso, and Johnson (2007) model of object perception, which provides an alternative explanation to Johnson's (1990) hypothesis on the role of cortical maturation of the frontal eye field for the acquisition of visual attention skills in early infancy.

9.1.3 Embodied and Situated Development

This is an area where significant achievements have been demonstrated in a variety of developmental robotics models and cognitive areas. Due to the intrinsically embodied nature of robotics, the validation of theories through experiments with robotic agents and their specific sensorimotor apparatus, it is natural that the great majority of developmental models emphasize the role of physical, embodied interaction with the environment.

Evidence of the role of embodiment in the very early (prenatal) stages of development comes from the fetus and neonatal model of Mori and Kuniyoshi (2010). Through their simulation model of fetuses and neonatal robots, they test the embodiment hypothesis that tactile sensation induces motions in the fetus. Using a fetus/infant body model with a humanlike distribution of tactile sensors, they investigate the development of the two types of reactive movements following the embryo's general movements, in other words, the isolated arm/leg movements (i.e., jerky movements independent from other body parts observed in human fetuses from around ten weeks of gestation) and hand/face contacts (i.e., when the fetus's hands touch the face slowly, observed from the eleventh gestation week).

In the developmental robotics models of motor development, the simulation of reaching behavior in agents by Schlesinger, Parisi, and Langer (2000) demonstrated that the freezing strategy (i.e., the solving of redundant DOFs by locking [freezing] the shoulder joint and reaching by rotating the body axis and elbow joint) does not need to be programmed into the model. On the contrary, it can emerge for free as a result of the learning process, and as a mechanism of morphological computation coming from the coupling of the body's properties and the environmental constraints.

Developmental models of language learning also show the importance of the embodiment biases in early word learning. The Modi Experiment by Morse, Belpaeme, et al. (2010) provides a demonstration of the embodiment bases of language learning. Smith (2005) observed that children use their body relationship with the objects (e.g., spatial location, shape of the object) to learn new object-word associations and expand their lexicon. In the ERA cognitive architecture used by Morse to model Smith's embodiment bias (see boxes 7.1 and 7.2) the robot's neural controller uses information on the body posture as a "hub" connecting information from other sensory streams and modalities. This hub allows for the spreading of activation and the priming of information across modalities, thus facilitating the embodied acquisition of the names of objects and actions.

In the field of abstract knowledge, evidence for the role of embodiment has been reported both for numerical cognition and for abstract word experiments. Rucinski, Cangelosi, and Belpaeme (2011; 2012) has proposed two developmental models of number embodiment. One model simulates the SNARC effect, the space-number association that links the number cardinality with a left/right spatial map (see section 8.2 and box 8.2). The other model investigates the role of gestures in learning to count (Rucinski, Cangelosi, and Belpaeme 2012), when the iCub is trained to count by either looking at objects being counted (visual condition) or pointing at objects while counting them (gesture condition). The comparison of the visual vs. gesture performance shows an advantage of the gesture condition in the counting ability, mirroring the developmental psychology data (Alibali and DiRusso 1999) on gestures and counting experiments. As for the role of embodiment in the acquisition of words with an abstract meaning component, the two models by Cangelosi and Riga (2006) and by Stramandinoli, Marocco, and Cangelosi (2012) explore the role of the symbol-grounding transfer mechanism in the continuum between concrete and abstract concepts. In the first model, the robot is taught higher-order complex actions via linguistic demonstrations, where the grounding transfer mechanism combines a lower-level action concept with higher-level actions such as "grab" and "carry. In subsequent work by Stramandinoli et al. (Ibid.), these complex concepts gradually become less concrete, with the learning of more abstract words such as "accept," "use" and "make."

All these examples of the strict coupling between embodiment and various other cognitive skills show the importance of grounded and situated learning in development.

9.1.4 Intrinsic Motivation and Social Learning

One other area of developmental robotics that has achieved important results is that of intrinsic motivation and social learning.

In chapter 4 we reviewed a wide set of studies that highlight the progress on the modeling of intrinsic motivation and artificial curiosity. Such studies show that intrinsically motivated robots are not specialized for solving a particular problem or task, but rather are capable of focusing on the process of learning itself and on artificial curiosity leading to exploring the environment. A major result in developmental robotics has been the establishment of a taxonomy of algorithms for modeling intrinsic motivation along the two classes of knowledge-based and competence-based frameworks (Oudeyer and Kaplan 2007; section 3.1.2). The knowledge-based view focuses on the properties of the environment, and how the organism gradually comes to know and understand these properties, objects, and events. Such a view includes novelty-based intrinsic motivation (i.e., when novel situations produce a mismatch or incongruity between an ongoing experience and stored knowledge) and prediction-based motivation (i.e., when the organism implicitly predicts how objects or events will respond to its actions). The competence-based view of intrinsic motivation focuses instead on the organism and the particular abilities or skills it possesses. Competence-based motivation promotes skill development by leading the agent to seek out challenging experiences and to discover what it can do. An example of this is Piaget's functional assimilation mechanism, that is, the tendency for infants and young children to systematically practice or repeat a newly emerging skill.

A significant achievement in the modeling of intrinsic motivation has come from the use of reinforcement learning (Sutton and Barto 1998; Oudeyer and Kaplan 2007). This learning approach is particularly suitable on the modeling of internal or intrinsic reward factors that influence behavior and their interaction with the environment's external reward.

This volume reviewed examples of knowledge-based novelty intrinsic motivation (section 3.3.2), as with the Vieira-Neto and Nehmzow (2007) model of visual exploration and habituation in a mobile robot, and the integration of both exploration and novelty-detection behaviors. Huang and Weng (2002) investigate novelty and habituation with the SAIL (Self-organizing, Autonomous, Incremental Learner) architecture and mobile-robot platform. This model combines sensory signals across several modalities (i.e., visual, auditory, and tactile) and implements novelty detection and habituation using the reinforcement learning framework.

In the field of knowledge-based prediction intrinsic motivation (section 3.3.3), we have seen the "Formal Theory of Creativity" proposed by Schmidhuber (1991). In this framework, the intrinsic reward is based on changes in the prediction error over time, in other words, on the learning progress. In a similar fashion, the prediction-based approach by Oudeyer, Kaplan and Hafner (2007), called intelligent adaptive curiosity (IAC), is centered on prediction learning and the mechanisms by which the "Meta Machine learner" module learns to predict the error and circumstances under which the "Classic Machine learner" forward model predictions are more or less accurate. The IAC framework has been successfully tested in the Playground Experiments with the AIBO mobile robot.

Examples of developmental models directly implementing the competence-based framework of intrinsic motivations are those based on the contingency perception phenomenon in early infancy with the child's ability to detect the influence that their actions have on events (section 3.3.4). This has been linked to neuroscientific investigations on dopamine burst in the superior colliculus for intrinsic motivation and contingency detection (Redgrave and Gurney 2006; Mirolli and Baldassarre 2013). These dopamine bursts have been proposed to serve as a contingency signal and intrinsic reinforcement signal that rewards the corresponding action. Computational models implementing such mechanisms include the study by Fiore et al. (2008), on the role of perceptually salient events in a simulated robot rat experiment on conditioning learning, and the model by Schembri, Mirolli, and Baldassarre (2007) on mobile robots with a childhood phase, when the reward signal is internally generated, and an adult phase with externally generated reward during adulthood.

Overall, such a variety of models has provided a series of computational algorithms and mechanisms for the operationalization of the key concepts of habituation, exploration, novelty-detection, and prediction in developmental robotics models of intrinsic motivation.

Significant progress has also been achieved in the field of social learning and the imitation "instinct," as seen in chapter 6. In developmental psychology, newborn babies show from the very first day of life the instinct to imitate the behavior of others and complex facial expressions (Meltzoff and Moore 1983). In comparative psychology studies with human infants and primates, evidence exists that eighteen- to twenty-four-month-old children have a tendency to cooperate altruistically, a capacity not observed in chimpanzees (Warneken, Chen, and Tomasello 2006). These empirical findings have directly inspired developmental robotics models of social learning, imitation, and cooperation. For example, imitation skills are acquired in robots via the implementation of the HAMMER architecture, directly based on the AIM child psychology imitation model of Demiris and Meltzoff (2008). The HAMMER architecture includes a top-down attentional system and bottom-up attentional processes, such as those depending on the salience properties of the stimulus itself. It also consists of a set of paired inverse/forward models, where the acquisition of forward models is similar to the motor babbling stage of infant development (association between random movements and their visual, proprioceptive, or environmental consequences). These learned forward associations can then be inverted to create approximations to the basic inverse models by observing and imitating others. This allows the robot to learn how certain goals correspond to input states.

The modeling of the social skills of cooperation and shared intention is demonstrated in the work of Dominey and Warneken (2011). They carry out experiments in which a robotic arm with a gripper jointly constructs a shared plan with a human participant to play a game with goals such as "Put the dog next to the rose" or "the horse chases the dog." These cooperative skills are implemented through a cognitive architecture based on the Action Sequence Storage and Retrieval System. This is used to represent the steps of the joint game as a sequential shared plan, where each action is associated with either the human or the robot agent. Such an ability to store a sequence of actions as a shared plan constitutes the core of collaborative cognitive representations, which Warneken and colleagues claim is a unique ability of human primates (Warneken, Chen, and Tomasello; see also box 6.1).

Finally, the experiment of the acquisition of the concept of negation and of the word "no" shows the importance of social interaction between the caregiver and the child, as with the negative intent interpretation of rejection where the adult caregiver provides the child with a linguistic interpretation of her negative motivational state.

9.1.5 Nonlinear, Stage-Like Development

The existence of developmental stages in cognitive development permeates child psychology, starting from Piaget's sensorimotor stage theory. Developmental stages are typically characterized by qualitative changes in the strategy and skills used by the child, and in nonlinear progression between stages. The (inverted) U-shaped phenomenon is an example of such nonlinearity, as a stage of good performance and low errors is followed by an unexpected decrease in performance, which is subsequently recovered to show high performance.

One of the main examples of developmental robotics studies demonstrating the emergence of nonlinear, qualitative changes from situated developmental interactions is Nagai et al.'s (2003) experiment on gaze and joint attention. Results from the human-robot experiments with a robotic head showed the emergence and transition of the three qualitative joint attention stages described by Butterworth (1991): ecological (stage I), geometrical (stage II), and representational (stage III). In the initial ecological stage, the robot can only look at the objects visible within its own view, thus only achieving joint attention at a chance level. In the second stage, the robot head can also achieve joint attention by gazing at a location outside its view. In the final stage, the robot shows joint attention in practically all trials and positions (see figure 6.6). The transition between stages is the result of developmental changes in the robot's neural and learning architecture and the history of interaction with the user, rather than top-down manipulations of the robot's attention strategies.

Another developmental robotics model replicating the stage-like development of social skills is the one based on Demiris's imitation experiments. In his experiments, the HAMMER model of imitation learning follows the developmental stages of imitation behavior in infants, such as the initial imitation of the surface behavior of the demonstrator, followed by the later understanding of the underlying intentions and goal-imitation skill using different behavioral strategies. Similarly, in Scassellatti's (2002) model of the theory of mind in robots, this implements Leslie's and Baron-Cohen's stages of the theory of mind in infants.

Other models have also directly addressed the modeling of nonlinear, U-shaped phenomena. This is the case for example of the Morse et al. (2011) model of error patterns in phonetic processing, built on the ERA architecture for early word learning, and Mayor and Plunkett's (2010) vocabulary spurt simulation.

9.1.6 Online, Open-Ended, Cumulative Learning

The principle that focuses on the simultaneous and cumulative learning of cognitive skills from the online, open-ended interaction with the world is the one where progress has been more limited. The great majority of models reviewed in this volume typically focus on the simulation, or experiment, of single, isolated sensorimotor or cognitive skills. For example in chapter 5, we have reviewed numerous models on motor development that look at the four key motor skills of reaching, grasping, crawling, and walking. Unfortunately, the full integrated spectrum of locomotion and manipulation behaviors has not yet been captured by existing studies.

Some models simulate the use of multiple cognitive capabilities to contribute to the acquisition of a specific skill (e.g., frequent integration of visual skills with motor knowledge), though there is no real accumulation, over time, of different skills leading to cognitive bootstrapping and the further development of complex skills. In this book, however, we have seen some attempts to look at the cumulative role of acquiring multiple skills for the subsequent learning of higher-order cognitive skills. For example, in their model of reaching Caligiore et al. (2008) propose that motor babbling provides a bootstrap for learning to grasp (section 5.3). Fitzpatrick et al. (2008), Natale et al. (2005a; 2007), and Nori et al. (2007) use motor skill development (reaching, grasping, and object exploration) as a bootstrap for the development of object perception and object segmentation. Moreover, in the experiments on the embodied nature of number learning and the SNARC effect, the previous training of the robot on simple motor babbling tasks allows its neural control system to develop a later association between left/right spatial representations and numbers of different size (section 8.2).

Specifically for the property of open-ended learning, some significant progress has been achieved in the field of intrinsic motivation. In the numerous studies in chapter 3 we have seen that once an intrinsically motivated robot achieves mastery in one area, it can subsequently shift its focus toward new features of the environment or new skills that it has not yet learned. This is possible through the robot's own general-purpose exploration, novelty-detection, and prediction capabilities.

The major potential contribution to the modeling of online, open-ended, continuous learning comes from the proposal of emergentist cognitive architectures (section 8.5). As cognitive architectures aim at the design of a broadly scoped, general-purpose computational model capable of capturing the essential structure and process of cognition, they offer a tool to model the simultaneous and cumulative acquisition of sensorimotor and cognitive skills. This constitutes, however, a potential contribution, as most of the existing developmental robotics cognitive architectures are specialized for a subset of cognitive skills. On the one hand, we have specialized architectures as the ERA (Morse, de Greeff, et al. 2010) and LESA (Lyon, Nehaniv, and Saunders 2012) architectures focusing on language learning tasks, though they permit the simultaneous consideration of sensorimotor, speech, semantic, and pragmatic phenomena. The shared gaze (Nagai et al. 2003), the HAMMER (Demiris and Meltzoff 2008), and the cooperation (Dominey and Warneken 2011) architectures focus on the integration of skills related to social learning. On the other hand, some general developmental cognitive architectures, including work on the iCub (Vernon, von Hofsten, and Fadiga 2010), aim at a more comprehensive consideration of most functional capabilities (see table 8.4 and figure 8.10) and therefore offer the most promising approach to the modeling of online, open-ended, cumulative learning in robots.

9.2 Additional Achievements

Further achievements related to methodological and technological issues have been achieved, in addition to the progress on the preceding developmental robotics principles. Specifically, we will look at attainments in (1) the direct modeling of child development data, (2) the development and access to benchmarking robot platforms and simulators, and (3) applications of developmental robotics to assistive robotics.

9.2.1 Modeling of Child Development Data

One of the key aims of developmental robotics is to take explicit inspiration from human developmental mechanisms to design cognitive skills in robots. In this volume we have seen numerous examples of how child psychology experiments and data have directly inspired developmental robotics. Specifically, we can distinguish two main types of relations between developmental psychology and developmental robotics. In the first case, the robot experiments are directly constructed to replicate specific child psychology experiments, even allowing a direct qualitative and quantitative comparison of empirical and modeling results. The other type of relation concerns a more generic, higher-level bio-inspired link between the broad developmental mechanism studied in child experiments and the general developmental aspects of the robotic algorithm.

While most of the studies reviewed in this book employ a general, higher-level bio-inspired approach, a few allow a direct comparison between empirical and modeling experiments. For example, in chapter 7 on visual development, Schlesinger, Amso, and Johnson (2007) propose a model of perceptual completion directly simulating Amso and Johnson's (2006) experiments with infants on the unity perception task. This computational investigation permits the comparison of the performance of the simulation model results (simulated distribution of scans) with three-month-olds infants' eye tracking data (figure 4.13). The computational model provides further evidence, and an operational hypothesis on the neural and developmental mechanisms, in support of Amso and Johnson's (ibid.) hypothesis that the development of perceptual completion is due to progressive improvements in oculomotor skill and visual selective attention. In the same chapter, the developmental robotics study by Hiraki, Sashima, and Phillips (1998) investigates the performance of a mobile robot on the search task described by Acredolo, Adams, and Goodwyn (1984). This is the experiment in which a participant is asked to find a hidden object, after being moved to a new location. In addition to using the experimental paradigm comparable to the task studied by Acredolo and colleagues, Hiraki's work also proposes a model specifically designed to test Acredolo's hypothesis that self-produced locomotion facilitates the transition from egocentric to allocentric spatial perception.

Dominey and Warneken (2011) used a direct comparison of robotics and child psychology studies in their investigation of social cooperation and shared plans. This is detailed in the comparison between the Warneken, Chen, and Tomasello (2006) child/animal psychology experiments in box 6.1 and the Dominey and Warneken (2011) model setup in box 6.2. In this case, the developmental robotics study comprises seven experimental conditions that extend in various ways the original 2×2 experimental design of the psychology study (two conditions for complimentary/parallel collaboration roles, and two for the problem solving and social game tasks).

In language acquisition models, the Morse, Belpaeme, et al. (2010) simulation of Smith and Samuelson's (2010) "Modi Experiment" is another clear example of the direct comparison between child psychology and robot experimental conditions and results. Specifically in boxes 7.1 and 7.2 we provide details on the child psychology and robot experiments. Moreover, figure 7.6 directly compares quantitative data from the four conditions of the children's and the robot's experiments. The close match between the empirical and modeling results is used to support the hypothesis that body posture and embodiment bias are central to early word learning.

A similar approach is shown in chapter 8 with the work by Rucinski, Cangelosi, and Belpaeme (2011; 2012) on the embodiment biases in numerical cognition. Section 8.2 details the direct comparison between robots' and human participants' data on the SNARC effect. Figures 8.3 and 8.4 demonstrate that the iCub's reaction time increases with larger numbers and for smaller distances between numbers, thus showing a space-number response code association similar to that shown with human participants reported in Chen and Verguts (2010). Moreover, Rucinski's developmental robot simulation on the role of gestures in the acquisition of numbers (two experimental conditions with/without the use of the pointing gestures when counting) provides quantitative evidence that there is a statistical increase in the number set size when the pointing gesture accompanies the number recitation. This directly models the observed advantage of the gesture condition in the child psychology study by Alibali and DiRusso (1999).

The great majority of developmental robotics studies, however, take on a more general, higher-level inspiration strategy from child psychology experiments. A typical example of a developmental robotics study that follows such a general inspiration approach is the work on negation by Förster (2013). In this case, two general hypotheses on the emergence of negation are considered as a starting point to investigate the acquisition of the use of the word "no" in human-robot interaction experiments: (1) the negative intent interpretation of rejection hypothesis by Pea (1980), and (2) the physical and linguistic prohibition hypothesis by Spitz (1957). As a consequence, two different experiments were designed. The first experiment used a rejection scenario, where the iCub robot elicits the participant's rejection interpretation with nonverbal behavior, as frowning and head turning, toward a disliked object. The second experiment used the prohibition scenario, where two of the objects are declared as forbidden and the human participant is instructed to restrain the robot from touching these two prohibited objects. This approach does not permit a direct comparison between experimental data and modeling results, but rather a higher-level validation of the two hypotheses. This is what Förster (2013) does when he uses numerous analyses of the human participants' and the robot's utterances and behavior to demonstrate that both the rejection and the prohibition strategies play a role in the development of negation capabilities.

Similarly, Demiris and Meltzoff's (2008) HAMMER model provides a higher-level computational implementation of the Active Intermodal Matching (AIM) theoretical model (figure 6.2). This computational architecture then inspired various robotic experiments by Demiris and collaborators (Demiris and Hayes 2002, Demiris and Johnson 2003; Demiris and Khadhouri 2006) to test different cognitive mechanisms involved in imitation behavior as top-down and bottom-up attention and limited memory capabilities.

Both the direct empirical-robotic comparison studies and the higher-level developmental-inspired works show the benefits of modeling the gradual acquisition of cognitive skills, which is a key tenet of developmental robotics. Moreover, both approaches benefit from the direct collaboration between roboticists and developmental psychologists. This is the case for most of the examples described above. For example, Schlesinger provided complementary computational expertise to the psychology work by Amso and Johnson on object perception in infants, Morse collaborates with the child psychologist Smith on the embodied bias in word learning, and the roboticist Demiris collaborates with Meltzoff to propose and test the robot computational architecture of the AIM model. The direct roboticist-psychologist collaboration has the significant advantage to make sure that the robotic-computational version of the model is robustly grounded on child psychology theories, experimental methodologies and data. Moreover, such a close collaboration allows a two-way interaction between the two communities. For example, in both the Schlesinger, Amso, and Johnson (1997) and the Morse, Belpaeme, et al. (2010) computational studies we have seen situations in which the robotics experiment can provides predictions and insights for human development mechanisms, leading to further experimental investigations in child psychology. Specifically, the embodiment and early word learning robot experiments of Morse and colleagues predicted novel phenomena on the change of full body posture during the "modi" experiments, which have been verified and confirmed in subsequent experiments at Smith's Babylab (Morse et al. in preparation).

9.2.2 Access to Benchmark Robot Platforms and Software Simulation Tools

Another important methodological and technological achievement in developmental robotics has been the design and dissemination of benchmark robot platforms and simulators. Some use the open source approach, while others are based on commercial platforms and software simulation suites.

In chapter 2, we looked at over ten "baby robot" platforms and three main software simulators used in this field. Of these robot platforms, three have had a significant impact in developmental robotics, contributing to its origins and subsequent growth: the AIBO mobile robot, as well as the iCub and the NAO humanoid platforms.

The mobile platform AIBO robot (see full details in section 2.4) was one of the very first platforms used in the pioneering developmental robotics studies. This was a commercial platform (now out of production) widely available in late 1990 in robotics and computer science labs, due to its standard use in the RoboCup league and its affordability. Seminal studies using the AIBO for developmental robot models are Oudeyer et al. (Oudeyer, Kaplan, and Hafner 2007; Oudeyer and Kaplan 2006, 2007) on artificial curiosity and intrinsic motivation, Kaplan and Hafner (2006a, 2006b) on joint attention, and Steels and Kaplan (2002) on word learning.

The NAO robot (cf. section 2.3.2) has more recently become one of the two main benchmark humanoid platforms for developmental modeling. This is again due to its affordability and wide availability in research labs, as this robot has become the new standard in the RoboCup competition (directly replacing the AIBO). The NAO in particular is suitable for studies on intrinsic motivation, navigation, and locomotion, and on social interaction involving action imitation given its robust motor capabilities. In this volume we have for example seen the use of the NAO for crawling and walking development (Li et al. 2011) and pointing gestures in social interaction (Hafner and Schillaci 2011).

The iCub robot (cf. section 2.3.1) is the most recent success story in developmental robotics. More than twenty-five laboratories worldwide have access to this open source platform (2013 data). This extensive dissemination is the result of the European Union Framework Programme investment in cognitive systems and robotics, and the Italian Institute of Technology's support for this open source platform. This volume has seen numerous developmental robotics studies with the iCub as in studies on motor development (Savastano and Nolfi 2012; Caligiore et al. 2008; Righetti and Ijspeert 2006a; Lu et al. 2012), social interaction and cooperation (Lallée et al. 2010), language acquisition (Morse, Belpaeme, et al. 2010; Lyon, Nehaniv, and Saunders 2012), number learning (Rucinski, Cangelosi, and Belpaeme 2011, 2012), negation (Förster 2013), and cognitive architecture (Vernon, von Hofsten, and Fadiga 2010).

The success and widespread use of platforms like the iCub and NAO have also been supported by robot software simulators that further facilitate research on the platforms. For example, the open-source iCub simulation software (section 2.4) has allowed hundreds of students and staff to participate in the yearly "Veni Vidi Vici" iCub Summer School (running from 2006) to learn to use the simulator robot, extending its use to labs with no access to the physical platform. As for the NAO simulator, its default availability in the Webots simulator has also facilitated wider use of this platform.

The enhanced access to developmental robot benchmark platforms (both hardware and simulators), accompanied by the continuous development of specialized platforms such as the ones reviewed in chapter 2, are key factors in the uptake and success of developmental robotics.

9.2.3 Applications in Assistive Robotics and Child Human-Robot Interaction (cHRI)

A further and important achievement of developmental robotics has been the translation of robot modeling research, especially the studies on social interaction, into applications of social assistive robotics for children. This started with the extension of the pioneering developmental robotics work on modeling social interaction and theory of mind to experiments on children with social skill disabilities, as in the autism spectrum disorders (Weir and Emanuel 1976; Dautenhahn 1999; Scassellati 2002; Kozima et al. 2004; Kozima, Nakagawa, and Yano 2005; Dautenhahn and Werry 2004; Thill et al. 2012). These applications have more recently encompassed other assistive robot areas, such as with children with Down syndrome and hospitalized child patients.

Autism spectrum disorders (ASDs; also referred to as the autism spectrum continuum) include a variety of chronic, life-long disabilities that affect children's capacities to communicate and interact with peers and adults and to understand social cues. Two of the most common types of ASDs are autistic disorder and Asperger's syndrome, which in addition to the social communication and social interaction deficits include symptoms such as restricted, repetitive patterns of behavior (Scassellati, Admoni, and Matarić 2012).

Dautenhahn (1999) was the first researcher to propose the use of physical robots as assistive social companions for ASD children, with the experiments of the AURORA project (AUtonomous RObotic platform as a Remedial tool for children with Autism; Dautenhahn and Werry 2004; www.aurora-project.com). The AURORA team has used various mobile and humanoid robots in different interaction studies with children with autism, including a study showing that the AIBO robot can adapt in real-time to the ways autistic children play with the robot (Francois, Dautenhahn, and Polani 2009a). In one of their pioneering studies, Wainer et al. (2013) investigated the effects of a session on autistic children's play interaction with the KASPAR humanoid robot on the subsequent improvement of social interaction with human adults. Comparison of the children's social behavior before and after the robot play session showed that autistic children were more invested in the game and collaborated better with the human partners after playing with KASPAR. The AURORA project led to numerous experimental investigations on autism and social robots by Dautenhahn and colleagues (e.g., Robins et al. 2012b; Dickerson, Robins, and Dautenhahn 2013), including a study by Wood et al. (2013) that shows the suitability of KASPAR as a tool for conducting robot-mediated interviews with children.

Another pioneering study on autistic children and robot interaction was presented in section 2.3.7, when we described the upper-torso humanoid platform Infanoid. Kozima et al. (2004; Kozima, Nakagawa, and Yano 2005) carried out child-robot interaction experiments with five- and six-year-olds, which compared typically developing children with autistic children. Their main interest was the children's perception of the robot and the demonstration of the robot's three stages of "ontological understanding": the neophobia stage marked by embarrassment and staring; the exploration stage featuring poking at the robot and showing toys; and finally, the interaction stage for reciprocal social exchanges. The comparison between autistic children and a control group showed similar responses in all three phases, with the only difference that autistic children enjoyed longer interactions, without losing interest as typically developing participants. Kozima has used both the Infanoid humanoid robot and the Keepon toy-like creature robot (Kozima, Nakagawa, and Yano 2005) to further explore these issues.

Scassellati (2005, 2007) also contributed to early studies on robot therapy for ASD children, given his interest in social development and theory of mind (section 6.5). His main contribution has been the use of social robots as diagnostic tools of ASD, in addition to their treatment role. For example he used the commercial robot face ESRA in two experimental conditions: (1) pre-scripted, noncontingent one-way dialogue with children, and (2) contingent play through Wizard of Oz control of the triggering of the robot's behavior in response to the child's reaction. Results show that autistic children are not sensitive to the differences in the two conditions, and did not show any reduction of interest in the last part of the interaction sessions (typically developing children's interest weaned during the last parts of the interactions). The manipulation of conditions that discriminate between children with different social behavior deficits can contribute to the definition of quantitative, objective measurements of social response and ASD symptom assessment. Scassellati proposes the implementation of structured interactions between ASD children and robots to create standardized social manipulation tasks designed to elicit particular social responses. The responses to monitor and quantify include gaze direction, focus of attention, position tracking, and vocal prosody.

A recent review of work on assistive social robots and ASD by Scassellati, Admoni, and Matarić (2012) gives an overview of the achievements in this field. This review for example points at consistent evidence that the use of these robots as autism therapy tools provides: (1) increased social engagement, (2) boosted levels of attention, (3) joint attention, (4) spontaneous imitation, (5) turn taking with another child, (6) manifestations of empathy, and (7) initiation of physical contact with the experimenter. Scassellati and colleagues claim that these improved social skills and behaviors are the consequences of the fact that robots provide novel sensory stimuli to the ASD child. Moreover, these social robots play an intermediate, novel role that sits between interactions with inanimate toys and animate social beings. On the one hand, inanimate toys do not elicit novel social behaviors, while people can be a source of confusion and distress for children with autism. Animated social robots, on the other hand, create new sensory situations facilitating novel behaviors as joint attention and empathy with the experimenters.

A similar approach to the use of social robots as therapy for ASD children has been applied to other disabilities. Lehmann et al. (2014) report an exploratory case study concerning a child with Down syndrome. These children have quite good nonverbal communication and social interaction skills, in contrast with the much more limited social interaction skills associated with ASDs. This study used two different robot platforms, the humanoid, doll-like KASPAR robot and the mobile IROMEC robot, for educational interactive games such as the "Move the robot" game and an "Imitation."

The case study results showed that the child was more interactive with the human experimenter and the robot in the games with the humanoid platform KASPAR for most behaviors such as looking at experimenter, pointing at the robot, vocalization, imitation of robot and prompted imitation of experimenter. The IROMEC mobile platform only showed advantages for the behavior of touching the robot. The advantage of the KASPAR platform is explained by the fact that its humanoid, robotic, and childlike appearance stimulates social behavior, in which Down syndrome children excel.

All these studies with varying developmental disorders like ASD and Down syndrome demonstrate the utility of the use of social assistive robots for different social and cognitive deficits. Moreover, other application areas of social robotics have been extended to their use as companions for children in the hospital (Belpaeme et al. 2012; Carlson and Demiris 2012; Sarabia and Demiris 2013). For example, Belpaeme et al. (2012) in the ALIZ-e project use the NAO humanoid platform as a long-term social companion for children with chronic disease such as diabetes. Specifically, the NAO robot has been used both in hospital environments and summer camps to support children to cope with the long term issues of diabetes and to help children understand dietary restrictions, medication and physical exercise requirements. This project includes action and dance imitation games, to encourage children to do regular physical exercises and help improve their self-image, and interactive tutoring and quiz sessions, to support the patient's learning about dietary restrictions and identify (and avoid) foods with high carbohydrate and sugar levels. The work on diabetes has also led to the extension of the NAO hospital companion role for different pathologies. A clinical case study involving an eight-year-old cerebral stroke patient specifically investigated the advantages of using a social robot to assist in physiotherapy for stroke recovery (Belpaeme et al. 2013).

The growing trend on studies on robots interacting with healthy, ill, and disabled children more generally contributes to the field of child human-robot interaction (cHRI) (Belpaeme et al. 2013). cHRI has distinctive features when compared with standard, adult HRI. This is due to the fact that because the child's neural, cognitive, and social development has not reached full maturity, they will thus respond to different interaction strategies employed by the robot. For example, in cHRI with two- to three-year-old children, the interaction will be facilitated with the use of nonverbal communication strategies and error-tolerant linguistic communication as the toddlers' linguistic skills have not fully developed. Moreover, cHRI benefits from the fact that young children readily engage in social play and pretend play, and have a tendency to anthropomorphize and attribute lifelike qualities or personalities to toys (and robots). Therefore, the design of cHRI studies will benefit from the utilization of play interaction strategies and the exploitation of the child's attribution of empathy and lifelike characteristics to the robot (Turkle et al. 2004; Rosenthal-von der Pütten et al. 2013). Given the joint interests and overlapping research issues between developmental robotics and cHRI, these two approaches will benefit from the sharing of methodologies on human-child interaction experiments and the general investigation of sociocognitive developmental mechanisms.

9.3 Open Challenges

We have analyzed in the previous sections a broad set of theoretical and technological achievements, which demonstrate the significant advances in the first ten to fifteen years since the beginning of developmental robotics. However, the aim of understanding developmental mechanisms in natural cognitive systems and their operationalization and implementation in artificial robotic cognitive systems is implicitly a very complex task, with a long-term timescale spanning well beyond the "infant" stage of this discipline. This leaves open a wide set of scientific and technological challenges. In this concluding section we are going to highlight some of the key open challenges, specifically future work on the cumulative, long-term learning of integrated skills, on the evolutionary and developmental changes in body and brain morphology, and on cHCI and child-robot interaction ethics.

9.3.1 Cumulative Learning and Skills Integration

As we saw in the previous section on achievements in the general developmental robotics principle of open-ended cumulative learning, this is an area where the level of scientific advances has been limited. Although we have seen the design and testing of an impressive number of models of the developmental learning of isolated sensory, motor, social, linguistic, and reasoning skills, in only a very few cases have we seen these integrated into a single cognitive agent.

One way to address this issue is the incremental training of a robot controller to learn skills of increasing complexity. For example in language development models, the same robot controller should be used for cumulative adding of phonetic skills and single-word learning, followed by two-word acquisition and the acquisition of simple grammar constructs, further followed by the gradual development of complex adultlike syntactic skills. In parallel, the same robot should be able to integrate lexical, semantic, and constructivist grammar skills with sensorimotor representations, along the lines of the embodied view of language development. A specific example of a research roadmap for language and action learning has been proposed by Cangelosi et al. (2010). Following a twenty-year research perspective, this roadmap proposes the following six milestones in linguistic skill acquisition research: (1) grounded acquisition, decomposition, and generalization of simple transitive holophrases in learning by demonstration tasks; (2) grounded acquisition, decomposition, and generalization of the five basic argument constructions of English from holophrastic instances and the event types that are associated with their prototypical uses; (3) grounded interactive language-learning games in simple joint attention scenarios based on the implementation of elementary sociocognitive/pragmatic capabilities; (4) learning from increasingly complex and diversified linguistic input within progressively less restricted learner-tutor interactions; (5) progressively more humanlike cooperative ostensive-inferential communication based on the implementation of more advanced sociocognitive and pragmatic capabilities; and (6) learning progressively more complex grammars from quantitatively naturalistic input. These language-specific milestones are accompanied by three parallel sets of research milestones respectively in action learning, social development, and action-language integration. Vernon, von Hofsten, and Fadiga (2010) have proposed similar exercises for a research plan covering all areas in a robot's cognitive development.

Another approach to the issue of cumulative learning is the use of developmental cognitive architectures. Such architectures encompass the integration of multiple skills and behaviors, and as such allow the modeling of cumulative skill development. In section 8.5 we discussed extensively the different types of emergentist cognitive architectures and their consideration of multiple cognitive mechanisms and behavior.

Recall also that one of the primary long-term goals of research on intrinsic motivation is to demonstrate hierarchical, cumulative learning. While work on this goal is still at an early stage, the conceptual framework for skill transfer across contexts and domains is relatively well established. In particular, the competence-based approach to IM that we discussed in chapter 3 proposes that many important skills are initially acquired "simply for fun or play," or in the service of "exploring the environment," but that these same skills are later recruited and reused as components within more complex, survival-related behaviors (Baldassarre 2011; Oudeyer and Kaplan 2007). Preliminary support for this approach is provided by studies such as the Fiore et al. (2008) model, which demonstrates an important element of the competence-based approach, that is, how contingency perception can provide an essential learning signal for the organism when there is no external reinforcement to drive behavior.

A further approach to open-ended, cumulative development and learning comes from long-term human-robot interaction experiments. Accumulating "life-long" experience means raising an infant robot into early childhood, if not longer, in an artificial "robot kindergarten." A recent contribution to this is the study by Araki, Nakamura, and Nagai (2013) on the interactive learning framework for long-term acquisition of concepts and words by robots. In this experiment a learning robot interacts for a full week with an experimenter, who teaches the robot the names of 200 objects during numerous online learning sessions. The analyses of the experiment revealed that the teacher produced 1,055 utterances, with the robot acquiring 924 words in total. Of these only a small set of 58 words are meaningful: 4 functional words, 10 adjectives, 40 nouns, and 4 verbs.

The approach of a "virtual school student" in a robot kindergarten/school scenario has been proposed by Adams et al. (2012) within the field of Artificial General

Intelligence. This method foresees the implementation of a virtual student robot growing in both "preschool learning" and "school learning" environments. The preschool learning scenario involves continuous, long-term HRI experiments for the practice and development of sensorimotor skills and basic cognitive capabilities. The School Learning scenario continues the virtual preschool scenario, but with a focus on the long-term practice of higher cognitive (symbolic) abilities. The design of methodologies for long-term robot learning and interaction experiments will be a key step in addressing the challenge of integrating ever more complex skills and the resulting bootstrapping of the cognitive system.

9.3.2 Evolutionary and Developmental Changes in Body and Brain Morphology

The interaction between evolutionary and ontogenetic mechanisms is another area where limited efforts have been dedicated to date, and that remains a key challenge in this field. The combination of both evolutionary and developmental algorithms permits the investigation of coevolutionary adaptation of brain-body systems, and the modeling of morphological changes in ontogenesis. As we saw in section 1.3.2, this includes maturational changes in the anatomy and physiology of both the infant robot's body and neural control system.

Evolutionary computation and modeling approaches such as the fields of evolutionary robotics (Nolfi and Floreano 2000) and of evo-devo models (Stanley and Miikkulainen 2003) can already directly address the interaction between brain and body and between phylogenetic and ontogenetic changes. Future research combining both evolutionary and developmental robotics models can further our understanding of the adaptive value of maturational changes. Simulation approaches like the fetus models by Kuniyoshi (see section 2.5.3) can already provide a methodology to investigate body-brain coadaptation. Another area of potential advances in this field can come from the recent development of modular, reconfigurable robots. Specifically, modular, self-reconfigurable robots are able to change their own shape by rearranging the connectivity of their body parts to respond to environmental changes and requirements (Yim et al. 2007). Such robots typically have a neural control system to deal with the nonlinearity and complexity of behavioral control, and strategies for dynamically varying body configurations. This type of robot could be used to model maturational changes in the baby robot's morphology, modeling anatomical maturational changes. Other advantages of reconfigurable robot research are the modeling of self-repair (Christensen 2006).

Another area of research that can contribute to the understanding of the complex and dynamic nature of body/brain adaptation and morphological changes is that of soft robots. As discussed in the conclusions of chapter 2, on baby robot platforms, recent advances in the field of material sciences (e.g., pneumatic artificial muscle actuators, rigid materials with soft dynamics as with the electromagnetic, piezoactive, or thermal actuators) have facilitated the production of new soft materials that can be used as robot sensors and effectors, and that are contributing to the production of soft robots prototypes (Pfeifer, Lungarella, and Iida 2012; Albu-Schaffer et al. 2008). Moreover, anthropomimetic robot platforms based on compliant musculoskeletal material and effectors, such as the ECCE robot platform (Holland and Knight 2006), provide a research tool to look at self-organization and emergent principles in robot control, and morphological computation strategies, which can be used to model the early stages of sensorimotor development in epigenetic robots.

9.3.3 cHRI, Robot Appearance, and Ethics

The scientific and technological progress in designing developmental robots has important implications for the design of intelligent interactive robots in a variety of domains. In particular, as we saw in the previous section on methodological and technological achievements, developmental robot studies have led to the design and testing of assistive robotics applications for children's use, and the general area of cHRI (Belpaeme et al. 2013). Increasing use of assistive and companion robots has important implications for robot platforms types and their level of autonomy and assistive capabilities. This raises critical issues on ethics principles in the use of robots, especially with children.

In the introductory chapter we discussed the uncanny valley phenomenon. This applies to the cases when users interact with android robots with appearances increasingly similar to the human body and face. If there is a mismatch between a humanlike robot whose appearance is indistinguishable from that of humans, but whose limited behavior capabilities do not correspond to the expected full capability of a person, the user can have a sense of revulsion and eeriness. The correct handling of the uncanny valley phenomenon is particularly important in cHRI. In addition to the general discussion on the uncanny valley (MacDorman and Ishiguro 2006), various researchers have looked at the role of the robot's appearance in HRI. Specifically, experimental investigations have considered the comparison between interactions with physical vs. simulated robots (Bainbridge et al. 2008) and with remote/present robots (Kiesler et al. 2008), the issue of proximity between the robot and the user (Walters et al. 2009), and the role of appearance (Walters et al. 2008). Such studies have however focused on adult participants, with little focus on the reaction of child users. Future work in cHRI should therefore look at children's expectations and reactions to the robot's appearance and physical and behavioral attributes. A direct open question is whether the humanoid design should focus on robot-like platforms, like the iCub and the NAO, or instead move toward a more humanlike appearance, as in the doll-like KASPAR robot.

In more general terms, research and practical applications of human-robot interaction, especially with child users or disabled and ill users, raise wider issues related to ethics and robotics. This has recently led to a more careful consideration of the legal, medical, and social-ethical rules and principles in robot use (van Wynsberghe 2012;

Gunkel, Bryson, and Torrance 2012; Lin, Abney, and Bekey 2011; Wallach and Allenn 2008; Veruggio and Operto 2008). For example, van Wynsberghe (2012) has looked specifically at ethical implications in assistive care robots. She proposes a framework where the assistive robot designers must be explicit about values, uses, and contexts throughout the whole design process. This should be achieved through a direct dialogue with all stakeholders, throughout the whole process from the conceptual design of the robot platform and its application, right to the point of implementation and testing. Specifically, in the context of assistive robot applications, the recommendation is to empower all users (doctors, caretakers, patients, as well as roboticists) to maintain full responsibility for the human-care providers, rather than the technology. Given the specific requirement of cHRI, and the additional constraints of the use of robots with disabled and child patients, fundamental questions regarding the ethical principles of developmental assistive robots remain open for future research.

Robot ethics implications go beyond the specific needs of child assistive robotics. For example, the ongoing progress in the design of increasingly complex motivational, behavioral, cognitive, and social skills in (baby) robots paves the way to research ethics considerations related to the issue of autonomy. In particular, the modeling of intrinsic motivation in robots, as seen in chapter 3, is a first step toward the design of fully autonomous robots with their own motivational and curiosity system. This, together with the implementation of developmental-inspired mechanisms allowing the autonomous learning of sensorimotor, cognitive, and reasoning capabilities, can result in the design of robots that no longer require continuous control by the human user and can apply their own decision making. Furthermore, research on increasingly complex and autonomous robots could lead to the potential design of self-awareness and consciousness capabilities in robots and machines (Aleksander 2005; Chella and Manzotti 2007). While progress on robot autonomy has reached advanced technological stages, the design of conscious machines and robots is a longer-term research issue. However, both have important ethical implications. Therefore, scientific and technological advances in autonomous developmental robotics necessitate the investigation, understanding, and definition of ethics principles constraining robot research and development.

The three open research issues discussed in this chapter constitute the key general scientific, technological, and ethical challenges that span the integration of a wide set of behavioral and cognitive skills. More specific open challenges and directions for future research on individual cognitive mechanisms have been discussed in the various central chapters (chapters 3–8) of this book. Through the review of the seminal models and experiments respectively covering motivational, perceptual, motor, social, linguistic, and reasoning capabilities, we have analyzed the achievements and limitations of current research, and identified the need for more work in these key areas of developmental robotics. The highly interdisciplinary nature of developmental robotics, ranging from robotics and computer science to cognitive and neural sciences, and reaching philosophical and ethics disciplines, can together contribute to the understanding of developmental principles and mechanisms in children and their operationalization for the autonomous design of behavioral and cognitive capabilities in artificial agents and robots.

To conclude, we can confidently state that developmental robotics has reached the end of its infancy, and now begins the years of early childhood! Therefore we expect in the next ten to fifteen years we will see child robots that can go completely from crawling to walking, that can speak in two- and three-word sentences, that can engage in pretend play and deceive others through their own theory of mind, that are beginning to develop a sense of gender and a sense of morality.

References

Abitz, M., R. D. Nielsen, E. G. Jones, H. Laursen, N. Graem, and B. Pakkenberg. 2007. "Excess of Neurons in the Human Newborn Mediodorsal Thalamus Compared with that of the Adult." *Cerebral Cortex* 17 (11) (Nov.): 2573–2578.

Acredolo, L. P. 1978. "Development of Spatial Orientation in Infancy." *Developmental Psychology* 14 (3): 224–234.

Acredolo, L. P., A. Adams, and S. W. Goodwyn. 1984. "The Role of Self-Produced Movement and Visual Tracking in Infant Spatial Orientation." *Journal of Experimental Child Psychology* 38 (2): 312–327.

Acredolo, L. P., and D. Evans. 1980. "Developmental-Changes in the Effects of Landmarks on Infant Spatial-Behavior." *Developmental Psychology* 16 (4): 312–318.

Adams, S. S., I. Arel, J. Bach, R. Coop, R. Furlan, B. Goertzel, J. S. Hall, A. Samsonovich, M. Scheutz, M. Schlesinger, S. C. Shapiro, and J. F. Sowa. 2012."Mapping the Landscape of Human-Level Artificial General Intelligence." *AI Magazine* 33 (1) (Spring): 25–41.

Adler, S. A., M. M. Haith, D. M. Arehart, and E. C. Lanthier. 2008. "Infants' Visual Expectations and the Processing of Time." *Journal of Cognition and Development* 9 (1) (Jan.–Mar.): 1–25.

Adolph, K. E. 2008. "Learning to Move." *Current Directions in Psychological Science* 17 (3) (June): 213–218.

Adolph, K. E. 1997. "Learning in the Development of Infant Locomotion." *Monographs of the Society for Research in Child Development* 62 (3): i–vi, 1–158.

Adolph, K. E., B. Vereijken, and M. A. Denny. 1998. "Learning to Crawl." *Child Development* 69 (5) (Oct.): 1299–1312.

Albu-Schaffer, A., O. Eiberger, M. Grebenstein, S. Haddadin, C. Ott, T. Wimbock, S. Wolf, and G. Hirzinger. 2008. "Soft Robotics." *IEEE Robotics & Automation Magazine* 15 (3): 20–30.

Aleksander, I. 2005. *The World in My Mind, My Mind in the World: Key Mechanisms of Consciousness in Humans, Animals and Machines*. Exeter, UK: Imprint Academic.

Aleotti, J., S. Caselli, and G. Maccherozzi. 2005. "Trajectory Reconstruction with NURBS Curves for Robot Programming by Demonstration." Paper presented at the 2005 IEEE International Symposium on Computational Intelligence in Robotics and Automation, New York, June 27–30.

Alibali, M. W., and A. A. DiRusso. 1999. "The Function of Gesture in Learning to Count: More than Keeping Track." *Cognitive Development* 14 (1) (Jan.–Mar.): 37–56.

Amso, D., and S. P. Johnson. 2006. "Learning by Selection: Visual Search and Object Perception in Young Infants." *Developmental Psychology* 42 (6) (Nov.): 1236–1245.

Anderson, J. R., D. Bothell, M. D. Byrne, S. Douglass, C. Lebiere, and Y. L. Qin. 2004. "An Integrated Theory of the Mind." *Psychological Review* 111 (4) (Oct.): 1036–1060.

Andres, M., X. Seron, and E. Olivier. 2007. "Contribution of Hand Motor Circuits to Counting." *Journal of Cognitive Neuroscience* 19 (4) (Apr.): 563–576.

Andry, P., A. Blanchard, and P. Gaussier. 2011. "Using the Rhythm of Nonverbal Human-Robot Interaction as a Signal for Learning." *IEEE Transactions on Autonomous Mental Development* 3 (1): 30–42.

Araki, T., T. Nakamura, and T. Nagai. 2013. "Long-Term Learning of Concept and Word by Robots: Interactive Learning Framework and Preliminary Results." Paper presented at the IEEE/RSJ International Conference on Intelligent Robots and Systems, Tokyo Big Sight, Tokyo, Japan, November 3–7.

Arena, P. 2000. "The Central Pattern Generator: A Paradigm for Artificial Locomotion." *Soft Computing* 4 (4): 251–266.

Arkin, R. C. 1998. *Behavior-Based Robotics*. Cambridge, MA: MIT Press.

Asada, M., K. Hosoda, Y. Kuniyoshi, H. Ishiguro, T. Inui, Y. Yoshikawa, M. Ogino, and C. Yoshida. 2009. "Cognitive Developmental Robotics: A Survey." *IEEE Transactions on Autonomous Mental Development* 1 (1) (May): 12–34.

Asada, M., K. F. MacDorman, H. Ishiguro, and Y. Kuniyoshi. 2001. "Cognitive Developmental Robotics as a New Paradigm for the Design of Humanoid Robots." *Robotics and Autonomous Systems* 37 (2–3) (Nov.): 185–193.

Asada, M., E. Uchibe, and K. Hosoda. 1999. "Cooperative Behavior Acquisition for Mobile Robots in Dynamically Changing Real Worlds Via Vision-Based Reinforcement Learning and Development." *Artificial Intelligence* 110 (2) (June): 275–292.

Ashmead, D. H., M. E. McCarty, L. S. Lucas, and M. C. Belvedere. 1993. "Visual Guidance in Infants Reaching toward Suddenly Displaced Targets." *Child Development* 64 (4) (Aug.): 1111–1127.

Austin, J. L. 1975. *How to Do Things with Words*. Vol. 1955. Oxford, UK: Oxford University Press.

Bahrick, L. E., L. Moss, and C. Fadil. 1996. "Development of Visual Self-Recognition in Infancy." *Ecological Psychology* 8 (3): 189–208.

Bahrick, L. E., and J. S. Watson. 1985. "Detection of Intermodal Proprioceptive Visual Contingency as a Potential Basis of Self-Perception in Infancy." *Developmental Psychology* 21 (6): 963–973.

Bainbridge, W. A., J. Hart, E. S. Kim, and B. Scassellati. 2008. "The Effect of Presence on Human-Robot Interaction." Paper presented at the 17th IEEE International Symposium on Robot and Human Interactive Communication, Technische Universitat, Munich, Germany, August 1–3.

Baldassarre, G. 2011. "What Are Intrinsic Motivations? A Biological Perspective." Paper presented at the 2011 IEEE International Conference on Development and Learning (ICDL), Frankfurt Biotechnology Innovation Center, Frankfurt, Germany, August 24–27.

Baldassarre, G., and M. Mirolli, eds. 2013. *Intrinsically Motivated Learning in Natural and Artificial Systems*. Berlin: Springer-Verlag.

Baldwin, D. A. 1993. "Early Referential Understanding—Infants Ability to Recognize Referential Acts for What They Are." *Developmental Psychology* 29 (5) (Sept.): 832–843.

Baldwin, D., and M. Meyer. 2008. "How Inherently Social Is Language?" In *Blackwell Handbook of Language Development*, ed. Erika Hoff and Marilyn Shatz, 87–106. Oxford, UK: Blackwell Publishing.

Balkenius, C., J. Zlatev, H. Kozima, K. Dautenhahn, and C. Breazeal, eds. 2001. Paper presented at the 1st International Workshop on Epigenetic Robotics: Modeling Cognitive Development in Robotic Systems, Lund, Sweden, September 17–18.

Ballard, D. H. 1991. "Animate Vision." *Artificial Intelligence* 48 (1) (Feb.): 57–86.

Bandura, A. 1986. *Social Foundations of Thought and Action: A Social Cognitive Theory*. Englewood Cliffs, NJ: Prentice Hall.

Baranes, A., and P. Y. Oudeyer. 2013. "Active Learning of Inverse Models with Intrinsically Motivated Goal Exploration in Robots." *Robotics and Autonomous Systems* 61 (1): 49–73.

Baranes, A., and P. Y. Oudeyer. 2009. "R-Iac: Robust Intrinsically Motivated Exploration and Active Learning." *IEEE Transactions on Autonomous Mental Development* 1 (3) (June): 155–169.

Barborica, A., and V. P. Ferrera. 2004. "Modification of Saccades Evoked by Stimulation of Frontal Eye Field during Invisible Target Tracking." *Journal of Neuroscience* 24 (13) (Mar.): 3260–3267.

Baron-Cohen, S. 1995. *Mindblindness: An Essay on Autism and Theory of Mind*. Cambridge, MA: MIT Press.

Barrett, M. 1999. *The Development of Language*. New York: Psychology Press.

Barsalou, L. W. 2008. "Grounded Cognition." *Annual Review of Psychology* 59: 617–645.

Barsalou, L. W. 1999. "Perceptual Symbol Systems." *Behavioral and Brain Sciences* 22 (4) (Aug.): 577–609.

Barsalou, L. W., and K. Wiemer-Hastings. 2005. "Situating Abstract Concepts." In *Grounding Cognition: The Role of Perception and Action in Memory, Language, and Thought*, ed. D. Pecher and R. Zwaan, 129–163. New York: Cambridge University Press.

Barto, A. G., S. Singh, and N. Chentanez. 2004. "Intrinsically Motivated Learning of Hierarchical Collections of Skills." Paper presented at the 3rd International Conference on Development and Learning (ICDL 2004), La Jolla, CA, October 20–22.

Bates, E., L. Benigni, I. Bretherton, L. Camaioni, and V. Volterra. 1979. *The Emergence of Symbols: Cognition and Communication in Infancy*. New York: Academic Press.

Bates, E., and J. L. Elman. 1993. "Connectionism and the Study of Change." In *Brain Development and Cognition: A Reader*, ed. Mark Johnson, 623–642. Oxford, UK: Blackwell Publishers.

Baxter, P., R. Wood, A. Morse, and T. Belpaeme. 2011. "Memory-Centred Architectures: Perspectives on Human-Level Cognitive Competencies." Paper presented at the AAAI Fall 2011 Symposium on Advances in Cognitive Systems, Arlington, VA, November 4–6.

Bednar, J. A., and R. Miikkulainen. 2003. "Learning Innate Face Preferences." *Neural Computation* 15 (7): 1525–1557.

Bednar, J. A., and R. Miikkulainen. 2002. "Neonatal Learning of Faces: Environmental and Genetic Influences." Paper presented at the 24th Annual Conference of the Cognitive Science Society, George Mason University, Fairfax, VA, August 7–10.

Beer, R. D. 2000. "Dynamical Approaches to Cognitive Science." *Trends in Cognitive Sciences* 4 (3): 91–99.

Behnke, S. 2008. "Humanoid Robots—From Fiction to Reality?" *Kunstliche Intelligenz Heft* 22 (4): 5–9.

Bellas, F., R. J. Duro, A. Faina, and D. Souto. 2010. "Multilevel Darwinist Brain (Mdb): Artificial Evolution in a Cognitive Architecture for Real Robots." *IEEE Transactions on Autonomous Mental Development* 2 (4) (Dec.): 340–354.

Belpaeme, T., P. Baxter, J. de Greeff, J. Kennedy, R. Read, R. Looije, M. Neerincx, I. Baroni, and M. C. Zelati. 2013. "Child-Robot Interaction: Perspectives and Challenges." In *Social Robotics*, ed. G. Herrmann et al., 452–459. New York: Springer International Publishing.

Belpaeme, T., P. E. Baxter, R. Read, R. Wood, H. Cuayáhuitl, B. Kiefer, S. Racioppa, I. Kruijff-Korbayová, G. Athanasopoulos, and V. Enescu. 2012. "Multimodal Child-Robot Interaction: Building Social Bonds." *Journal of Human-Robot Interaction* 1 (2): 33–53.

Benjamin, D. P., D. Lyons, and D. Lonsdale. 2004. "Adapt: A Cognitive Architecture for Robotics." Paper presented at the 6th International Conference on Cognitive Modeling: Integrating Models, Carnegie Mellon University, University of Pittsburgh, Pittsburgh, PA, July 30–August 1.

Berk, L. 2003. *Child Development*. 6th ed. Boston: Allyn & Bacon.

Berlyne, D. E. 1960. *Conflict, Arousal, and Curiosity*. New York: McGraw-Hill.

Bernstein, N. A. 1967 *The Co-ordination and Regulation of Movements*. New York: Pergamon Press.

Berrah, A.-R., H. Glotin, R. Laboissière, P. Bessière, and L.-J. Boë. 1996. "From Form to Formation of Phonetic Structures: An Evolutionary Computing Perspective." Paper presented at the International Conference on Machine Learning, Workshop on Evolutionary Computing and Machine Learning, Bari, Italy.

Berthier, N. E. 2011. "The Syntax of Human Infant Reaching." Paper presented at the 8th International Conference on Complex Systems, Cambridge, MA.

Berthier, N. E. 1996. "Learning to Reach: A Mathematical Model." *Developmental Psychology* 32 (5) (Sept.): 811–823.

Berthier, N. E., B. I. Bertenthal, J. D. Seaks, M. R. Sylvia, R. L. Johnson, and R. K. Clifton. 2001. "Using Object Knowledge in Visual Tracking and Reaching." *Infancy* 2 (2): 257–284.

Berthier, N. E., and R. Keen. 2005. "Development of Reaching in Infancy." *Experimental Brain Research* 169 (4) (Mar.): 507–518.

Berthier, N. E., M. T. Rosenstein, and A. G. Barto. 2005. "Approximate Optimal Control as a Model for Motor Learning." *Psychological Review* 112 (2) (Apr.): 329–346.

Berthouze, L., and M. Lungarella. 2004. "Motor Skill Acquisition under Environmental Perturbations: On the Necessity of Alternate Freezing and Freeing of Degrees of Freedom." *Adaptive Behavior* 12 (1): 47–64.

Berthouze, L., and T. Ziemke. 2003. "Epigenetic Robotics—Modelling Cognitive Development in Robotic Systems." *Connection Science* 15 (4) (Dec.): 147–150.

Billard, A., and K. Dautenhahn. 1999. "Experiments in Learning by Imitation—Grounding and Use of Communication in Robotic Agents." *Adaptive Behavior* 7 (3–4) (Winter): 415–438.

Billard, A., and M. J. Matarić. 2001. "Learning Human Arm Movements by Imitation: Evaluation of a Biologically Inspired Connectionist Architecture." *Robotics and Autonomous Systems* 37 (2–3) (Nov.): 145–160.

Bisanz, H., J. L. Sherman, C. Rasmussen, and E. Ho. 2005. "Development of Arithmetic Skills and Knowledge in Preschool Children." In *Handbook of Mathematical Cognition*, ed. J. I. D. Campbell, 143–162. New York: Psychology Press.

Bjorklund, D. F., and A. D. Pellegrini. 2002. *Evolutionary Developmental Psychology*. Washington, DC: American Psychological Association.

Bloom, L. 1973. *One Word at a Time: The Use of Single Word Utterances before Syntax*. Vol. 154. The Hague: Mouton.

Bloom, L. 1970. *Language Development; Form and Function in Emerging Grammars*. Cambridge, MA: MIT Press.

Bloom, P. 2000. *How Children Learn the Meaning of Words*. Cambridge, MA: MIT Press.

Bolland, S., and S. Emami. 2007. "The Benefits of Boredom: An Exploration in Developmental Robotics." Paper presented at the 2007 IEEE Symposium on Artificial Life, New York.

Bongard, J. C., and R. Pfeifer. 2003. "Evolving Complete Agents Using Artificial Ontogeny." In *Morpho-Functional Machines: The New Species: Designing Embodied Intelligence*, ed. F. Hara and R. Pfeifer, 237–258. Berlin: Springer-Verlag.

Borenstein, E., and E. Ruppin. 2005. "The Evolution of Imitation and Mirror Neurons in Adaptive Agents." *Cognitive Systems Research* 6 (3) (Sept.): 229–242.

Borghi, A. M., and F. Cimatti. 2010. "Embodied Cognition and Beyond: Acting and Sensing the Body." *Neuropsychologia* 48 (3) (Feb.): 763–773.

Borghi, A. M., A. Flumini, F. Cimatti, D. Marocco, and C. Scorolli. 2011. "Manipulating Objects and Telling Words: A Study on Concrete and Abstract Words Acquisition." *Frontiers in Psychology* 2 (15): 1–14.

Borisyuk, R., Y. Kazanovich, D. Chik, V. Tikhanoff, and A. Cangelosi. 2009. "A Neural Model of Selective Attention and Object Segmentation in the Visual Scene: An Approach Based on Partial Synchronization and Star-Like Architecture of Connections." *Neural Networks* 22 (5–6) (July–Aug.): 707–719.

Bortfeld, H., J. L. Morgan, R. M. Golinkoff, and K. Rathbun. 2005. "Mommy and Me—Familiar Names Help Launch Babies into Speech-Stream Segmentation." *Psychological Science* 16 (4): 298–304.

Bower, T. G. R., J. M. Broughton, and M. K. Moore. 1970. "Demonstration of Intention in Reaching Behaviour of Neonate Humans." *Nature* 228 (5272) (Nov.): 679–681.

Bradski, G., and A. Kaehler. 2008. *Learning Opencv: Computer Vision with the Opencv Library*. Sebastopol, CA: O'Reilly.

Braine, M. D. 1976. *Children's First Word Combinations*. Chicago: University of Chicago.

Brandl, H., B. Wrede, F. Joublin, and C. Goerick. 2008. "A Self-Referential Childlike Model to Acquire Phones, Syllables and Words from Acoustic Speech." Paper presented at the IEEE 7th International Conference on Development and Learning, Monterey, CA, August 9–12.

Breazeal, C. 2003. "Emotion and Sociable Humanoid Robots." *International Journal of Human-Computer Studies* 59 (1–2) (July): 119–155.

Breazeal, C., D. Buchsbaum, J. Gray, D. Gatenby, and B. Blumberg. 2005. "Learning from and About Others: Towards Using Imitation to Bootstrap the Social Understanding of Others by Robots." *Artificial Life* 11 (1–2): 31–62.

Breazeal, C., and B. Scassellati. 2002. "Robots That Imitate Humans." *Trends in Cognitive Sciences* 6 (11): 481–487.

Bril, B., and Y. Breniere. 1992. "Postural Requirements and Progression Velocity in Young Walkers." *Journal of Motor Behavior* 24 (1) (Mar.): 105–116.

Bromberg-Martin, E. S., and O. Hikosaka. 2009. "Midbrain Dopamine Neurons Signal Preference for Advance Information About Upcoming Rewards." *Neuron* 63 (1): 119–126.

Bronson, G. W. 1991. "Infant Differences in Rate of Visual Encoding." *Child Development* 62 (1) (Feb.): 44–54.

Brooks, R. A. 1991. "Intelligence without Representation." *Artificial Intelligence* 47 (1–3): 139–159.

Brooks, R. A. 1990. "Elephants Don't Play Chess." *Robotics and Autonomous Systems* 6 (1): 3–15.

Brooks, R. A. 1986. "A Robust Layered Control-System for a Mobile Robot." *IEEE Journal on Robotics and Automation* 2 (1) (Mar.): 14–23.

Brooks, R. A., C. Breazeal, R. Irie, C. C. Kemp, M. Marjanovic, B. Scassellati, and M. M. Williamson. 1998 "Alternative Essences of Intelligence." Paper presented at the Fifteenth National Conference on Artificial Intelligence, Menlo Park, CA.

Brooks, R. A., C. Breazeal, M. Marjanović, B. Scassellati, and M. M. Williamson. 1999. "The Cog Project: Building a Humanoid Robot." In *Computation for Metaphors, Analogy, and Agents*, ed. C. L. Nehaniv, 52–87. Berlin: Springer-Verlag.

Browatzki, B., V. Tikhanoff, G. Metta, H. H. Bulthoff, and C. Wallraven. 2012. "Active Object Recognition on a Humanoid Robot." Paper presented at the IEEE International Conference on Robotics and Automation, St. Paul, MN, May 14–18.

Browman, C. P., and L. Goldstein. 2000. "Competing Constraints on Intergestural Coordination and Self-Organization of Phonological Structures." *Les Cahiers de l'IC, Bulletin de la Communication Parlée* 5:25–34.

Bruner, J. S., and H. Haste, eds. 1987. *Making Sense: The Child's Construction of the World*. New York: Methuen & Co.

Bucciarelli, M., and P. N. Johnson-Laird. 1999. "Strategies in Syllogistic Reasoning." *Cognitive Science* 23 (3) (July–Sept.): 247–303.

Bullock, D., S. Grossberg, and F. H. Guenther. 1993. "A Self-Organizing Neural Model of Motor Equivalent Reaching and Tool Use by a Multijoint Arm." *Journal of Cognitive Neuroscience* 5 (4) (Fall): 408–435.

Burghart, C., R. Mikut, R. Stiefelhagen, T. Asfour, H. Holzapfel, P. Steinhaus, and R. Dillmann. 2005. "A Cognitive Architecture for a Humanoid Robot: A First Approach." Paper presented at the 5th IEEE-RAS International Conference on Humanoid Robots, New York.

Bushnell, E. W. 1985. "The Decline of Visually Guided Reaching During Infancy." *Infant Behavior and Development* 8 (2): 139–155.

Bushnell, E. W., and J. P. Boudreau. 1993. "Motor Development and the Mind—the Potential Role of Motor Abilities as a Determinant of Aspects of Perceptual Development." *Child Development* 64 (4): 1005–1021.

Bushnell, I. W. R. 2001. "Mother's Face Recognition in Newborn Infants: Learning and Memory." *Infant and Child Development* 10:67–74.

Bushnell, I. W. R., F. Sai, and J. T. Mullin. 1989. "Neonatal Recognition of the Mother's Face." *British Journal of Developmental Psychology* 7 (1): 3–15.

Butler, R. A. 1953. "Discrimination Learning by Rhesus Monkeys to Visual-Exploration Motivation." *Journal of Comparative and Physiological Psychology* 46 (2): 95–98.

Butterworth, G. 1992. "Origins of Self-Perception in Infancy." *Psychological Inquiry* 3 (2): 103–111.

Butterworth, G. 1991. "The Ontogeny and Phylogeny of Joint Visual Attention." In *Natural Theories of Mind*, ed. A. Whiten, 223–232. Oxford, UK: Blackwell Publishers.

Butterworth, G., and N. Jarrett. 1991. "What Minds Have in Common Is Space—Spatial Mechanisms Serving Joint Visual-Attention in Infancy." *British Journal of Developmental Psychology* 9 (Mar.): 55–72.

Byrne, R. M. J. 1989. "Suppressing Valid Inferences with Conditionals." *Cognition* 31 (1) (Feb.): 61–83.

Caligiore, D., A. M. Borghi, D. Parisi, and G. Baldassarre. 2010. "TRoPICALS: A Computational Embodied Neuroscience Model of Compatibility Effects." *Psychological Review* 117 (4) (Oct.): 1188–1228.

Caligiore, D., A. M. Borghi, D. Parisi, R. Ellis, A. Cangelosi, and G. Baldassarre. 2013. "How Affordances Associated with a Distractor Object Affect Compatibility Effects: A Study with the Computational Model TRoPICALS." *Psychological Research-Psychologische Forschung* 77 (1): 7–19.

Caligiore, D., T. Ferrauto, D. Parisi, N. Accornero, M. Capozza, and G. Baldassarre. 2008. "Using Motor Babbling and Hebb Rules for Modeling the Development of Reaching with Obstacles and Grasping." Paper presented at the International Conference on Cognitive Systems, University of Karlsruhe, Karlsruhe, Germany, April 2–4.

Call, J., and M. Carpenter. 2002. "Three Sources of Information in Social Learning." In *Imitation in Animals and Artifacts*, ed. K. Dautenhahn and C. L. Nehaniv, 211–228. Cambridge, MA: MIT Press.

Call, J., and M. Tomasello. 2008. "Does the Chimpanzee Have a Theory of Mind? 30 Years Later." *Trends in Cognitive Sciences* 12 (5): 187–192.

Campbell, J. I. D. 2005. *Handbook of Mathematical Cognition*. New York: Psychology Press.

Campos, J. J., B. I. Bertenthal, and R. Kermoian. 1992. "Early Experience and Emotional Development: The Emergence of Wariness of Heights." *Psychological Science* 3 (1): 61–64.

Campos, J. J., A. Langer, and A. Krowitz. 1970. "Cardiac Responses on Visual Cliff in Prelocomotor Human Infants." *Science* 170 (3954): 196–197.

Canfield, R. L., and M. M. Haith. 1991. "Young Infants Visual Expectations for Symmetrical and Asymmetric Stimulus Sequences." *Developmental Psychology* 27 (2) (Mar.): 198–208.

Cangelosi, A. 2010. "Grounding Language in Action and Perception: From Cognitive Agents to Humanoid Robots." *Physics of Life Reviews* 7 (2) (Jun.): 139–151.

Cangelosi, A. 1999. "Heterochrony and Adaptation in Developing Neural Networks." Paper presented at the Genetic and Evolutionary Computation Conference, San Francisco, CA.

Cangelosi, A., A. Greco, and S. Harnad. 2000. "From Robotic Toil to Symbolic Theft: Grounding Transfer from Entry-Level to Higher-Level Categories." *Connection Science* 12 (2) (Jun.): 143–162.

Cangelosi, A., G. Metta, G. Sagerer, S. Nolfi, C. Nehaniv, K. Fischer, J. Tani, et al. 2010. "Integration of Action and Language Knowledge: A Roadmap for Developmental Robotics." *IEEE Transactions on Autonomous Mental Development* 2 (3) (Sept.): 167–195.

Cangelosi, A., and D. Parisi, eds. 2002. *Simulating the Evolution of Language*. London: Springer.

Cangelosi, A., and T. Riga. 2006. "An Embodied Model for Sensorimotor Grounding and Grounding Transfer: Experiments with Epigenetic Robots." *Cognitive Science* 30 (4) (July–Aug.): 673–689.

Cangelosi, A., V. Tikhanoff, J. F. Fontanari, and E. Hourdakis. 2007. "Integrating Language and Cognition: A Cognitive Robotics Approach." *IEEE Computational Intelligence Magazine* 2 (3) (Aug.): 65–70.

Cannata, G., M. Maggiali, G. Metta, and G. Sandini. 2008. "An Embedded Artificial Skin for Humanoid Robots." Paper presented at the International Conference on Multisensor Fusion and Integration for Intelligent Systems, Seoul, Korea, August 20–22.

Čapek, K. 1920. *R.U.R. (Rossumovi Univerzální Roboti)*, Prague, Aventinum (English transaltion by Claudia Novack, 2004, *R.U.R. Rossum's Universal Robots*, New York: Penguin Books)

Caramelli, N., A. Setti, and D. D. Maurizzi. 2004. "Concrete and Abstract Concepts in School Age Children." *Psychology of Language and Communication* 8 (2): 19–34.

Carey, S. 2009. *The Origin of Concepts*. Oxford, UK: Oxford University Press.

Carlson, E., and J. Triesch. 2004. *A Computational Model of the Emergence of Gaze Following. Connectionist Models of Cognition and Perception II*. Vol. 15. Ed. H. Bowman and C. Labiouse. Singapore: World Scientific Publishing.

Carlson, T., and Y. Demiris. 2012. "Collaborative Control for a Robotic Wheelchair: Evaluation of Performance, Attention, and Workload." *IEEE Transactions on Systems, Man, and Cybernetics. Part B, Cybernetics* 42 (3) (Jun.): 876–888.

Carpenter, M. 2009. "Just How Joint Is Joint Action in Infancy?" *Topics in Cognitive Science* 1 (2) (Apr.): 380–392.

Carpenter, M., K. Nagell, and M. Tomasello. 1998. "Social Cognition, Joint Attention, and Communicative Competence from 9 to 15 Months of Age." *Monographs of the Society for Research in Child Development* 63 (4): 1–143.

Carpenter, M., M. Tomasello, and T. Striano. 2005. "Role Reversal Imitation and Language in Typically Developing Infants and Children with Autism." *Infancy* 8 (3): 253–278.

Chaminade, T., and G. Cheng. 2009. "Social Cognitive Neuroscience and Humanoid Robotics." *Journal of Physiology, Paris* 103 (3–5) (May–Sept.): 286–295.

Charlesworth, W. R. 1969. "The Role of Surprise in Cognitive Development." In *Studies in Cognitive Development. Essays in Honor of Jean Piaget*, ed. D. Elkind and J. Flavell, 257–314. Oxford, UK: Oxford University Press.

Chaudhuri, A. 2011. *Fundamentals of Sensory Perception*. Don Mills, Ontario, Canada: Oxford University Press.

Chella, A., and R. Manzotti. 2007. *Artificial Consciousness*. Exeter, UK: Imprint Academic.

Chen, Q., and T. Verguts. 2010. "Beyond the Mental Number Line: A Neural Network Model of Number-Space Interactions." *Cognitive Psychology* 60 (3) (May): 218–240.

Chen, Y., and J. Weng. 2004. "Developmental Learning: A Case Study in Understanding 'Object Permanence.'" Paper presented at the 4th International Workshop on Epigenetic Robotics: Modeling Cognitive Development in Robotic Systems, Genoa, Italy, August 25–27.

Cheng, G., N. A. Fitzsimmons, J. Morimoto, M. A. Lebedev, M. Kawato, and M. A. Nicolelis. 2007a. "Bipedal Locomotion with a Humanoid Robot Controlled by Cortical Ensemble Activity." Paper presented at the Society for Neuroscience 37th Annual Meeting, San Diego, CA.

Cheng, G., S. H. Hyon, J. Morimoto, A. Ude, J. G. Hale, G. Colvin, W. Scroggin, and S. C. Jacobsen. 2007b. "CB: A Humanoid Research Platform for Exploring Neuroscience." *Advanced Robotics* 21 (10): 1097–1114.

Choi, S. 1988. "The Semantic Development of Negation—A Cross-Linguistic Longitudinal-Study." *Journal of Child Language* 15 (3) (Oct.): 517–531.

Chomsky, N. 1965. *Aspects of the Theory of Syntax*. Vol. 11. Cambridge, MA: MIT Press.

Chomsky, N. 1957. *Syntactic Structures*. The Hague, Netherlands: Mouton.

Christensen, D. J. 2006. "Evolution of Shape-Changing and Self-Repairing Control for the Atron Self-Reconfigurable Robot." Paper presented at the IEEE International Conference on Robotics and Automation, New York.

Christensen, W. D., and C. A. Hooker. 2000. "An Interactivist-Constructivist Approach to Intelligence: Self-Directed Anticipative Learning." *Philosophical Psychology* 13 (1) (Mar): 5–45.

Christiansen, M. H., and N. Chater. 2001. "Connectionist Psycholinguistics: Capturing the Empirical Data." *Trends in Cognitive Sciences* 5 (2): 82–88.

Clark, A. 1997. *Being There: Putting Brain, Body, and World Together Again*. Cambridge, MA: MIT Press.

Clark, E. V. 1993. *The Lexicon in Acquisition*. Cambridge, UK: Cambridge University Press.

Clark, J. E., and S. J. Phillips. 1993. "A Longitudinal-Study of Intralimb Coordination in the 1st Year of Independent Walking—a Dynamical-Systems Analysis." *Child Development* 64 (4) (Aug.): 1143–1157.

Clifton, R. K., D. W. Muir, D. H. Ashmead, and M. G. Clarkson. 1993. "Is Visually Guided Reaching in Early Infancy a Myth?" *Child Development* 64 (4) (Aug.): 1099–1110.

CMU. 2008. "The CMU Pronouncing Dictionary." http://www.speech.cs.cmu.edu/cgi-bin/cmudict.

Collett, T. H. J., B. A. MacDonald, and B. P. Gerkey. 2005. "Player 2.0: Toward a Practical Robot Programming Framework." Paper presented at the Australasian Conference on Robotics and Automation (ACRA 2005), Sydney, Australia, December 5–7.

Collins, S., A. Ruina, R. Tedrake, and M. Wisse. 2005. "Efficient Bipedal Robots Based on Passive-Dynamic Walkers." *Science* 307 (5712) (Feb.): 1082–1085.

Colombo, J., and C. L. Cheatham. 2006. "The Emergence and Basis of Endogenous Attention in Infancy and Early Childhood." In *Advances in Child Development and Behavior*, ed. R. V. Kail, 283–322. New York: Academic Press.

Colombo, J., and D. W. Mitchell. 2009. "Infant Visual Habituation." *Neurobiology of Learning and Memory* 92 (2) (Sept.): 225–234.

Cook, G., and J. Littlefield Cook. 2014. *The World of Children*. 3rd ed. Upper Saddle River, NJ: Pearson Education, Inc.

Cordes, S., and R. Gelman. 2005. "The Young Numerical Mind: When Does It Count?" In *Handbook of Mathematical Cognition*, ed. J. Campbell, 127–142. New York: Psychology Press.

Cos-Aguilera, I., L. Cañamero, and G. M. Hayes. 2003. "Motivation-Driven Learning of Object Affordances: First Experiments Using a Simulated Khepera Robot." Paper presented at the 9th International Conference in Cognitive Modelling, Bamberg, Germany.

Courage, M. L., and M. L. Howe. 2002. "From Infant to Child: The Dynamics of Cognitive Change in the Second Year of Life." *Psychological Bulletin* 128 (2) (Mar.): 250–277.

Coventry, K. R., A. Cangelosi, S. Newstead, A. Bacon, and R. Rajapakse. 2005. "Grounding Natural Language Quantifiers in Visual Attention." Paper presented at the 27th Annual Meeting of the Cognitive Science Society, Stresa, Italy.

Cowan, R., A. Dowker, A. Christakis, and S. Bailey. 1996. "Even More Precisely Assessing Children's Understanding of the Order-Irrelevance Principle." *Journal of Experimental Child Psychology* 62 (1) (June): 84–101.

Cox, I. J., and S. L. Hingorani. 1996. "An Efficient Implementation of Reid's Multiple Hypothesis Tracking Algorithm and Its Evaluation for the Purpose of Visual Tracking." *IEEE Transactions on Pattern Analysis and Machine Intelligence* 18 (2): 138–150.

Croker, S. 2012. *The Development of Cognition*. Andover, UK: Cengage Learning EMEA.

Daelemans, W., and A. Van den Bosch. 2005. *Memory-Based Language Processing*. Cambridge, UK: Cambridge University Press.

Dannemiller, J. L. 2000. "Competition in Early Exogenous Orienting between 7 and 21 Weeks." *Journal of Experimental Child Psychology* 76 (4) (Aug.): 253–274.

Dautenhahn, K. 1999. "Robots as Social Actors: Aurora and the Case of Autism." Paper presented at the Proceedings of the Third International Cognitive Technology Conference, August, San Francisco.

Dautenhahn, K., C. L. Nehaniv, M. L. Walters, B. Robins, H. Kose-Bagci, N. A. Mirza, and M. Blow. 2009. "KASPAR—a Minimally Expressive Humanoid Robot for Human–Robot Interaction Research." *Applied Bionics and Biomechanics* 6 (3–4): 369–397.

Dautenhahn, K., and I. Werry. 2004. "Towards Interactive Robots in Autism Therapy: Background, Motivation and Challenges." *Pragmatics & Cognition* 12 (1): 1–35.

Davies, M., and T. Stone. 1995. *Mental Simulation: Evaluations and Applications*. Oxford, UK: Blackwell Publishers.

de Boer, B. 2010. "Investigating the Acoustic Effect of the Descended Larynx with Articulatory Models." *Journal of Phonetics* 38 (4) (Oct.): 679–686.

de Boer, B. 2001. *The Origins of Vowel Systems: Studies in the Evolution of Language*. Oxford, UK: Oxford University Press.

de Charms, R. 1968. *Personal Causation: The Internal Affective Determinants of Behavior*. New York: Academic Press.

de Haan, M., O. Pascalis, and M. H. Johnson. 2002. "Specialization of Neural Mechanisms Underlying Face Recognition in Human Infants." *Journal of Cognitive Neuroscience* 14 (2): 199–209.

de Pina Filho, A. C., ed. 2007. *Humanoid Robots: New Developments*. Vienna: I-Tech Education and Publishing.

Deci, E. L., and R. M. Ryan. 1985. *Intrinsic Motivation and Self-Determination in Human Behavior*. New York: Plenum Press.

Degallier, S., L. Righetti, and A. Ijspeert. 2007. "Hand Placement During Quadruped Locomotion in a Humanoid Robot: A Dynamical System Approach." Paper presented at the IEEE/RSJ International Conference on Intelligent Robots and Systems, San Diego, CA, October 29–November 2.

Degallier, S., L. Righetti, L. Natale, F. Nori, G. Metta, and A. Ijspeert. 2008. "A Modular Bio-Inspired Architecture for Movement Generation for the Infant-Like Robot iCub." Paper presented at the 2nd IEEE RAS & EMBS International Conference on Biomedical Robotics and Biomechatronics, New York, October 19–22.

Dehaene, S. 1997. *The Number Sense: How the Mind Creates Mathematics*. Oxford, UK: Oxford University Press.

Dehaene, S., S. Bossini, and P. Giraux. 1993. "The Mental Representation of Parity and Number Magnitude." *Journal of Experimental Psychology: General* 122 (3) (Sept.): 371–396.

Della Rosa, P. A., E. Catricala, G. Vigliocco, and S. F. Cappa. 2010. "Beyond the Abstract-Concrete Dichotomy: Mode of Acquisition, Concreteness, Imageability, Familiarity, Age of Acquisition,

Context Availability, and Abstractness Norms for a Set of 417 Italian Words." *Behavior Research Methods* 42 (4) (Nov.): 1042–1048.

Demiris, Y., and G. Hayes. 2002. "Imitation as a Dual-Route Process Featuring Predictive and Learning Components: A Biologically Plausible Computational Model." In *Imitation in Animals and Artifacts*, ed. K. Dautenhahn and C. L. Nehaniv, 327–361. Cambridge, MA: MIT Press.

Demiris, Y., S. Rougeaux, G. M. Hayes, L. Berthouze, and Y. Kuniyoshi. 1997. "Deferred Imitation of Human Head Movements by an Active Stereo Vision Head." Paper presented at the 6th IEEE International Workshop on Robot and Human Communication, New York, September 29–October 1.

Demiris, Y., and A. Dearden. 2005. "From Motor Babbling to Hierarchical Learning by Imitation: A Robot Developmental Pathway." In *Proceedings of the 5th International Workshop on Epigenetic Robotics: Modeling Cognitive Development in Robotic Systems*, ed. L. Berthouze et al., 31–37. Vol. 123. Lund University Cognitive Studies. http://cogprints.org/4961.

Demiris, Y., and M. Johnson. 2003. "Distributed, Predictive Perception of Actions: A Biologically Inspired Robotics Architecture for Imitation and Learning." *Connection Science* 15 (4) (Dec.): 231–243.

Demiris, Y., and B. Khadhouri. 2006. "Hierarchical Attentive Multiple Models for Execution and Recognition of Actions." *Robotics and Autonomous Systems* 54 (5): 361–369.

Demiris, Y., and A. Meltzoff. 2008. "The Robot in the Crib: A Developmental Analysis of Imitation Skills in Infants and Robots." *Infant and Child Development* 17 (1) (Jan.–Feb.): 43–53.

Demiris, Y., and G. Simmons. 2006. "Perceiving the Unusual: Temporal Properties of Hierarchical Motor Representations for Action Perception." *Neural Networks* 19 (3): 272–284.

Dickerson, P., B. Robins, and K. Dautenhahn. 2013. "Where the Action Is: A Conversation Analytic Perspective on Interaction between a Humanoid Robot, a Co-Present Adult and a Child with an ASD." *Interaction Studies: Social Behaviour and Communication in Biological and Artificial Systems* 14 (2): 297–316.

Dittes, B., M. Heracles, T. Michalke, R. Kastner, A. Gepperth, J. Fritsch, and C. Goerick. 2009. "A Hierarchical System Integration Approach with Application to Visual Scene Exploration for Driver Assistance." In *Proceedings of the 7th International Conference on Computer Vision Systems: Computer Vision Systems*, ed. M. Fritz, B. Schiele, and J. H. Piater, 255–264. Berlin: Springer-Verlag.

Dominey, P. F., and J. D. Boucher. 2005a. "Developmental Stages of Perception and Language Acquisition in a Perceptually Grounded Robot." *Cognitive Systems Research* 6 (3): 243–259.

Dominey, P. F., and J. D. Boucher. 2005b. "Learning to Talk about Events from Narrated Video in a Construction Grammar Framework." *Artificial Intelligence* 167 (1–2) (Dec.): 31–61.

Dominey, P. F., M. Hoen, and T. Inui. 2006. "A Neurolinguistic Model of Grammatical Construction Processing." *Journal of Cognitive Neuroscience* 18 (12): 2088–2107.

Dominey, P. F., and F. Warneken. 2011. "The Basis of Shared Intentions in Human and Robot Cognition." *New Ideas in Psychology* 29 (3) (Dec.): 260–274.

Driesen, J., L. ten Bosch, and H. van Hamme. 2009. "Adaptive Non-negative Matrix Factorization in a Computational Model of Language Acquisition." Paper presented at the 10th Annual Conference of the International Speech Communication Association, Brighton, UK, September 6–10.

Eimas, P. D., E. R. Siquelan, P. Jusczyk, and J. Vigorito. 1971 "Speech Perception in Infants." *Science* 171 (3968): 303–306.

Elman, J. L., E. A. Bates, M. H. Johnson, A. Karmiloff-Smith, D. Parisi, and K. Plunkett. 1996. *Rethinking Innateness: A Connectionist Perspective on Development.* Vol. 10. Cambridge, MA: MIT Press.

Ennouri, K., and H. Bloch. 1996. "Visual Control of Hand Approach Movements in New-Borns. *British Journal of Developmental Psychology* 14 (Sept.): 327–338.

Erhardt, R. P. 1994. *Developmental Hand Dysfunction: Theory, Assessment, and Treatment.* Tuscon, AZ: Therapy Skill Builders.

Fadiga, L., L. Fogassi, G. Pavesi, and G. Rizzolatti. 1995. "Motor Facilitation During Action Observation: A Magnetic Stimulation Study." *Journal of Neurophysiology* 73 (6): 2608–2611.

Fantz, R. L. 1956. "A Method for Studying Early Visual Development." *Perceptual and Motor Skills* 6: 13–15.

Fasel, I., G. O. Deak, J. Triesch, and J. Movellan. 2002. "Combining Embodied Models and Empirical Research for Understanding the Development of Shared Attention." Paper presented at the 2nd International Conference on Development and Learning, Massachusetts Institute of Technology, Cambridge, MA, June 12–15.

Fenson, L., P. S. Dale, J. S. Reznick, E. Bates, D. J. Thal, and S. J. Pethick. 1994. "Variability in Early Communicative Development." *Monographs of the Society for Research in Child Development* 59 (5): 1–173, discussion 74–85.

Ferrari, P. F., E. Visalberghi, A. Paukner, L. Fogassi, A. Ruggiero, and S. J. Suomi. 2006. "Neonatal Imitation in Rhesus Macaques." *PLoS Biology* 4 (9): 1501–1508.

Ferrera, V. P., and A. Barborica. 2010. "Internally Generated Error Signals in Monkey Frontal Eye Field During an Inferred Motion Task." *Journal of Neuroscience* 30 (35) (Sept.): 11612–11623.

Field, J. 1977. "Coordination of Vision and Prehension in Young Infants." *Child Development* 48 (1): 97–103.

Field, T. M., R. Woodson, D. Cohen, R. Greenberg, R. Garcia, and K. Collins. 1983. "Discrimination and Imitation of Facial Expressions by Term and Preterm Neonates." *Infant Behavior and Development* 6 (4): 485–489.

Fiore, V., F. Mannella, M. Mirolli, K. Gurney, and G. Baldassarre. 2008. "Instrumental Conditioning Driven by Neutral Stimuli: A Model Tested with a Simulated Robotic Rat." Paper presented at the 8th International Workshop on Epigenetic Robotics: Modeling Cognitive Development in Robotic Systems, Brighton, UK, July 30–31.

Fischer, K. W. 1980. "A Theory of Cognitive-Development: The Control and Construction of Hierarchies of Skills." *Psychological Review* 87 (6): 477–531.

Fischer, M. H. 2008. "Finger Counting Habits Modulate Spatial-Numerical Associations." *Cortex* 44 (4) (Apr.): 386–392.

Fischer, M. H., A. D. Castel, M. D. Dodd, and J. Pratt. 2003. "Perceiving Numbers Causes Spatial Shifts of Attention." *Nature Neuroscience* 6 (6): 555–556.

Fitzpatrick, P., and A. Arsenio. 2004. "Feel the Beat: Using Cross-Modal Rhythm to Integrate Perception of Objects, Others, and Self." Paper presented at the 4th International Workshop on Epigenetic Robotics: Modeling Cognitive Development in Robotic Systems, Genoa, Italy, August 25–27.

Fitzpatrick, P. M., and G. Metta. 2002."Towards Manipulation-Driven Vision." Paper presented at the IEEE/RSJ International Conference on Intelligent Robots and Systems, Lausanne, Switzerland, September 30–October 4.

Fitzpatrick, P., A. Needham, L. Natale, and G. Metta. 2008. "Shared Challenges in Object Perception on for Robots and Infants." *Infant and Child Development* 17 (1) (Jan.–Feb.): 7–24.

Flege, J. E. 1987. "A Critical Period for Learning to Pronounce Foreign-Languages." *Applied Linguistics* 8 (2) (Summer): 162–177.

Fontaine, R. 1984. "Imitative Skills between Birth and Six Months." *Infant Behavior and Development* 7 (3): 323–333.

Förster, F. 2013. "Robots that Say 'No': Acquisition of Linguistic Behaviour in Interaction Games with Humans." PhD Thesis. Hertfordshire University, UK.

Förster, F., C. L. Nehaniv, and J. Saunders. 2011. "Robots That Say 'No.'" In *Advances in Artificial Life: Darwin Meets Von Neumann*, ed. G. Kampis, I. Karsai, and E. Szathmáry, 158–166. Berlin: Springer-Verlag.

François, D., K. Dautenhahn, and D. Polani. 2009a. "Using Real-Time Recognition of Human-Robot Interaction Styles for Creating Adaptive Robot Behaviour in Robot-Assisted Play." Paper presented at the 2nd IEEE Symposium on Artificial Life, Nashville, TN.

François, D., S. Powell, and K. Dautenhahn. 2009b. "A Long-Term Study of Children with Autism Playing with a Robotic Pet Taking Inspirations from Non-Directive Play Therapy to Encourage Children's Proactivity and Initiative-Taking." *Interaction Studies: Social Behaviour and Communication in Biological and Artificial Systems* 10 (3): 324–373.

Franklin, S., T. Madl, S. D'Mello, and J. Snaider. 2014. "Lida: A Systems-Level Architecture for Cognition, Emotion, and Learning." *IEEE Transactions on Autonomous Mental Development* 6 (1) (Mar.): 19–41.

Franz, A., and J. Triesch. 2010. "A Unified Computational Model of the Development of Object Unity, Object Permanence, and Occluded Object Trajectory Perception." *Infant Behavior and Development* 33 (4) (Dec.): 635–653.

Freedland, R. L., and B. I. Bertenthal. 1994. "Developmental-Changes in Interlimb Coordination—Transition to Hands-and-Knees Crawling." *Psychological Science* 5 (1): 26–32.

Fritz, G., L. Paletta, R. Breithaupt, E. Rome, and G. Dorffner. 2006. "Learning Predictive Features in Affordance Based Robotic Perception Systems." Paper presented at the IEEE/RSJ International Conference on Intelligent Robots and Systems, Beijing, China, October 9–15.

Fujita, M. 2001. "AIBO: Toward the Era of Digital Creatures." *International Journal of Robotics Research* 20 (10): 781–794.

Fujita, M., Y. Kuroki, T. Ishida, and T. T. Doi. 2003. "A Small Humanoid Robot Sdr-4x for Entertainment Applications." Paper presented at the IEEE/ASME International Conference on Advanced Intelligent Mechatronics (AIM2003), Kobe, Japan, July 20–24.

Fuke, S., M. Ogino, and M. Asada. 2007. "Visuo-Tactile Face Representation through Self-Induced Motor Activity." Paper presented at the 1st International Conference on Epigenetic Robotics: Modeling Cognitive Development in Robotic Systems, Lund, Sweden, September 17–18.

Furl, N., P. J. Phillips, and A. J. O'Toole. 2002. "Face Recognition Algorithms and the Other-Race Effect: Computational Mechanisms for a Developmental Contact Hypothesis." *Cognitive Science* 26 (6) (Nov.–Dec.): 797–815.

Gallup, G. G. 1970. "Chimpanzees: Self-Recognition." *Science* 167 (3914) (Jan. 2): 86–87.

Ganger, J., and M. R. Brent. 2004. "Reexamining the Vocabulary Spurt." *Developmental Psychology* 40 (4) (July): 621–632.

Gaur, V., and B. Scassellati. 2006. "A Learning System for the Perception of Animacy." Paper presented at the 6th International Conference on Development and Learning, Bloomington, IN.

Gelman, R. 1990. "First Principles Organize Attention to and Learning About Relevant Data: Number and the Animate-Inanimate Distinction as Examples." *Cognitive Science* 14 (1): 79–106.

Gelman, R., E. Meck, and S. Merkin. 1986. "Young Childrens Numerical Competence." *Cognitive Development* 1 (1): 1–29.

Gelman, R., and M. F. Tucker. 1975. "Further Investigations of the Young Child's Conception of Number." *Child Development* 46 (1): 167–175.

Gentner, D. 2010. "Bootstrapping the Mind: Analogical Processes and Symbol Systems." *Cognitive Science* 34 (5) (July): 752–775.

Geppert, L. 2004. "Qrio the Robot That Could." *IEEE Spectrum* 41 (5): 34–36.

Gerber, R. J., T. Wilks, and C. Erdie-Lalena. 2010. "Developmental Milestones: Motor Development." *Pediatrics* 31 (7) (July): 267–277.

Gergely, G. 2003. "What Should a Robot Learn from an Infant? Mechanisms of Action Interpretation and Observational Learning in Infancy." *Connection Science* 15 (4) (Dec.): 191–209.

Gergely, G., and J. S. Watson. 1999. "Early Socio-Emotional Development: Contingency Perception and the Social-Biofeedback Model." In *Early Social Cognition: Understanding Others in the First Months of Life*, ed. P. Rochat, 101–136. Mahwah, NJ: Erlbaum.

Gesell, A. 1946. "The Ontogenesis of Infant Behavior" In *Manual of Child Psychology*, ed. L. Carmichael, 295–331. New York: Wiley.

Gesell, A. 1945. *The Embryology of Behavior*. New York: Harper and Row.

Gibson, E. J., and R. D. Walk. 1960. "The 'Visual Cliff.'" *Scientific American* 202: 64–71.

Gibson, J. J. 1986. *The Ecological Approach to Visual Perception*. Hillsdale, NJ: Lawrence Erlbaum Associates.

Gilmore, R. O., and H. Thomas. 2002. "Examining Individual Differences in Infants' Habituation Patterns Using Objective Quantitative Techniques." *Infant Behavior and Development* 25 (4): 399–412.

Ginsburg, H. P., and S. Opper. 1988. *Piaget's Theory of Intellectual Development*. Englewood Cliffs, NJ: Prentice-Hall, Inc.

Gläser, C., and F. Joublin. 2010. "A Computational Model for Grounding Words in the Perception of Agents." Paper presented at the 9th IEEE International Conference on Development and Learning, Ann Arbor, MI, August 18–21.

Gleitman, L. 1990. "The Structural Sources of Verb Meanings." *Language Acquisition* 1 (1): 3–55.

Goerick, C., B. Bolder, H. Janssen, M. Gienger, H. Sugiura, M. Dunn, I. Mikhailova, T. Rodemann, H. Wersing, and S. Kirstein. 2007. "Towards Incremental Hierarchical Behavior Generation for Humanoids." Paper presented at the 7th IEEE-RAS International Conference on Humanoid Robots, New York.

Goerick, C., J. Schmudderich, B. Bolder, H. Janssen, M. Gienger, A. Bendig, M. Heckmann, et al. 2009. "Interactive Online Multimodal Association for Internal Concept Building in Humanoids." Paper presented at the 9th IEEE-RAS International Conference on Humanoid Robots, Paris, France, December 7–10.

Gold, K., and B. Scassellati. 2009. "Using Probabilistic Reasoning over Time to Self-Recognize." *Robotics and Autonomous Systems* 57 (4): 384–392.

Goldberg, A. 2006. *Constructions at Work: The Nature of Generalization in Language*. Oxford, UK: Oxford University Press.

Goldfield, E. C. 1989. "Transition from Rocking to Crawling—Postural Constraints on Infant Movement." *Developmental Psychology* 25 (6) (Nov.): 913–919.

Goldfield, E. C., B. A. Kay, and W. H. Warren. 1993. "Infant Bouncing—the Assembly and Tuning of Action Systems." *Child Development* 64 (4) (Aug.): 1128–1142.

Golinkoff, R. M., C. B. Mervis, and K. Hirshpasek. 1994. "Early Object Labels—the Case for a Developmental Lexical Principles Framework." *Journal of Child Language* 21 (1) (Feb.): 125–155.

Gordon, S. M., K. Kawamura, and D. M. Wilkes. 2010. "Neuromorphically Inspired Appraisal-Based Decision Making in a Cognitive Robot." *IEEE Transactions on Autonomous Mental Development* 2 (1) (Mar.): 17–39.

Gori, I., U. Pattacini, F. Nori, G. Metta, and G. Sandini. 2012. "Dforc: A Real-Time Method for Reaching, Tracking and Obstacle Avoidance in Humanoid Robots." Paper presented at the IEEE/RSJ International Conference on Intelligent Robots and Systems, Osaka, Japan, November 29–December 1.

Gottlieb, J., P. Y. Oudeyer, M. Lopes, and A. Baranes. 2013. "Information-Seeking, Curiosity, and Attention: Computational and Neural Mechanisms." *Trends in Cognitive Sciences* 17 (11): 585–593.

Gouaillier, D., V. Hugel, P. Blazevic, C. Kilner, J. Monceaux, P. Lafourcade, B. Marnier, J. Serre, and B. Maisonnier. 2008. "The NAO Humanoid: A Combination of Performance and Affordability." *CoRR* abs/0807.3223.

Graham, T. A. 1999. "The Role of Gesture in Children's Learning to Count." *Journal of Experimental Child Psychology* 74 (4) (Dec.): 333–355.

Greenough, W. T., and J. E. Black. 1999. "Experience, Neural Plasticity, and Psychological Development." Paper presented at The Role of Early Experience in Infant Development, Pediatric Round Table, New York, January.

Guerin, F., and D. McKenzie. 2008. "A Piagetian Model of Early Sensorimotor Development." Paper presented at the 8th International Workshop on Epigenetic Robotics: Modeling Cognitive Development in Robotic Systems, Brighton, UK, July 30–31.

Guizzo, E. 2010. "The Robot Baby Reality Matrix." *IEEE Spectrum* 47 (7) (July): 16.

Gunkel, D. J., J. J. Bryson, and S. Torrance. 2012. "The Machine Question: AI, Ethics and Moral Responsibility." Paper presented at the Symposium of the AISB/IACAP World Congress 2012, Birmingham, UK.

Hafner, V. V., and F. Kaplan. 2008. "Interpersonal Maps: How to Map Affordances for Interaction Behaviour." In *Towards Affordance-Based Robot Control*, ed. E. Rome, J. Hertzberg, and G. Dorffner, 1–15. Berlin: Springer-Verlag.

Hafner, V. V., and F. Kaplan. 2005. "Learning to Interpret Pointing Gestures: Experiments with Four-Legged Autonomous Robots." In *Biomimetic Neural Learning for Intelligent Robots: Intelligent Systems, Cognitive Robotics, and Neuroscience*, ed. S. Wermter, G. Palm, and M. Elshaw, 225–234. Berlin: Springer-Verlag.

Hafner, V. V., and G. Schillaci. 2011. "From Field of View to Field of Reach—Could Pointing Emerge from the Development of Grasping?" Paper presented at the IEEE Conference on Development and Learning and Epigenetic Robotics, Frankfurt, Germany, August 24–27.

Haith, M. M. 1980. *Rules That Babies Look By: The Organization of Newborn Visual Activity*. Hillsdale, NJ: Erlbaum.

Haith, M. M., C. Hazan, and G. S. Goodman. 1988. "Expectation and Anticipation of Dynamic Visual Events by 3.5-Month-Old Babies." *Child Development* 59 (2) (Apr.): 467–479.

Haith, M. M., N. Wentworth, and R. L. Canfield. 1993. "The Formation of Expectations in Early Infancy." *Advances in Infancy Research* 8:251–297.

Hannan, M. W., and I. D. Walker. 2001. "Analysis and Experiments with an Elephant's Trunk Robot." *Advanced Robotics* 15 (8): 847–858.

Harlow, H. F. 1950. "Learning and Satiation of Response in Intrinsically Motivated Complex Puzzle Performance by Monkeys." *Journal of Comparative and Physiological Psychology* 43 (4): 289–294.

Harlow, H. F., M. K. Harlow, and D. R. Meyer. 1950. "Learning Motivated by a Manipulation Drive." *Journal of Experimental Psychology* 40 (2): 228–234.

Harnad, S. 1990. "The Symbol Grounding Problem." *Physica D. Nonlinear Phenomena* 42 (1–3) (June): 335–346.

Hase, K., and N. Yamazaki. 1998. "Computer Simulation of the Ontogeny of Bipedal Walking." *Anthropological Science* 106 (4) (Oct.): 327–347.

Hashimoto, T., S. Hitramatsu, T. Tsuji, and H. Kobayashi. 2006. "Development of the Face Robot Saya for Rich Facial Expressions." Paper presented at the SICE-ICASE International Joint Conference, New York.

Hawes, N., and J. Wyatt. 2008. "Developing Intelligent Robots with CAST." Paper presented at the IEEE/RSJ International Conference on Intelligent Robots and Systems, Nice, France, September, 22–26.

Higgins, C. I., J. J. Campos, and R. Kermoian. 1996. "Effect of Self-Produced Locomotion on Infant Postural Compensation to Optic Flow." *Developmental Psychology* 32 (5) (Sept.): 836–841.

Hikita, M., S. Fuke, M. Ogino, T. Minato, and M. Asada. 2008. "Visual Attention by Saliency Leads Cross-Modal Body Representation." Paper presented at the 7th IEEE International Conference on Development and Learning, Monterey, CA, August 9–12.

Hinton, G. E., and S. J. Nowlan. 1987. "How Learning Can Guide Evolution." *Complex Systems* 1 (3): 495–502.

Hiolle, A., and L. Cañamero. 2008. "Why Should You Care?—An Arousal-Based Model of Exploratory Behavior for Autonomous Robot." Paper presented at the 11th International Conference on Artificial Life, Cambridge, MA.

Hiraki, K., A. Sashima, and S. Phillips. 1998. "From Egocentric to Allocentric Spatial Behavior: A Computational Model of Spatial Development." *Adaptive Behavior* 6 (3–4) (Winter–Spring): 371–391.

Hirose, M., and K. Ogawa. 2007. "Honda Humanoid Robots Development." Philosophical Transactions of the Royal Society—Mathematical Physical and Engineering Sciences 365 (1850) (Jan.): 11–19.

Ho, W. C., K. Dautenhahn, and C. L. Nehaniv. 2006. "A Study of Episodic Memory-Based Learning and Narrative Structure for Autobiographic Agents." Paper presented at the AISB 2006: Adaptation in Artificial and Biological Systems Conference, University of Bristol, Bristol, UK, April 3–6.

Hofe, R., and R. K. Moore. 2008. "Towards an Investigation of Speech Energetics Using 'Anton': An Animatronic Model of a Human Tongue and Vocal Tract." *Connection Science* 20 (4): 319–336.

Hoff, E. 2009. *Language Development*. Belmont, CA: Wadsworth Thomson Learning.

Holland, O., and R. Knight. 2006."The Anthropomimetic Principle." Paper presented at the AISB 2006: Adaptation in Artificial and Biological Systems Conference, University of Bristol, Bristol, UK, April 3–6.

Hollerbach, J. M. 1990. "Planning of Arm Movements." In *An Invitation to Cognitive Science: Visual Cognition and Action*, ed. D. N. Osherson and S. M. Kosslyn, 183–211. Cambridge, MA: MIT Press.

Hornstein, J., and J. Santos-Victor. 2007. "A Unified Approach to Speech Production and Recognition Based on Articulatory Motor Representations." Paper presented at the IEEE/RSJ International Conference on Intelligent Robots and Systems, San Diego, CA, October 29–November 2.

Horswill, I. 2002. "Cerebus: A Higher-Order Behavior-Based System." *AI Magazine* 23 (1) (Spring): 27.

Horvitz, J. C. 2000. "Mesolimbocortical and Nigrostriatal Dopamine Responses to Salient Non-Reward Events." *Neuroscience* 96 (4): 651–656.

Hsu, F. H. 2002. *Behind Deep Blue: Building the Computer That Defeated the World Chess Champion*. Princeton, NJ: Princeton University Press.

Huang, X., and J. Weng. 2002. "Novelty and Reinforcement Learning in the Value System of Developmental Robots." Paper presented at the 2nd International Workshop on Epigenetic Robotics: Modeling Cognitive Development in Robotic Systems, Edinburgh, Scotland, August 10–11.

Hubel, D. H., and T. N. Wiesel. 1970. "Period of Susceptibility to Physiological Effects of Unilateral Eye Closure in Kittens." *Journal of Physiology* 206 (2): 419–436.

Hugues, L., and N. Bredeche. 2006."Simbad: An Autonomous Robot Simulation Package for Education and Research." Paper presented at the International Conference on the Simulation of Adaptive Behavior, Rome, Italy, September 25–29.

Hull, C. L. 1943. *Principles of Behavior: An Introduction to Behavior Theory*. New York: Appleton -Century-Croft.

Hulse, M., S. McBride, J. Law, and M. H. Lee. 2010. "Integration of Active Vision and Reaching from a Developmental Robotics Perspective." *IEEE Transactions on Autonomous Mental Development* 2 (4) (Dec.): 355–367.

Hulse, M., S. McBride, and M. H. Lee. 2010. "Fast Learning Mapping Schemes for Robotic Hand-Eye Coordination." *Cognitive Computation* 2 (1) (Mar.): 1–16.

Hunt, J. M. 1965. "Intrinsic Motivation and Its Role in Psychological Development." Paper presented at the Nebraska Symposium on Motivation, Lincoln, NE.

Hunt, J. M. 1970. "Attentional Preference and Experience." *Journal of Genetic Psychology* 117 (1): 99–107.

Iacoboni, M., R. P. Woods, M. Brass, H. Bekkering, J. C. Mazziotta, and G. Rizzolatti. 1999. "Cortical Mechanisms of Human Imitation." *Science* 286 (5449) (Dec.): 2526–2528.

Ieropoulos, I. A., J. Greenman, C. Melhuish, and I. Horsfield. 2012. "Microbial Fuel Cells for Robotics: Energy Autonomy through Artificial Symbiosis." *ChemSusChem* 5 (6): 1020–1026.

Ijspeert, A. J. 2008. "Central Pattern Generators for Locomotion Control in Animals and Robots: A Review." *Neural Networks* 21 (4) (May): 642–653.

Ijspeert, A. J., J. Nakanishi, and S. Schaal. 2002. "Movement Imitation with Nonlinear Dynamical Systems in Humanoid Robots." Paper presented at the IEEE International Conference on Robotics and Automation, New York.

Ikemoto, S., H. Ben Amor, T. Minato, B. Jung, and H. Ishiguro. 2012. "Physical Human-Robot Interaction Mutual Learning and Adaptation." *IEEE Robotics & Automation Magazine* 19 (4) (Dec.): 24–35.

Ikemoto, S., T. Minato, and H. Ishiguro. 2009. "Analysis of Physical Human–Robot Interaction for Motor Learning with Physical Help." *Applied Bionics and Biomechanics* 5 (4): 213–223.

Imai, M., T. Ono, and H. Ishiguro. 2003. "Physical Relation and Expression: Joint Attention for Human-Robot Interaction." *IEEE Transactions on Industrial Electronics* 50 (4) (Aug.): 636–643.

Iriki, A., M. Tanaka, and Y. Iwamura. 1996. "Coding of Modified Body Schema During Tool Use by Macaque Postcentral Neurones." *Neuroreport* 7 (14): 2325–2330.

Ishiguro, H., T. Ono, M. Imai, T. Maeda, T. Kanda, and R. Nakatsu. 2001. "Robovie: An Interactive Humanoid Robot." *Industrial Robot* 28 (6): 498–503.

Ishihara, H., and M. Asada. 2013. "'Affetto': Towards a Design of Robots Who Can Physically Interact with People, Which Biases the Perception of Affinity (Beyond 'Uncanny')." Paper presented at the IEEE International Conference on Robotics and Automation, Karlsruhe, Germany, May 10.

Ishihara, H., Y. Yoshikawa, and M. Asada. 2011. "Realistic Child Robot 'Affetto' for Understanding the Caregiver-Child Attachment Relationship That Guides the Child Development." Paper presented at the IEEE International Conference on Development and Learning (ICDL), Frankfurt Biotechnology Innovation Center, Frankfurt, Germany, August 24–27.

Ishihara, H., Y. Yoshikawa, K. Miura, and M. Asada. 2009. "How Caregiver's Anticipation Shapes Infant's Vowel through Mutual Imitation." *IEEE Transactions on Autonomous Mental Development* 1 (4) (Dec.): 217–225.

Isoda, M., and O. Hikosaka. 2008. "A Neural Correlate of Motivational Conflict in the Superior Colliculus of the Macaque." *Journal of Neurophysiology* 100 (3) (Sept.): 1332–1342.

Ito, M., and J. Tani. 2004. "On-Line Imitative Interaction with a Humanoid Robot Using a Dynamic Neural Network Model of a Mirror System." *Adaptive Behavior* 12 (2): 93–115.

Itti, L., and C. Koch. 2001. "Computational Modelling of Visual Attention." *Nature Reviews. Neuroscience* 2 (3): 194–203.

Itti, L., and C. Koch. 2000. "A Saliency-Based Search Mechanism for Overt and Covert Shifts of Visual Attention." *Vision Research* 40 (10–12): 1489–1506.

Iverson, J. M., and S. Goldin-Meadow. 2005. "Gesture Paves the Way for Language Development." *Psychological Science* 16 (5): 367–371.

James, W. 1890. *The Principles of Psychology*. Cambridge, MA: Harvard University Press.

Jamone, L., G. Metta, F. Nori, and G. Sandini. 2006. "James: A Humanoid Robot Acting over an Unstructured World." Paper presented at the 6th IEEE-RAS International Conference on Humanoid Robots (HUMANOIDS2006), Genoa, Italy, December 4–6.

Jasso, H., J. Triesch, and G. Deak. 2008. "A Reinforcement Learning Model of Social Referencing." Paper presented at the IEEE 7th International Conference on Development and Learning, Monterey, CA, August 9–12.

Joh, A. S., and K. E. Adolph. 2006. "Learning from Falling." *Child Development* 77 (1): 89–102.

Johnson, C. P., and P. A. Blasco. 1997. "Infant Growth and Development." *Pediatrics in Review/American Academy of Pediatrics* 18 (7): 224–242.

Johnson, J. S., and E. L. Newport. 1989. "Critical Period Effects in Second Language Learning: The Influence of Maturational State on the Acquisition of English as a Second Language." *Cognitive Psychology* 21 (1): 60–99.

Johnson, M., and Y. Demiris. 2004. "Abstraction in Recognition to Solve the Correspondence Problem for Robot Imitation." *Proceedings of TAROS*, 63–70.

Johnson, M. H. 1990. "Cortical Maturation and the Development of Visual Attention in Early Infancy." *Journal of Cognitive Neuroscience* 2 (2): 81–95.

Johnson, S. P. 2004. "Development of Perceptual Completion in Infancy." *Psychological Science* 15 (11): 769–775.

Johnson, S. P., D. Amso, and J. A. Slemmer. 2003a. "Development of Object Concepts in Infancy: Evidence for Early Learning in an Eye-Tracking Paradigm." *Proceedings of the National Academy of Sciences of the United States of America* 100 (18) (Sept.): 10568–10573.

Johnson, S. P., J. G. Bremner, A. Slater, U. Mason, K. Foster, and A. Cheshire. 2003b. "Infants' Perception of Object Trajectories." *Child Development* 74 (1): 94–108.

Jurafsky, D., J. H. Martin, A. Kehler, L. K. Vander, and N. Ward. 2000. *Speech and Language Processing: An Introduction to Natural Language Processing, Computational Linguistics, and Speech Recognition*. Vol. 2. Cambridge, MA: MIT Press.

Jusczyk, P. W. 1999. "How Infants Begin to Extract Words from Speech." *Trends in Cognitive Sciences* 3 (9): 323–328.

Kagan, J. 1972. "Motives and Development." *Journal of Personality and Social Psychology* 22 (1): 51–66.

Kaipa, K. N., J. C. Bongard, and A. N. Meltzoff. 2010. "Self Discovery Enables Robot Social Cognition: Are You My Teacher?" *Neural Networks* 23 (8–9) (Oct.–Nov.): 1113–1124.

Kaplan, F., and V. V. Hafner. 2006a. "Information-Theoretic Framework for Unsupervised Activity Classification." *Advanced Robotics* 20 (10): 1087–1103.

Kaplan, F., and V. V. Hafner. 2006b. "The Challenges of Joint Attention." *Interaction Studies: Social Behaviour and Communication in Biological and Artificial Systems* 7 (2): 135–169.

Kaplan, F., and P. Y. Oudeyer. 2007. "The Progress-Drive Hypothesis: An Interpretation of Early Imitation." In *Models and Mechanisms of Imitation and Social Learning: Behavioural, Social and Communication Dimensions*, ed. K. Dautenhahn and C. Nehaniv, 361–377. Cambridge, UK: Cambridge University Press.

Kaplan, F., and P. Y. Oudeyer. 2003 "Motivational Principles for Visual Know-How Development." Paper presented at the 3rd International Workshop on Epigenetic Robotics: Modeling Cognitive Development in Robotic Systems, Boston, MA, August 4–5.

Kaplan, F., P. Y. Oudeyer, E. Kubinyi, and A. Miklósi. 2002. "Robotic Clicker Training." *Robotics and Autonomous Systems* 38 (3–4) (Mar.): 197–206.

Karmiloff-Smith, A. 1995. *Beyond Modularity: A Developmental Perspective on Cognitive Science*. Cambridge, MA: MIT Press.

Kaufman, E. L., M. W. Lord, T. W. Reese, and J. Volkmann. 1949. "The Discrimination of Visual Number." *American Journal of Psychology* 62 (4): 498–525.

Kawamura, K., S. M. Gordon, P. Ratanaswasd, E. Erdemir, and J. F. Hall. 2008. "Implementation of Cognitive Control for a Humanoid Robot." *International Journal of Humanoid Robotics* 5 (4) (Dec.): 547–586.

Kawato, M. 2008. "Brain Controlled Robots." *HFSP Journal* 2 (3): 136–142.

Keil, Frank C. 1989. *Concepts, Kinds, and Cognitive Development*. Cambridge, MA: MIT Press.

Kellman, P. J., and E. S. Spelke. 1983. "Perception of Partly Occluded Objects in Infancy." *Cognitive Psychology* 15 (4): 483–524.

Kermoian, R., and J. J. Campos. 1988. "Locomotor Experience: A Facilitator of Spatial Cognitive-Development." *Child Development* 59 (4) (Aug.): 908–917.

Kestenbaum, R., N. Termine, and E. S. Spelke. 1987. "Perception of Objects and Object Boundaries by 3-Month-Old Infants." *British Journal of Developmental Psychology* 5 (4): 367–383.

Kiesler, S., A. Powers, S. R. Fussell, and C. Torrey. 2008. "Anthropomorphic Interactions with a Robot and Robot-Like Agent." *Social Cognition* 26 (2) (Apr.): 169–181.

Kirby, S. 2001. "Spontaneous Evolution of Linguistic Structure: An Iterated Learning Model of the Emergence of Regularity and Irregularity." *IEEE Transactions on Evolutionary Computation* 5 (2) (Apr.): 102–110.

Kirstein, S., H. Wersing, and E. Körner. 2008. "A Biologically Motivated Visual Memory Architecture for Online Learning of Objects." *Neural Networks* 21 (1): 65–77.

Kish, G. B., and J. J. Antonitis. 1956. "Unconditioned Operant Behavior in Two Homozygous Strains of Mice." *Journal of Genetic Psychology* 88 (1): 121–129.

Kisilevsky, B. S., and J. A. Low. 1998. "Human Fetal Behavior: 100 Years of Study." *Developmental Review* 18 (1) (Mar.): 1–29.

Konczak, J., and J. Dichgans. 1997. "The Development toward Stereotypic Arm Kinematics during Reaching in the First Three Years of Life." *Experimental Brain Research* 117 (2): 346–354.

Kose-Bagci, H., K. Dautenhahn, and C. L. Nehaniv. 2008. "Emergent Dynamics of Turn-Taking Interaction in Drumming Games with a Humanoid Robot." Paper presented at the 17th IEEE International Symposium on Robot and Human Interactive Communication (ROMAN 2008), Munich, Germany, August 1–3.

Kousta, S. T., G. Vigliocco, D. P. Vinson, M. Andrews, and E. del Campo. 2011. "The Representation of Abstract Words: Why Emotion Matters." *Journal of Experimental Psychology. General* 140 (1) (Feb.): 14–34.

Kozima, H. 2002. "Infanoid—a Babybot that Explores the Social Environment." In *Socially Intelligent Agents: Creating Relationships with Computers and Robots*, ed. K. Dautenhahn, A. H. Bond, L. Cañamero, and B. Edmonds, 157–164. Amsterdam: Kluwer Academic Publishers.

Kozima, H., C. Nakagawa, N. Kawai, D. Kosugi, and Y. Yano. 2004. "A Humanoid in Company with Children." Paper presented at the 4th IEEE/RAS International Conference on Humanoid Robots, Santa Monica, CA, November 10–12.

Kozima, H., C. Nakagawa, and H. Yano. 2005. "Using Robots for the Study of Human Social Development." Paper presented at the AAAI Spring Symposium on Developmental Robotics, Stanford University, Stanford, CA, March 21–23.

Kozima, H., and H. Yano. 2001. "A Robot That Learns to Communicate with Human Caregivers." Paper presented at the 1st International Workshop on Epigenetic Robotics: Modeling Cognitive Development in Robotic Systems, Lund, Sweden, September 17–18.

Kraft, D., E. Baseski, M. Popovic, A. M. Batog, A. Kjær-Nielsen, N. Krüger, R. Petrick, C. Geib, N. Pugeault, and M. Steedman. 2008. "Exploration and Planning in a Three-Level Cognitive Architecture." Paper presented at the International Conference on Cognitive Systems, University of Karlsruhe, Karlsruhe, Germany, April 2–4.

Krichmar, J. L., and G. M. Edelman. 2005. "Brain-Based Devices for the Study of Nervous Systems and the Development of Intelligent Machines." *Artificial Life* 11 (1–2) (Winter): 63–77.

Krüger, V., D. Herzog, S. Baby, A. Ude, and D. Kragic. 2010. "Learning Actions from Observations." *IEEE Robotics & Automation Magazine* 17 (2): 30–43.

Kubinyi, E., A. Miklosi, F. Kaplan, M. Gacsi, J. Topal, and V. Csanyi. 2004. "Social Behaviour of Dogs Encountering Aibo, an Animal-Like Robot in a Neutral and in a Feeding Situation." *Behavioural Processes* 65 (3): 231–239.

Kumar, S., and P. J. Bentley. 2003. *On Growth, Form and Computers*. Amsterdam: Elsevier.

Kumaran, D., and E. A. Maguire. 2007. "Which Computational Mechanisms Operate in the Hippocampus During Novelty Detection?" *Hippocampus* 17 (9): 735–748.

Kuniyoshi, Y., N. Kita, S. Rougeaux, and T. Suehiro. 1995. "Active Stereo Vision System with Foveated Wide Angle Lenses." Paper presented at the Invited Session Papers from the Second Asian Conference on Computer Vision: Recent Developments in Computer Vision, Singapore, December 5–8.

Kuniyoshi, Y., and S. Sangawa. 2006. "Early Motor Development from Partially Ordered Neural-Body Dynamics: Experiments with a Cortico-Spinal-Musculo-Skeletal Model." *Biological Cybernetics* 95 (6): 589–605.

Kuperstein, M. 1991. "Infant Neural Controller for Adaptive Sensory Motor Coordination." *Neural Networks* 4 (2): 131–145.

Kuperstein, M. 1988. "Neural Model of Adaptive Hand-Eye Coordination for Single Postures." *Science* 239 (4845) (Mar.): 1308–1311.

Kuroki, Y., M. Fujita, T. Ishida, K. Nagasaka, and J. Yamaguchi. 2003. "A Small Biped Entertainment Robot Exploring Attractive Applications." Paper presented at the IEEE International Conference on Robotics and Automation (ICRA 03), Taipei, Taiwan, September 14–19.

Laird, J. E., A. Newell, and P. S. Rosenbloom. 1987. "Soar: An Architecture for General Intelligence." *Artificial Intelligence* 33 (1) (Sept.): 1–64.

Lakoff, G., and M. Johnson. 1999. *Philosophy in the Flesh: The Embodied Mind and Its Challenge to Western Thought*. New York: Basic Books.

Lakoff, G., and R. E. Núñez. 2000. *Where Mathematics Comes From: How the Embodied Mind Brings Mathematics into Being*. New York: Basic Books.

Lallée, S., U. Pattacini, S. Lemaignan, A. Lenz, C. Melhuish, L. Natale, S. Skachek, et al. 2012. "Towards a Platform-Independent Cooperative Human Robot Interaction System: III. An Architecture for Learning and Executing Actions and Shared Plans." *IEEE Transactions on Autonomous Mental Development* 4 (3) (Sept.): 239–253.

Lallée, S., E. Yoshida, A. Mallet, F. Nori, L. Natale, G. Metta, F. Warneken, and P. F. Dominey. 2010. "Human-Robot Cooperation Based on Interaction Learning." In *Motor Learning to Interaction Learning in Robot.*, ed. O. Sigaud and J. Peters, 491–536. Berlin: Springer-Verlag.

Langacker, R. W. 1987. *Foundations of Cognitive Grammar: Theoretical Prerequisites*. Vol. 1. Stanford, CA: Stanford University Press.

Langley, P., and D. Choi. 2006. "A Unified Cognitive Architecture for Physical Agents." In *Proceedings of the 21st National Conference on Artificial Intelligence, Volume 2*, 1469–1474. Boston: AAAI Press.

Langley, P., J. E. Laird, and S. Rogers. 2009. "Cognitive Architectures: Research Issues and Challenges." *Cognitive Systems Research* 10 (2) (June): 141–160.

Laschi, C., M. Cianchetti, B. Mazzolai, L. Margheri, M. Follador, and P. Dario. 2012. "Soft Robot Arm Inspired by the Octopus." *Advanced Robotics* 26 (7): 709–727.

Laurent, R., C. Moulin-Frier, P. Bessière, J. L. Schwartz, and J. Diard. 2011. "Noise and Inter-Speaker Variability Improve Distinguishability of Auditory, Motor and Perceptuo-Motor Theories of Speech Perception: An Exploratory Bayesian Modeling Study." Paper presented at the 9th International Seminar on Speech Production, Montreal, Canada, June 20–23.

Law, J., M. H. Lee, M. Hulse, and A. Tomassetti. 2011. "The Infant Development Timeline and Its Application to Robot Shaping." *Adaptive Behavior* 19 (5) (Oct.): 335–358.

Lee, G., R. Lowe, and T. Ziemke. 2011. "Modelling Early Infant Walking: Testing a Generic Cpg Architecture on the NAO Humanoid." Paper presented at the IEEE Joint Conference on Development and Learning and on Epigenetic Robotics, Frankfurt, Germany.

Lee, M. H., Q. G. Meng, and F. Chao. 2007. "Staged Competence Learning in Developmental Robotics." *Adaptive Behavior* 15 (3): 241–255.

Lee, R., R. Walker, L. Meeden, and J. Marshall. 2009 "Category-Based Intrinsic Motivation." Paper presented at the 9th International Conference on Epigenetic Robotics: Modeling Cognitive Development in Robotic Systems, Venice, Italy, November 12–14.

Lehmann, H., I. Iacono, K. Dautenhahn, P. Marti, and B. Robins. 2014. "Robot Companions for Children with Down Syndrome: A Case Study." *Interaction Studies* 15 (1) (May): 99–112.

Leinbach, M. D., and B. I. Fagot. 1993. "Categorical Habituation to Male and Female Faces—Gender Schematic Processing in Infancy." *Infant Behavior and Development* 16 (3) (July–Sept.): 317–332.

Lenneberg, E. H. 1967. *Biological Foundations of Language*. New York: Wiley.

Leslie, A. M. 1994. "Tomm, Toby, and Agency: Core Architecture and Domain Specificity." In *Mapping the Mind: Domain Specificity in Cognition and Culture*, ed. L. A. Hirschfeld and S. A. Gelman, 119–148. Cambridge, UK: Cambridge University Press.

Lewis, M., and J. Brooks-Gunn. 1979. *Social Cognition and the Acquisition of Self*. New York: Plenum Press.

Li, C., R. Lowe, B. Duran, and T. Ziemke. 2011. "Humanoids that Crawl: Comparing Gait Performance of iCub and NAO Using a CPG Architecture." Paper presented at the International Conference on Computer Science and Automation Engineering (CSAE), Shanghai.

Li, C., R. Lowe, and T. Ziemke. 2013. "Humanoids Learning to Crawl Based on Natural CPG-Actor-Critic and Motor Primitives." Paper presented at the Workshop on Neuroscience and Robotics (IROS 2013): Towards a Robot-Enabled, Neuroscience-Guided Healthy Society, Tokyo, Japan, November 3.

Lin, P., K. Abney, and G. A. Bekey. 2011. *Robot Ethics: The Ethical and Social Implications of Robotics*. Cambridge, MA: MIT Press.

Lockman, J. J., D. H. Ashmead, and E. W. Bushnell. 1984. "The Development of Anticipatory Hand Orientation during Infancy." *Journal of Experimental Child Psychology* 37 (1): 176–186.

Lopes, L. S., and A. Chauhan. 2007. "How Many Words Can My Robot Learn? An Approach and Experiments with One-Class Learning." *Interaction Studies: Social Behaviour and Communication in Biological and Artificial Systems* 8 (1): 53–81.

Lovett, A., and B. Scassellati. 2004. "Using a Robot to Reexamine Looking Time Experiments." Paper presented at the 3rd International Conference on Development and Learning, San Diego, CA.

Lu, Z., S. Lallee, V. Tikhanoff, and P. F. Dominey. 2012. "Bent Leg Walking Gait Design for Humanoid Robotic Child-iCub Based on Key State Switching Control." Paper presented at the IEEE Symposium on Robotics and Applications (ISRA), Kuala Lumpur, June 3–5.

Lungarella, M., and L. Berthouze. 2004. "Robot Bouncing: On the Synergy between Neural and Body-Environment Dynamics." In *Embodied Artificial Intelligence*, ed. F. Iida, R. Pfeifer, L. Steels, and Y. Kuniyoshi, 86–97. Berlin: Springer-Verlag.

Lungarella, M., and L. Berthouze. 2003 "Learning to Bounce: First Lessons from a Bouncing Robot." Paper presented at the 2nd International Symposium on Adaptive Motion in Animals and Machines, Kyoto, Japan, March 4–8.

Lungarella, M., G. Metta, R. Pfeifer, and G. Sandini. 2003. "Developmental Robotics: A Survey." *Connection Science* 15 (4) (Dec.): 151–190.

Lyon, C., C. L. Nehaniv, and A. Cangelosi. 2007. *Emergence of Communication and Language*. London: Springer-Verlag.

Lyon, C., C. L. Nehaniv, and J. Saunders. 2012. "Interactive Language Learning by Robots: The Transition from Babbling to Word Forms." *PLoS ONE* 7 (6): 1–16.

Lyon, C., C. L. Nehaniv, and J. Saunders. 2010. "Preparing to Talk: Interaction between a Linguistically Enabled Agent and a Human Teacher." Paper presented at the Dialog with Robots AAAI Fall Symposium Series, Arlington, VA, November 11–13.

MacDorman, K. F., and H. Ishiguro. 2006a. "Toward Social Mechanisms of Android Science: A Cogsci 2005 Workshop." *Interaction Studies: Social Behaviour and Communication in Biological and Artificial Systems* 7 (2): 289–296.

MacDorman, K. F., and H. Ishiguro. 2006b. "The Uncanny Advantage of Using Androids in Cognitive and Social Science Research." *Interaction Studies: Social Behaviour and Communication in Biological and Artificial Systems* 7 (3): 297–337.

Macura, Z., A. Cangelosi, R. Ellis, D. Bugmann, M. H. Fischer, and A. Myachykov. 2009. "A Cognitive Robotic Model of Grasping." Paper presented at the 9th International Conference on Epigenetic Robotics: Modeling Cognitive Development in Robotic Systems, Venice, Italy, November 12–14.

MacWhinney, B. 1998. "Models of the Emergence of Language." *Annual Review of Psychology* 49: 199–227.

Mangin, O., and P. Y. Oudeyer. 2012."Learning to Recognize Parallel Combinations of Human Motion Primitives with Linguistic Descriptions Using Non-Negative Matrix Factorization." Paper presented at the IEEE/RSJ International Conference on Intelligent Robots and Systems, New York.

Mareschal, D., M. H. Johnson, S. Sirois, M. Spratling, M. S. C. Thomas, and G. Westermann. 2007. *Neuroconstructivism: How the Brain Constructs Cognition.* Vol. 1. Oxford, UK: Oxford University Press.

Mareschal, D., and S. P. Johnson. 2002. "Learning to Perceive Object Unity: A Connectionist Account." *Developmental Science* 5 (2) (May): 151–172.

Markman, E. M., and J. E. Hutchinson. 1984. "Children's Sensitivity to Constraints on Word Meaning: Taxonomic Versus Thematic Relations." *Cognitive Psychology* 16 (1): 1–27.

Markman, E. M., and G. F. Wachtel. 1988. "Children's Use of Mutual Exclusivity to Constrain the Meanings of Words." *Cognitive Psychology* 20 (2) (Apr.): 121–157.

Marocco, D., A. Cangelosi, K. Fischer, and T. Belpaeme. 2010. "Grounding Action Words in the Sensorimotor Interaction with the World: Experiments with a Simulated iCub Humanoid Robot." *Frontiers in Neurorobotics* 4 (7) (May 31). doi:10.3389/fnbot.2010.00007.

Marques, H. G., M. Jäntsch, S. Wittmeier, C. Alessandro, O. Holland, C. Alessandro, A. Diamond, M. Lungarella, and R. Knight. 2010. "Ecce1: The First of a Series of Anthropomimetic Musculoskelal Upper Torsos." Paper presented at the International Conference on Humanoids, Nashville, TN, December 6–8.

Marr, D. 1982. *Vision: A Computational Investigation into the Human Representation and Processing of Visual Information.* San Francisco: Freeman.

Marshall, J., D. Blank, and L. Meeden. 2004. "An Emergent Framework for Self-Motivation in Developmental Robotics." Paper presented at the 3rd International Conference on Development and Learning (ICDL 2004), La Jolla, CA, October 20–22.

Martinez, R. V., J. L. Branch, C. R. Fish, L. H. Jin, R. F. Shepherd, R. M. D. Nunes, Z. G. Suo, and G. M. Whitesides. 2013. "Robotic Tentacles with Three-Dimensional Mobility Based on Flexible Elastomers." *Advanced Materials* 25 (2) (Jan.): 205–212.

Massera, G., A. Cangelosi, and S. Nolfi. 2007. "Evolution of Prehension Ability in an Anthropomorphic Neurorobotic Arm." *Frontiers in Neurorobotics* 1 (4): 1–9.

Massera, G., T. Ferrauto, O. Gigliotta, and S. Nolfi. 2013. "Farsa: An Open Software Tool for Embodied Cognitive Science." Paper presented at the 12th European Conference on Artificial Life, Cambridge, MA.

Matarić, M. J. 2007. *The Robotics Primer*. Cambridge, MA: MIT Press.

Matsumoto, M., K. Matsumoto, H. Abe, and K. Tanaka. 2007. "Medial Prefrontal Cell Activity Signaling Prediction Errors of Action Values." *Nature Neuroscience* 10 (5): 647–656.

Maurer, D., and M. Barrera. 1981. "Infants' Perception of Natural and Distorted Arrangements of a Schematic Face." *Child Development* 52 (1): 196–202.

Maurer, D., and P. Salapatek. 1976. "Developmental-Changes in Scanning of Faces by Young Infants." *Child Development* 47 (2): 523–527.

Mavridis, N., and D. Roy. 2006. "Grounded Situation Models for Robots: Where Words and Percepts Meet." Paper presented at the IEEE/RSJ International Conference on Intelligent Robots and Systems, Beijing, China, October 9–15.

Mayor, J., and K. Plunkett. 2010. "Vocabulary Spurt: Are Infants Full of Zipf?" Paper presented at the 32nd Annual Conference of the Cognitive Science Society, Austin, TX.

McCarty, M. E., and D. H. Ashmead. 1999. "Visual Control of Reaching and Grasping in Infants." *Developmental Psychology* 35 (3) (May): 620–631.

McCarty, M. E., R. K. Clifton, D. H. Ashmead, P. Lee, and N. Goubet. 2001a. How Infants Use Vision for Grasping Objects. *Child Development* 72 (4) (Jul–Aug.): 973–987.

McCarty, M. E., R. K. Clifton, and R. R. Collard. 2001b. "The Beginnings of Tool Use by Infants and Toddlers." *Infancy* 2 (2): 233–256.

McCarty, M. E., R. K. Clifton, and R. R. Collard. 1999. "Problem Solving in Infancy: The Emergence of an Action Plan." *Developmental Psychology* 35 (4) (Jul): 1091–1101.

McClelland, J. L., and D. E. Rumelhart. 1981. "An Interactive Activation Model of Context Effects in Letter Perception. 1. An Account of Basic Findings." *Psychological Review* 88 (5): 375–407.

McDonnell, P. M., and W. C. Abraham. 1979. "Adaptation to Displacing Prisms in Human Infants." *Perception* 8 (2): 175–185.

McGeer, T. 1990. "Passive Dynamic Walking." *International Journal of Robotics Research* 9 (2) (Apr.): 62–82.

McGraw, M. B. 1945. *The Neuro-Muscular Maturation of the Human Infant*. New York: Columbia University.

McGraw, M. B. 1941. "Development of Neuro-Muscular Mechanisms as Reflected in the Crawling and Creeping Behavior of the Human Infant." *Pedagogical Seminary and Journal of Genetic Psychology* 58 (1): 83–111.

McKinney, M. L., and K. J. McNamara. 1991. *Heterochrony: The Evolution of Ontogeny*. London: Plenum Press.

McMurray, B. 2007. "Defusing the Childhood Vocabulary Explosion." *Science* 317 (5838) (Aug.): 631.

Meissner, C. A., and J. C. Brigham. 2001. "Thirty Years of Investigating the Own-Race Bias in Memory for Faces—a Meta-Analytic Review." *Psychology, Public Policy, and Law* 7 (1) (Mar.): 3–35.

Meltzoff, A. N. 2007. "The 'Like Me' Framework for Recognizing and Becoming an Intentional Agent." *Acta Psychologica* 124 (1): 26–43.

Meltzoff, A. N. 1995. "Understanding the Intentions of Others: Re-Enactment of Intended Acts by Eighteen-Month-Old Children." *Developmental Psychology* 31 (5): 838–850.

Meltzoff, A. N. 1988. "Infant Imitation after a One-Week Delay—Long-Term-Memory for Novel Acts and Multiple Stimuli." *Developmental Psychology* 24 (4) (July): 470–476.

Meltzoff, A. N., and R. W. Borton. 1979. "Inter-Modal Matching by Human Neonates." *Nature* 282 (5737): 403–404.

Meltzoff, A. N., and M. K. Moore. 1997. "Explaining Facial Imitation: A Theoretical Model." *Early Development & Parenting* 6 (3–4) (Sept.–Dec.): 179–192.

Meltzoff, A. N., and M. K. Moore. 1989. "Imitation in Newborn-Infants—Exploring the Range of Gestures Imitated and the Underlying Mechanisms." *Developmental Psychology* 25 (6) (Nov.): 954–962.

Meltzoff, A. N., and M. K. Moore. 1983. "Newborn-Infants Imitate Adult Facial Gestures." *Child Development* 54 (3): 702–709.

Meltzoff, A. N., and M. K. Moore. 1977. "Imitation of Facial and Manual Gestures by Human Neonates." *Science* 198 (4312): 75–78.

Merrick, K. E. 2010. "A Comparative Study of Value Systems for Self-Motivated Exploration and Learning by Robots." *IEEE Transactions on Autonomous Mental Development* 2 (2) (June): 119–131.

Mervis, C. B. 1987."Child-Basic Object Categories and Early Lexical Development." In *Concepts and Conceptual Development: Ecological and Intellectural Factors in Categorization*, ed. U. Neisser, 201–233. Cambridge, UK: Cambridge University Press.

Metta, G., P. Fitzpatrick, and L. Natale. 2006. "YARP: Yet Another Robot Platform." *International Journal of Advanced Robotic Systems* 3 (1): 43–48.

Metta, G., A. Gasteratos, and G. Sandini. 2004. "Learning to Track Colored Objects with Log-Polar Vision." *Mechatronics* 14 (9) (Nov.): 989–1006.

Metta, G., L. Natale, F. Nori, G. Sandini, D. Vernon, L. Fadiga, C. von Hofsten, K. Rosander, J. Santos-Victor, A. Bernardino, and L. Montesano. 2010. "The iCub Humanoid Robot: An Open-Systems Platform for Research in Cognitive Development." *Neural Networks* 23: 1125–1134.

Metta, G., F. Panerai, R. Manzotti, and G. Sandini. 2000. "Babybot: An Artificial Developing Robotic Agent." Paper presented at From Animals to Animats: The 6th International Conference on the Simulation of Adaptive Behavior, Paris, France, September 11–16.

Metta, G., G. Sandini, and J. Konczak. 1999. "A Developmental Approach to Visually-Guided Reaching in Artificial Systems." *Neural Networks* 12 (10) (Dec.): 1413–1427.

Metta, G., G. Sandini, L. Natale, and F. Panerai. 2001. "Development and Robotics." Paper presented at the 2nd IEEE-RAS International Conference on Humanoid Robots, Tokyo, Japan, November 22–24.

Metta, G., G. Sandini, D. Vernon, L. Natale, and F. Nori. 2008. "The iCub Humanoid Robot: An Open Platform for Research in Embodied Cognition." Paper presented at the 8th IEEE Workshop on Performance Metrics for Intelligent Systems, Gaithersburg, MD, August 19–21.

Metta, G., D. Vernon, and G. Sandini. 2005. "The Robotcub Approach to the Development of Cognition: Implications of Emergent Systems for a Common Research Agenda in Epigenetic Robotics." Paper presented at the 5th International Workshop on Epigenetic Robotics : Modeling Cognitive Development in Robotic Systems, Nara, Japan, July 22–24.

Michalke, T., J. Fritsch, and C. Goerick. 2010. "A Biologically-Inspired Vision Architecture for Resource-Constrained Intelligent Vehicles." *Computer Vision and Image Understanding* 114 (5): 548–563.

Michel, O. 2004. "Webotstm: Professional Mobile Robot Simulation." *International Journal of Advanced Robotic Systems* 1 (1): 39–42.

Michel, P., K. Gold, and B. Scassellati. 2004. "Motion-Based Robotic Self-Recognition." Paper presented at the IEEE/RSJ International Conference on Intelligent Robots and Systems, Sendai, Japan, September 28–October 2.

Mikhailova, I., M. Heracles, B. Bolder, H. Janssen, H. Brandl, J. Schmüdderich, and C. Goerick. 2008. "Coupling of Mental Concepts to a Reactive Layer: Incremental Approach in System Design." Paper presented at the Proceedings of the 8th International Workshop on Epigenetic Robotics: Modeling Cognitive Development in Robotic Systems, Brighton, UK, July 30–31.

Minato, T., M. Shimada, H. Ishiguro, and S. Itakura. 2004. "Development of an Android Robot for Studying Human-Robot Interaction." Paper presented at the Seventeenth International Conference on Industrial and Engineering Applications of Artificial Intelligence and Expert Systems (IEA/AIE), Berlin, May.

Minato, T., Y. Yoshikawa, T. Noda, S. Ikemoto, H. Ishiguro, and M. Asada. 2007. "CB2: A Child Robot with Biomimetic Body for Cognitive Developmental Robotics." Paper presented at the 7th IEEE-RAS International Conference on Humanoid Robots, Pittsburgh, PA, November 29–December 1.

Mirolli, M., and G. Baldassarre. 2013. "Functions and Mechanisms of Intrinsic Motivations." In *Intrinsically Motivated Learning in Natural and Artificial Systems*, ed. G. Baldassarre and M. Mirolli, 49–72. Heidelberg: Springer-Verlag.

Mirza, N. A., C. L. Nehaniv, K. Dautenhahn, and R. T. Boekhorst. 2008. "Developing Social Action Capabilities in a Humanoid Robot Using an Interaction History Architecture." Paper presented at the 8th IEEE/RAS International Conference on Humanoid Robots, New York.

Mohan, V., P. Morasso, G. Metta, and G. Sandini. 2009. "A Biomimetic, Force-Field Based Computational Model for Motion Planning and Bimanual Coordination in Humanoid Robots." *Autonomous Robots* 27 (3) (Oct.): 291–307.

Mohan, V., P. Morasso, J. Zenzeri, G. Metta, V. S. Chakravarthy, and G. Sandini. 2011. "Teaching a Humanoid Robot to Draw 'Shapes.'" *Autonomous Robots* 31 (1) (July): 21–53.

Mondada, F., M. Bonani, X. Raemy, J. Pugh, C. Cianci, A. Klaptocz, S. Magnenat, J. C. Zufferey, D. Floreano, and A. Martinoli. 2009. "The E-Puck, a Robot Designed for Education in Engineering." Paper presented at the 9th Conference on Autonomous Robot Systems and Competitions, Polytechnic Institute of Castelo Branco, Castelo Branco, Portugal.

Montesano, L., M. Lopes, A. Bernardino, and J. Santos-Victor. 2008. "Learning Object Affordances: From Sensory-Motor Coordination to Imitation." *IEEE Transactions on Robotics* 24 (1) (Feb.): 15–26.

Montgomery, K. C. 1954. "The Role of the Exploratory Drive in Learning." *Journal of Comparative and Physiological Psychology* 47 (1): 60–64.

Mori, H., and Y. Kuniyoshi. 2010. "A Human Fetus Development Simulation: Self-Organization of Behaviors through Tactile Sensation." Paper presented at the IEEE 9th International Conference on Development and Learning, Ann Arbor, MI, August 18–21.

Mori, M. 1970/2012. "The Uncanny Valley." *IEEE Robotics and Automation* 19 (2): 98–100. (English trans. by K. F. MacDorman and N. Kageki.)

Morimoto, J., G. Endo, J. Nakanishi, S. H. Hyon, G. Cheng, D. Bentivegna, and C. G. Atkeson. 2006. "Modulation of Simple Sinusoidal Patterns by a Coupled Oscillator Model for Biped Walking." Paper presented at the IEEE International Conference on Robotics and Automation, Orlando, FL, May 15–19.

Morse, A. F., T. Belpaeme, A. Cangelosi, and C. Floccia. 2011. "Modeling U-Shaped Performance Curves in Ongoing Development." In *Expanding the Space of Cognitive Science: Proceedings of the 23rd Annual Meeting of the Cognitive Science Society*, ed. L. Carlson, C. Hoelscher, and T. F. Shipley, 3034–3039. Austin, TX: Cognitive Science Society.

Morse, A. F., T. Belpaeme, A. Cangelosi, and L. B. Smith. 2010. "Thinking with Your Body: Modelling Spatial Biases in Categorization Using a Real Humanoid Robot." Paper presented at the 32nd Annual Meeting of the Cognitive Science Society, Portland, OR, August 11–14.

Morse, A. F.,V. L. Benitez, T. Belpaeme, A. Cangelosi, and L. B. Smith. In preparation. "Posture Affects Word Learning in Robots and Infants."

Morse, A. F., J. de Greeff, T. Belpeame, and A. Cangelosi. 2010. "Epigenetic Robotics Architecture (ERA)." *IEEE Transactions on Autonomous Mental Development* 2 (4) (Dec.): 325–339.

Morse, A., R. Lowe, and T. Ziemke. 2008. "Towards an Enactive Cognitive Architecture." Paper presented at the 1st International Conference on Cognitive Systems, Karlsruhe, Germany, April 2–4.

Morton, J., and M. H. Johnson. 1991. "Conspec and Conlern—A 2-Process Theory of Infant Face Recognition." *Psychological Review* 98 (2) (Apr.): 164–181.

Moxey, L. M., and A. J. Sanford. 1993. *Communicating Quantities: A Psychological Perspective*. Hove, UK: Erlbaum.

Mühlig, M., M. Gienger, S. Hellbach, J. J. Steil, and C. Goerick. 2009. "Task-Level Imitation Learning Using Variance-Based Movement Optimization." Paper presented at the IEEE International Conference on Robotics and Automation, Kobe, Japan, May.

Nadel, J. Ed, and G. Ed Butterworth. 1999. *Imitation in Infancy. Cambridge Studies in Cognitive Perceptual Development*. New York: Cambridge University Press.

Nagai, Y., M. Asada, and K. Hosoda. 2006. "Learning for Joint Attention Helped by Functional Development." *Advanced Robotics* 20 (10): 1165–1181.

Nagai, Y., K. Hosoda, and M. Asada. 2003. "How Does an Infant Acquire the Ability of Joint Attention?: A Constructive Approach." Paper presented at the 3rd International Workshop on Epigenetic Robotics, Lund, Sweden.

Nagai, Y., K. Hosoda, A. Morita, and M. Asada. 2003. "A Constructive Model for the Development of Joint Attention." *Connection Science* 15 (4) (Dec.): 211–229.

Narioka, K., and K. Hosoda. 2008. "Designing Synergistic Walking of a Whole-Body Humanoid Driven by Pneumatic Artificial Muscles: An Empirical Study." *Advanced Robotics* 22 (10): 1107–1123.

Nagai, Y., and K. J. Rohlfing. 2009. "Computational Analysis of Motionese Toward Scaffolding Robot Action Learning." *IEEE Transactions on Autonomous Mental Development* 1 (1): 44–54.

Narioka, K., S. Moriyama, and K. Hosoda. 2011. "Development of Infant Robot with Musculoskeletal and Skin System." Paper presented at the 3rd International Conference on Cognitive Neurodynamics, Hilton Niseko Village, Hokkaido, Japan, June 9–13.

Narioka, K., R. Niiyama, Y. Ishii, and K. Hosoda. 2009. "Pneumatic Musculoskeletal Infant Robots." Paper presented at the IEEE/RSJ International Conference on Intelligent Robots and Systems Workshop.

Natale, L., G. Metta, and G. Sandini. 2005. "A Developmental Approach to Grasping." Paper presented at the the AAAI Spring Symposium on Developmental Robotics, Stanford University, Stanford, CA, March 21–23.

Natale, L., F. Nori, G. Sandini, and G. Metta. 2007. "Learning Precise 3D Reaching in a Humanoid Robot." Paper presented at the IEEE 6th International Conference on Development and Learning, London, July 11–13.

Natale, L., F. Orabona, F. Berton, G. Metta, and G. Sandini. 2005. "From Sensorimotor Development to Object Perception." Paper presented at the 5th IEEE-RAS International Conference on Humanoid Robots, Japan.

Nava, N. E., G. Metta, G. Sandini, and V. Tikhanoff. 2009. "Kinematic and Dynamic Simulations for the Design of Icub Upper-Body Structure." Paper presented at the 9th Biennial Conference on Engineering Systems Design and Analysis (ESDA)—2008, Haifa, Israel.

Nehaniv, C. L., and K. Dautenhahn, eds. 2007. *Imitation and Social Learning in Robots, Humans and Animals: Behavioural, Social and Communicative Dimensions*. Cambridge, UK: Cambridge University Press.

Nehaniv, C. L., and K. Dautenhahn. 2002. "The Correspondence Problem." In *Imitation in Animals and Artifacts*, ed. K. Dautenhahn and C. L. Nehaniv, 41–61. Cambridge, MA: MIT Press.

Nehaniv, C. L., C. Lyon, and A. Cangelosi. 2007. "Current Work and Open Problems: A Road-Map for Research into the Emergence of Communication and Language." In *Emergence of Communication and Language*, ed. C. L. Nehaniv, C. Lyon, and A. Cangelosi, 1–27. London: Springer.

Newell, K. M., D. M. Scully, P. V. McDonald, and R. Baillargeon. 1989. "Task Constraints and Infant Grip Configurations." *Developmental Psychobiology* 22 (8) (Dec.): 817–832.

Nishio, S., H. Ishiguro, and N. Hagita. 2007. "Geminoid: Teleoperated Android of an Existing Person." In *Humanoid Robots—New Developments*, ed. A. C. de Pina Filho, 343–352. Vienna: I-Tech Education and Publishing.

Nolfi, S., and D. Floreano. 2000. *Evolutionary Robotics: The Biology, Intelligence, and Technology of Self-Organizing Machines*. Vol. 26. Cambridge, MA: MIT Press.

Nolfi, S., and O. Gigliotta. 2010. "Evorobot*: A Tool for Running Experiments on the Evolution of Communication." In *Evolution of Communication and Language in Embodied Agents*, 297–301. Berlin: Springer-Verlag.

Nolfi, S., and D. Parisi. 1999. "Exploiting the Power of Sensory-Motor Coordination." In *Advances in Artificial Life*, ed. D. Floreano, J.-D. Nicoud, and F. Mondada, 173–182. Berlin: Springer-Verlag.

Nolfi, S., D. Parisi, and J. L. Elman. 1994. "Learning and Evolution in Neural Networks." *Adaptive Behavior* 3 (1) (Summer): 5–28.

Nori, F., L. Natale, G. Sandini, and G. Metta. 2007 "Autonomous Learning of 3D Reaching in a Humanoid Robot." Paper presented at the IEEE/RSJ International Conference on Intelligent Robots and Systems, San Diego, CA, October 29–November 2.

Nosengo, N. 2009. "The Bot That Plays Ball." *Nature* 460 (7259) (Aug.): 1076–1078.

Ogino, M., M. Kikuchi, and M. Asada. 2006. "How Can Humanoid Acquire Lexicon?—Active Approach by Attention and Learning Biases Based on Curiosity." Paper presented at the IEEE/RSJ International Conference on Intelligent Robots and Systems, Beijing, China, October 9–15.

Oller, D. K. 2000. *The Emergence of the Speech Capacity*. London, UK: Erlbaum.

Otero, N., J. Saunders, K., Dautenhahn, and C. L. Nehaniv. 2008. "Teaching Robot Companions: The Role of Scaffolding and Event Structuring." *Connection Science* 20 (2–3): 111–134.

Oudeyer, P. Y. 2011. "Developmental Robotics." In *Encyclopedia of the Sciences of Learning*, ed. N. M. Seel, 329. New York: Springer.

Oudeyer, P. Y. 2006. *Self-Organization in the Evolution of Speech*. Vol. 6. Oxford, UK: Oxford University Press.

Oudeyer, P. Y., and F. Kaplan. 2007. "What Is Intrinsic Motivation? A Typology of Computational Approaches." *Frontiers in Neurorobotics* 1:1–14.

Oudeyer, P. Y., and F. Kaplan. 2006. "Discovering Communication." *Connection Science* 18 (2) (June): 189–206.

Oudeyer, P. Y., F. Kaplan, and V. V. Hafner. 2007. "Intrinsic Motivation Systems for Autonomous Mental Development." *IEEE Transactions on Evolutionary Computation* 11 (2): 265–286.

Oudeyer, P. Y., F. Kaplan, V. V. Hafner, and A. Whyte. 2005."The Playground Experiment: Task-Independent Development of a Curious Robot." Paper presented at the AAAI Spring Symposium on Developmental Robotics.

Oztop, E., N. S. Bradley, and M. A. Arbib. 2004. "Infant Grasp Learning: A Computational Model." *Experimental Brain Research* 158 (4) (Oct.): 480–503.

Parisi, D., and M. Schlesinger. 2002. "Artificial Life and Piaget." *Cognitive Development* 17 (3–4) (Sept.–Dec.): 1301–1321.

Parmiggiani, A., M. Maggiali, L. Natale, F. Nori, A. Schmitz, N. Tsagarakis, J. S. Victor, F. Becchi, G. Sandini, and G. Metta. 2012. "The Design of the iCub Humanoid Robot." *International Journal of Humanoid Robotics* 9 (4): 1–24.

Pattacini, U., F. Nori, L. Natale, G. Metta, and G. Sandini. 2010. "An Experimental Evaluation of a Novel Minimum-Jerk Cartesian Controller for Humanoid Robots." Paper presented at the IEEE/RSJ International Conference on Intelligent Robots and Systems, Taipei, Taiwan, October 18–22.

Pea, R. D. 1980. "The Development of Negation in Early Child Language." In *The Social Foundations of Language and Thought: Essays in Honor of Jerome S. Bruner*, ed. D. R. Olson, 156–186. New York: W. W. Norton.

Pea, R. D. 1978. "The Development of Negation in Early Child Language." Unpublished diss., University of Oxford.

Pecher, D., and R. A. Zwaan, eds. 2005. *Grounding Cognition: The Role of Perception and Action in Memory, Language, and Thinking*. Cambridge, UK: Cambridge University Press.

Peelle, J. E., J. Gross, and M. H. Davis. 2013. "Phase-Locked Responses to Speech in Human Auditory Cortex Are Enhanced During Comprehension." *Cerebral Cortex* 23 (6): 1378–1387.

Peniak, M., A. Morse, C. Larcombe, S. Ramirez-Contla, and A. Cangelosi. 2011. "Aquila: An Open-Source GPU-Accelerated Toolkit for Cognitive and Neuro-Robotics Research." Paper presented at the International Joint Conference on Neural Networks, San Jose, CA.

Pezzulo, G., L. W. Barsalou, A. Cangelosi, M. H. Fischer, K. McRae, and M. J. Spivey. 2011. "The Mechanics of Embodiment: A Dialog on Embodiment and Computational Modeling." *Frontiers in Psychology* 2 (5): 1–21.

Pfeifer, R., and J. Bongard. 2007. *How the Body Shapes the Way We Think: A New View of Intelligence.* Cambridge, MA: MIT Press.

Pfeifer, R., M. Lungarella, and F. Iida. 2012. "The Challenges Ahead for Bio-Inspired 'Soft' Robotics." *Communications of the ACM* 55 (11): 76–87.

Pfeifer, R., and C. Scheier. 1999. *Understanding Intelligence.* Cambridge, MA: MIT Press.

Piaget, J. 1972. *The Psychology of Intelligence.* Totowa, NJ: Littlefields Adams.

Piaget, J. 1971. *Biology and Knowledge: An Essay on the Relation between Organic Regulations and Cognitive Processes,* trans. Beautrix Welsh. Edinburgh: Edinburgh University Press.

Piaget, J. 1952. *The Origins of Intelligence in Children.* New York: International Universities Press.

Piantadosi, S. T., J. B. Tenenbaum, and N. D. Goodman. 2012. "Bootstrapping in a Language of Thought: A Formal Model of Numerical Concept Learning." *Cognition* 123 (2) (May): 199–217.

Pierris, G., and T. S. Dahl. 2010. "Compressed Sparse Code Hierarchical Som on Learning and Reproducing Gestures in Humanoid Robots." Paper presented at the IEEE International Symposium in Robot and Human Interactive Communication (RO-MAN'10), Viareggio, Italy, September 13–15.

Pinker, S. 1994. *The Language Instinct: How the Mind Creates Language.* New York: HarperCollins.

Pinker, S. 1989. *Learnability and Cognition: The Acquisition of Argument Structure.* Cambridge, MA: MIT Press.

Pinker, S., and P. Bloom. 1990. "Natural Language and Natural Selection." *Behavioral and Brain Sciences* 13 (4) (Dec.): 707–726.

Pinker, S., and A. Prince. 1988. "On Language and Connectionism: Analysis of a Parallel Distributed Processing Model of Language Acquisition." *Cognition* 28 (1–2) (Mar.): 73–193.

Plunkett, K., and V. A. Marchman. 1996. "Learning from a Connectionist Model of the Acquisition of the English Past Tense." *Cognition* 61 (3) (Dec.): 299–308.

Plunkett, K., C. Sinha, M. F. Møller, and O. Strandsby. 1992. "Symbol Grounding or the Emergence of Symbols? Vocabulary Growth in Children and a Connectionist Net." *Connection Science* 4 (3–4): 293–312.

Pulvermüller, F. 2003. *The Neuroscience of Language: On Brain Circuits of Words and Serial Order.* Cambridge, UK: Cambridge University Press.

Quinn, P. C., J. Yahr, A. Kuhn, A. M. Slater, and O. Pascalis. 2002. "Representation of the Gender of Human Faces by Infants: A Preference for Female." *Perception* 31 (9): 1109–1121.

Rajapakse, R. K., A. Cangelosi, K. R. Coventry, S. Newstead, and A. Bacon. 2005. "Connectionist Modeling of Linguistic Quantifiers." In *Artificial Neural Networks: Formal Models and Their Applications—ICANN 2005*, ed.W. Duch et al., 679–684. Berlin: Springer-Verlag.

Rao, R. P. N., A. P. Shon, and A. N. Meltzoff. 2007. "A Bayesian Model of Imitation in Infants and Robots." In *Imitation and Social Learning in Robots, Humans, and Animals: Behavioural, Social and Communicative Dimensions*, ed. C. L. Nehaniv and K. Dautenhahn, 217–247. Cambridge, UK: Cambridge University Press.

Redgrave, P., and K. Gurney. 2006. "The Short-Latency Dopamine Signal: A Role in Discovering Novel Actions?" *Nature Reviews. Neuroscience* 7 (12): 967–975.

Regier, T. 1996. *The Human Semantic Potential: Spatial Language and Constrained Connectionism.* Cambridge, MA: MIT Press.

Reinhart, R. F., and J. J. Steil. 2009. "Reaching Movement Generation with a Recurrent Neural Network Based on Learning Inverse Kinematics for the Humanoid Robot iCub." Paper presented at the 9th IEEE-RAS International Conference on Humanoid Robots, Paris, France, December 7–10.

Riesenhuber, M., and T. Poggio. 1999. "Hierarchical Models of Object Recognition in Cortex." *Nature Neuroscience* 2 (11): 1019–1025.

Righetti, L., and A. J. Ijspeert. 2006a. "Design Methodologies for Central Pattern Generators: An Application to Crawling Humanoids." Paper presented at the Robotics: Science and Systems II, University of Pennsylvania, Philadelphia, PA, August 16–19.

Righetti, L., and A. J. Ijspeert. 2006b. "Programmable Central Pattern Generators: An Application to Biped Locomotion Control." Paper presented at the IEEE International Conference on Robotics and Automation, New York.

Rizzolatti, G., and M. A. Arbib. 1998. "Language within Our Grasp." *Trends in Neurosciences* 21 (5): 188–194.

Rizzolatti, G., and L. Craighero. 2004. "The Mirror-Neuron System." *Annual Review of Neuroscience* 27: 169–192.

Rizzolatti, G., L. Fogassi, and V. Gallese. 2001. "Neurophysiological Mechanisms Underlying the Understanding and Imitation of Action." *Nature Reviews. Neuroscience* 2 (9): 661–670.

Robins, B., K. Dautenhahn, and P. Dickerson. 2012a. "Embodiment and Cognitive Learning—Can a Humanoid Robot Help Children with Autism to Learn about Tactile Social Behaviour?" Paper presented at the International Conference on Social Robotics (ICSR 2012), Chengdu, China, October 29–31.

Robins, B., K. Dautenhahn, and P. Dickerson. 2009. "From Isolation to Communication: A Case Study Evaluation of Robot Assisted Play for Children with Autism with a Minimally Expressive

Humanoid Robot." Paper presented at the Second International Conference on Advances in Computer-Human Interactions (ACHI'09), Cancun, Mexico.

Robins, B., K. Dautenhahn, E. Ferrari, G. Kronreif, B. Prazak-Aram, P. Marti, I. Iacono, et al. 2012b. "Scenarios of Robot-Assisted Play for Children with Cognitive and Physical Disabilities." *Interaction Studies: Social Behaviour and Communication in Biological and Artificial Systems* 13 (2): 189–234.

Rochat, P. 1998. "Self-Perception and Action in Infancy." *Experimental Brain Research* 123 (1–2): 102–109.

Rochat, P., and T. Striano. 2002. "Who's in the Mirror? Self–Other Discrimination in Specular Images by Four- and Nine-Month-Old Infants." *Child Development* 73 (1): 35–46.

Roder, B. J., E. W. Bushnell, and A. M. Sasseville. 2000. "Infants' Preferences for Familiarity and Novelty During the Course of Visual Processing." *Infancy* 1 (4): 491–507.

Rodriguez, P., J. Wiles, and J. L. Elman. 1999. "A Recurrent Neural Network That Learns to Count." *Connection Science* 11 (1) (May): 5–40.

Rosenthal-von der Pütten, A. M., F. P. Schulte, S. C. Eimler, L. Hoffmann, S. Sobieraj, S. Maderwald, N. C. Kramer, and M. Brand. 2013. "Neural Correlates of Empathy towards Robots." Paper presented at the 8th ACM/IEEE International Conference on Human-Robot Interaction, New York.

Rothwell, A., C. Lyon, C. L. Nehaniv, and J. Saunders. 2011. "From Babbling towards First Words: The Emergence of Speech in a Robot in Real-Time Interaction." Paper presented at the Artificial Life (ALIFE) IEEE Symposium, Paris, France, April 11–15.

Rovee-Collier, C. K., and J. B. Capatides. 1979. "Positive Behavioral-Contrast in Three-Month-Old Infants on Multiple Conjugate Reinforcement Schedules." *Journal of the Experimental Analysis of Behavior* 32 (1): 15–27.

Rovee-Collier, C. K., and M. W. Sullivan. 1980. "Organization of Infant Memory." *Journal of Experimental Psychology. Human Learning and Memory* 6 (6) (Nov.): 798–807.

Rovee-Collier, C. K., M. W. Sullivan, M. Enright, D. Lucas, and J. W. Fagen. 1980. "Reactivation of Infant Memory." *Science* 208 (4448): 1159–1161.

Roy, D., K. Y. Hsiao, and N. Mavridis. 2004. "Mental Imagery for a Conversational Robot." *IEEE Transactions on Systems, Man, and Cybernetics. Part B, Cybernetics* 34 (3): 1374–1383.

Rucinski, M., A. Cangelosi, and T. Belpaeme. 2012. "Robotic Model of the Contribution of Gesture to Learning to Count." Paper presented at the IEEE International Conference on Development and Learning and Epigenetic Robotics, New York.

Rucinski, M., A. Cangelosi, and T. Belpaeme. 2011. "An Embodied Developmental Robotic Model of Interactions between Numbers and Space." Paper presented at the Expanding the Space of Cognitive Science: 23rd Annual Meeting of the Cognitive Science Society, Boston, MA, July 20–23.

Ruesch, J., M. Lopes, A. Bernardino, J. Hornstein, J. Santos-Victor, and R. Pfeifer. 2008. "Multimodal Saliency-Based Bottom-up Attention a Framework for the Humanoid Robot iCub." Paper presented at the IEEE International Conference on Robotics and Automation, New York, May 19–23.

Ryan, R. M., and E. L. Deci. 2000. "Intrinsic and Extrinsic Motivations: Classic Definitions and New Directions." *Contemporary Educational Psychology* 25 (1): 54–67.

Ryan, R. M., and E. L. Deci. 2008. "Self-Determination Theory and the Role of Basic Psychological Needs in Personality and the Organization of Behavior." In *Handbook of Personality: Theory and Research*, 3rd ed., ed. O. P. John, R. W. Robins, and L. A. Pervin, 654–678. New York: Guilford Press.

Saegusa, R., G. Metta, and G. Sandini. 2012. "Body Definition Based on Visuomotor Correlation." *IEEE Transactions on Industrial Electronics* 59 (8): 3199–3210.

Saffran, J. R., E. L. Newport, and R. N. Aslin. 1996. "Word Segmentation: The Role of Distributional Cues." *Journal of Memory and Language* 35 (4) (Aug.): 606–621.

Sahin, E., M. Cakmak, M. R. Dogar, E. Ugur, and G. Ucoluk. 2007. "To Afford or Not to Afford: A New Formalization of Affordances toward Affordance-Based Robot Control." *Adaptive Behavior* 15 (4): 447–472.

Sakagami, Y., R. Watanabe, C. Aoyama, S. Matsunaga, N. Higaki, and K. Fujimura. 2002. "The Intelligent Asimo: System Overview and Integration." Paper presented at the IEEE/RSJ International Conference on Intelligent Robots and Systems, Lausanne, Switzerland, September 30–October 4.

Sakamoto, D., T. Kanda, T. Ono, H. Ishiguro, and N. Hagita. 2007. "Android as a Telecommunication Medium with a Human-Like Presence." Paper presented at the 2nd ACM/IEEE International Conference on Human-Robot Interaction (HRI), Arlington, VA, March.

Samuelson, L. K., and L. B. Smith. 2005. "They Call It Like They See It: Spontaneous Naming and Attention to Shape." *Developmental Science* 8 (2) (Mar.): 182–198.

Sandini, G., G. Metta, and J. Konczak. 1997. "Human Sensori-Motor Development and Artificial Systems." Paper presented at the International Symposium on Artificial Intelligence, Robotics, and Intellectual Human Activity Support for Applications, Wakoshi, Japan.

Sandini, G., G. Metta, D. Vernon, D. Caldwell, N. Tsagarakis, R. Beira, J. Santos-Victor, et al. 2004. "Robotcub: An Open Framework for Research in Embodied Cognition." Paper presented at the 4th IEEE/RAS International Conference on Humanoid Robots, Santa Monica, CA, November 10–12.

Sandini, G., and V. Tagliasco. 1980. "An Anthropomorphic Retina-Like Structure for Scene Analysis." *Computer Graphics and Image Processing* 14 (4): 365–372.

Sann, C., and A. Streri. 2007. "Perception of Object Shape and Texture in Human Newborns: Evidence from Cross-Modal Transfer Tasks." *Developmental Science* 10 (3) (May): 399–410.

Sangawa, S., and Y. Kuniyoshi. 2006. "Body and Brain-Spinal Cord Model of Embryo and Neonate, Self-Organization of Somatic Sensory Area and Motor Area." Paper presented at the Robot Society of Japan, 24th Academic Lecture.

Santrock, J. W. 2011. *Child Development.* 13th ed. New York: McGraw Hill.

Sarabia, M., and Y. Demiris. 2013. "A Humanoid Robot Companion for Wheelchair Users." In *Social Robotics*, ed. Guido Herrmann, Martin J. Pearson, Alexander Lenz, Paul Bremner, Adam Spiers, and Ute Leonards, 432–441. Berlin: Springer International Publishing.

Sarabia, M., R. Ros, and Y. Demiris. 2011. "Towards an Open-Source Social Middleware for Humanoid Robots." Paper presented at the 11th IEEE-RAS International Conference on Humanoid Robots, Slovenia, October 26–28.

Saunders, J., C. Lyon, F. Forster, C. L. Nehaniv, and K. Dautenhahn. 2009. "A Constructivist Approach to Robot Language Learning via Simulated Babbling and Holophrase Extraction." Paper presented at the IEEE Symposium on Artificial Life, New York, March 3–April 2.

Saunders, J., C. L. Nehaniv, and C. Lyon. 2011. "The Acquisition of Word Semantics by a Humanoid Robot Via Interaction with a Human Tutor." In *New Frontiers in Human-Robot Interaction*, ed. K. Dautenhahn and J. Saunders, 211–234. Philadelphia: John Benjamins.

Sauser, E. L., B. D. Argall, G. Metta, and A. G. Billard. 2012. "Iterative Learning of Grasp Adaptation through Human Corrections." *Robotics and Autonomous Systems* 60 (1): 55–71.

Savastano, P., and S. Nolfi. 2012. "Incremental Learning in a 14 Dof Simulated iCub Robot: Modeling Infant Reach/Grasp Development." In *Biomimetic and Biohybrid Systems*, ed. T. J. Prescott et al., 250–261. Berlin: Springer-Verlag.

Saylor, M. M., M. A. Sabbagh, and D. A. Baldwin. 2002. "Children Use Whole-Part Juxtaposition as a Pragmatic Cue to Word Meaning." *Developmental Psychology* 38 (6): 993–1003.

Scassellati, B. 2007. "How Social Robots Will Help Us to Diagnose, Treat, and Understand Autism." In *Robotics Research*, ed. S. Thrun, R. Brooks, and H. DurrantWhyte, 552–563. Berlin: Springer-Verlag.

Scassellati, B. 2005. "Quantitative Metrics of Social Response for Autism Diagnosis." Paper presented at the 2005 IEEE International Workshop on Robot and Human Interactive Communication, New York.

Scassellati, B. 2002. "Theory of Mind for a Humanoid Robot." *Autonomous Robots* 12 (1): 13–24.

Scassellati, B. 1999. "Imitation and Mechanisms of Joint Attention: A Developmental Structure for Building Social Skills on a Humanoid Robot." In *Computation for Metaphors, Analogy, and Agents*, ed. C. L. Nehaniv, 176–195. Heidelberg, Germany: Springer-Verlag Berlin.

Scassellati, B. 1998. "Building Behaviors Developmentally: A New Formalism." Paper presented at the AAAI Spring Symposium on Integrating Robotics Research, Palo Alto, CA, March 23–25.

Scassellati, B., H. Admoni, and M. Matarić. 2012. "Robots for Use in Autism Research." In *Annual Review of Biomedical Engineering*, vol 14, ed. M. L. Yarmush,, 275–294. Palo Alto, CA: Annual Reviews.

Schaal, S. 1999. "Is Imitation Learning the Route to Humanoid Robots?" *Trends in Cognitive Sciences* 3 (6): 233–242.

Schembri, M., M. Mirolli, and G. Baldassarre. 2007. "Evolving Internal Reinforcers for an Intrinsically Motivated Reinforcement-Learning Robot." Paper presented at the IEEE 6th International Conference on Development and Learning, London, July 11–13.

Schlesinger, M., D. Amso, and S. P. Johnson. 2012. "Simulating the Role of Visual Selective Attention during the Development of Perceptual Completion." *Developmental Science* 15 (6) (Nov.): 739–752.

Schlesinger, M., D. Amso, and S. P. Johnson. 2007. "Simulating Infants' Gaze Patterns during the Development of Perceptual Completion." Paper presented at the 7th International Workshop on Epigenetic Robotics: Modeling Cognitive Development in Robotic Systems, Piscataway, NJ, November 5–7.

Schlesinger, M., and J. Langer. 1999. "Infants' Developing Expectations of Possible and Impossible Tool-Use Events between Ages Eight and Twelve Months." *Developmental Science* 2 (2) (May): 195–205.

Schlesinger, M., and D. Parisi. 2007. "Connectionism in an Artificial Life Perspective: Simulating Motor, Cognitive, and Language Development." In *Neuroconstructivism*, ed. D. Mareschal, S. Sirois, G. Westermann, and M. H. Johnson, 129–158. Oxford, UK: Oxford University Press.

Schlesinger, M., and D. Parisi. 2001. "The Agent-Based Approach: A New Direction for Computational Models of Development." *Developmental Review* 21 (1) (Mar.): 121–146.

Schlesinger, M., D. Parisi, and J. Langer. 2000. "Learning to Reach by Constraining the Movement Search Space." *Developmental Science* 3 (1) (Mar.): 67–80.

Schmidhuber, J. 2013. "Formal Theory of Creativity, Fun, and Intrinsic Motivation (1990–2010)." In *Intrinsically Motivated Learning in Natural and Artificial Systems*, ed. G. Baldassarre and M. Mirolli, 230–247. Heidelberg: Springer-Verlag.

Schmidhuber, J. 1991. "Curious Model-Building Control-Systems." Paper presented at the 1991 IEEE International Joint Conference on Neural Networks, Singapore.

Schwanenflugel, P. J., ed. 1991. *Why Are Abstract Concepts Hard to Understand. Psychology of Word Meanings*. Hillsdale, NJ: Erlbaum.

Schwarz, W., and F. Stein. 1998. "On the Temporal Dynamics of Digit Comparison Processes." *Journal of Experimental Psychology. Learning, Memory, and Cognition* 24 (5) (Sept.): 1275–1293.

Sebastián-Gallés, N., and L. Bosch. 2009. "Developmental Shift in the Discrimination of Vowel Contrasts in Bilingual Infants: Is the Distributional Account All There Is to It?" *Developmental Science* 12 (6) (Nov): 874–887.

Serre, T., and T. Poggio. 2010. "A Neuromorphic Approach to Computer Vision." *Communications of the ACM* 53 (10): 54–61.

Shadmehr, R., and S. P. Wise. 2004. *The Computational Neurobiology of Reaching and Pointing*. Cambridge, MA: MIT Press.

Shafii, N., L. P. Reis, and R. J. F. Rossetti. 2011. "Two Humanoid Simulators: Comparison and Synthesis." Paper presented at the 6th Iberian Conference on Information Systems and Technologies (CISTI), Chaves, Portugal, June 15–18.

Shamsuddin, S., H. Yussof, L. Ismail, F. A. Hanapiah, S. Mohamed, H. A. Piah, and N. I. Zahari. 2012. "Initial Response of Autistic Children in Human-Robot Interaction Therapy with Humanoid Robot NAO." Paper presented at the IEEE 8th International Colloquium on Signal Processing and Its Applications (CSPA).

Shanahan, M. 2006. "A Cognitive Architecture That Combines Internal Simulation with a Global Workspace." *Consciousness and Cognition* 15 (2) (June): 433–449.

Shapiro, L., and G. Stockman. 2002. *Computer Vision*. London: Prentice Hall.

Shapiro, S. C., and H. O. Ismail. 2003. "Anchoring in a Grounded Layered Architecture with Integrated Reasoning." *Robotics and Autonomous Systems* 43 (2–3) (May): 97–108.

Siciliano, B., and O. Khatib, eds. 2008. *Springer Handbook of Robotics*. Berlin and Heidelberg: Springer.

Simons, D. J., and F. C. Keil. 1995. "An Abstract to Concrete Shift in the Development of Biological Thought—the Insides Story." *Cognition* 56 (2) (Aug.): 129–163.

Sinha, P. 1996. "Perceiving and Recognizing Three-Dimensional Forms." PhD diss., Dept. of Electrical Engineering and Computer Science, Massachusetts Institute of Technology, Cambridge, MA.

Sirois, S., and D. Mareschal. 2004. "An Interacting Systems Model of Infant Habituation." *Journal of Cognitive Neuroscience* 16 (8): 1352–1362.

Siviy, S. M., and J. Panksepp. 2011. "In Search of the Neurobiological Substrates for Social Playfulness in Mammalian Brains." *Neuroscience and Biobehavioral Reviews* 35 (9) (Oct.): 1821–1830.

Slater, A., S. P. Johnson, E. Brown, and M. Badenoch. 1996. "Newborn Infant's Perception of Partly Occluded Objects." *Infant Behavior and Development* 19 (1): 145–148.

Smeets, J. B. J., and E. Brenner. 1999. "A New View on Grasping." *Motor Control* 3 (3) (July): 237–271.

Smilansky, S. 1968. *The Effects of Sociodramatic Play on Disadvantaged Preschool Children*. New York: Wiley.

Smith, L. B. 2005. "Cognition as a Dynamic System: Principles from Embodiment." *Developmental Review* 25 (3–4) (Sept.–Dec.): 278–298.

Smith, L. B., and L. K. Samuelson. 2010. "Objects in Space and Mind: From Reaching to Words." In The Spatial F*oundations of Language and Cognition*: *Thinking t*hrough Space, ed. K. S. Mix, L. B. Smith, and M. Gasser, 188–207. Oxford: Oxford University Press.

Smith, L. B., and E. Thelen. 2003. Development as a Dynamic System. *Trends in Cognitive Sciences* 7 (8): 343–348.

Sokolov, E. N. 1963. *Perception and the Conditioned Reflex*. New York: Pergamon.

Spelke, E. S. 1990. "Principles of Object Perception." *Cognitive Science* 14 (1): 29–56.

Spitz, R. A. 1957. *No and Yes: On the Genesis of Human Communication*. Madison, CT: International Universities Press.

Sporns, O., and G. M. Edelman. 1993. "Solving Bernstein Problem—a Proposal for the Development of Coordinated Movement by Selection." *Child Development* 64 (4) (Aug.): 960–981.

Spranger, M. 2012a. "A Basic Emergent Grammar for Space." In *Experiments in Cultural Language Evolution*, ed. L. Steels, 207–232. Amsterdam: John Benjamins.

Spranger, M. 2012b. "The Co-evolution of Basic Spatial Terms and Categories." In *Experiments in Cultural Language Evolution*, 111–141. Amsterdam: John Benjamins.

Stanley, K. O., and R. Miikkulainen. 2003, "A Taxonomy for Artificial Embryogeny." *Artificial Life* 9 (2) (Spring): 93–130.

Starkey, P., and R. G. Cooper. 1995. "The Development of Subitizing in Young-Children." *British Journal of Developmental Psychology* 13 (Nov.): 399–420.

Starkey, P., and R. G. Cooper. 1980. "Perception of Numbers by Human Infants." *Science* 210 (4473): 1033–1035.

Steels, L., ed. 2012. *Experiments in Cultural Language Evolution*. Vol. 3. Amsterdam: John Benjamins.

Steels, L. 2011. *Design Patterns in Fluid Construction Grammar*. Vol. 11. Amsterdam: John Benjamins.

Steels, L. 2003. "Evolving Grounded Communication for Robots." *Trends in Cognitive Sciences* 7 (7) (July): 308–312.

Steels, L., and J. de Beule. 2006. "A (Very) Brief Introduction to Fluid Construction Grammar." In *Proceedings of the Third Workshop on Scalable Natural Language Understanding*, 73–80. Stroudsburg, PA: Association for Computational Linguistics.

Steels, L., and F. Kaplan. 2002. "Aibos First Words: The Social Learning of Language and Meaning." *Evolution of Communication* 4 (1): 3–32.

Stewart, J., O. Gapenne, and E. A. Di Paolo. 2010. *Enaction: Toward a New Paradigm for Cognitive Science*. Cambridge, MA: MIT Press.

Stojanov, G. 2001. "Petitagé: A Case Study in Developmental Robotics." Paper presented at the 1st International Workshop on Epigenetic Robotics: Modeling Cognitive Development in Robotic Systems, Lund, Sweden, September 17–18.

Stoytchev, A. 2005. "Behavior-Grounded Representation of Tool Affordances." Paper presented at the 2005 IEEE International Conference on Robotics and Automation, New York.

Stoytchev, A. 2008. "Learning the Affordances of Tools Using a Behavior-Grounded Approach." In *Towards Affordance-Based Robot Control*, ed. E. Rome, J. Hertzberg, and G. Dorffner, 140–158. Berlin: Springer-Verlag Berlin.

Stoytchev, A. 2011. "Self-Detection in Robots: A Method Based on Detecting Temporal Contingencies." *Robotica* 29 (Jan.): 1–21.

Stramandinoli, F., D. Marocco, and A. Cangelosi. 2012. "The Grounding of Higher Order Concepts in Action and Language: A Cognitive Robotics Model." *Neural Networks* 32 (Aug.): 165–173.

Sturm, J., C. Plagemann, and W. Burgard. 2008. "Unsupervised Body Scheme Learning through Self-Perception." Paper presented at the IEEE International Conference on Robotics and Automation (ICRA), New York.

Sugita, Y., and J. Tani. 2005. "Learning Semantic Combinatoriality from the Interaction between Linguistic and Behavioral Processes." *Adaptive Behavior* 13 (1): 33–52.

Sun, R. 2007. "The Importance of Cognitive Architectures: An Analysis Based on Clarion." *Journal of Experimental & Theoretical Artificial Intelligence* 19 (2) (June): 159–193.

Sun, R., E. Merrill, and T. Peterson. 2001. "From Implicit Skills to Explicit Knowledge: A Bottom-up Model of Skill Learning." *Cognitive Science* 25 (2) (Mar.–Apr.): 203–244.

Sutton, R. S., and A. G. Barto. 1998. *Reinforcement Learning: An Introduction*. Cambridge, MA: MIT Press.

Szeliski, R. 2011. *Computer Vision: Algorithms and Applications*. London: Springer.

Taga, G. 2006. "Nonlinear Dynamics of Human Locomotion: From Real-Time Adaptation to Development." In *Adaptive Motion of Animals and Machines*, ed. H. Kimura, K. Tsuchiya, A. Ishiguro, and H. Witte, 189–204. Tokyo: Springer-Verlag.

Tallerman, M., and K. R. Gibson. 2012. *The Oxford Handbook of Language Evolution*. Oxford, UK: Oxford University Press.

Tanaka, F., A. Cicourel, and J. R. Movellan. 2007. "Socialization between Toddlers and Robots at an Early Childhood Education Center." *National Academy of Sciences of the United States of America* 104 (46): 17954–17958.

Tani, J. 2003. "Learning to Generate Articulated Behavior through the Bottom-Up and the Top-Down Interaction Processes." *Neural Networks* 16 (1): 11–23.

Tanz, J. 2011. "Kinect Hackers Are Changing the Future of Robotics." *Wired Magazine*, June 28. http://www.wired.com/2011/06/mf_kinect.

Tapus, A., M. J. Matarić, and B. Scassellati. 2007. "Socially Assistive Robotics—the Grand Challenges in Helping Humans through Social Interaction." *IEEE Robotics & Automation Magazine* 14 (1) (Mar.): 35–42.

ten Bosch, L., and L. Boves. 2008. "Unsupervised Detection of Words-Questioning the Relevance of Segmentation." Paper presented at ITRW on Speech Analysis and Processing for Knowledge Discovery Workshop, Aalborg, Denmark, June 4–6.

Thelen, E. 1986. "Treadmill-Elicited Stepping in 7-Month-Old Infants." *Child Development* 57 (6) (Dec.): 1498–1506.

Thelen, E., G. Schöner, C. Scheier, and L. B. Smith. 2001. "The Dynamics of Embodiment: A Field Theory of Infant Perseverative Reaching." *Behavioral and Brain Sciences* 24 (1) (Feb.): 1–86.

Thelen, E., and L. B. Smith. 1994. *A Dynamic Systems Approach to the Development of Cognition and Action*. Cambridge, MA: MIT Press.

Thelen, E., and B. D. Ulrich. 1991. "Hidden Skills: A Dynamic Systems Analysis of Treadmill Stepping during the First Year." *Monographs of the Society for Research in Child Development* 56 (1): 1–98, discussion 99–104.

Thelen, E., B. D. Ulrich, and D. Niles. 1987. "Bilateral Coordination in Human Infants—Stepping on a Split-Belt Treadmill." *Journal of Experimental Psychology. Human Perception and Performance* 13 (3) (Aug.): 405–410.

Thill, S., C. A. Pop, T. Belpaeme, T. Ziemke, and B. Vanderborght. 2012. "Robot-Assisted Therapy for Autism Spectrum Disorders with (Partially) Autonomous Control: Challenges and Outlook." *Paladyn Journal of Behavioral Robotics* 3 (4): 209–217.

Thomaz, A. L., M. Berlin, and C. Breazeal. 2005. "An Embodied Computational Model of Social Referencing." Paper presented at the 2005 IEEE International Workshop on Robot and Human Interactive Communication, New York.

Thornton, S. 2008. *Understanding Human Development*. Basingstoke, UK: Palgrave Macmillan.

Thrun, S., and J. J. Leonard. 2008. "Simultaneous Localization and Mapping." In *Springer Handbook of Robotics*, ed. B. Siciliano and O. Khatib, 871–889. Berlin: Springer.

Tikhanoff, V., A. Cangelosi, P. Fitzpatrick, G. Metta, L. Natale, and F. Nori. 2008. "An Open-Source Simulator for Cognitive Robotics Research: The Prototype of the iCub Humanoid Robot Simulator." Paper presented at the 8th Workshop on Performance Metrics for Intelligent Systems, Washington, DC.

Tikhanoff, V., A. Cangelosi, J. F. Fontanari, and L. I. Perlovsky. 2007. "Scaling up of Action Repertoire in Linguistic Cognitive Agents." Paper presented at the International Conference on Integration of Knowledge Intensive Multi-Agent Systems, New York.

Tikhanoff, V., A. Cangelosi, and G. Metta. 2011. "Integration of Speech and Action in Humanoid Robots: iCub Simulation Experiments." *IEEE Transactions on Autonomous Mental Development* 3 (1): 17–29.

Tomasello, M. 2009. *Why We Cooperate*. Vol. 206. Cambridge, MA: MIT Press.

Tomasello, M. 2008. *Origins of Human Communication*. Cambridge, MA: MIT Press.

Tomasello, M. 2003. *Constructing a Language: A Usage-Based Theory of Language Acquisition*. Cambridge, MA: Harvard University Press.

Tomasello, M. 1995. "Language Is Not an Instinct." *Cognitive Development* 10 (1): 131–156.

Tomasello, M. 1992. *First Verbs: A Case Study of Early Grammatical Development*. Cambridge, UK: Cambridge University Press.

Tomasello, M., and P. J. Brooks. 1999. *Early Syntactic Development: A Construction Grammar Approach*. New York: Psychology Press.

Tomasello, M., M. Carpenter, J. Call, T. Behne, and H. Moll. 2005. "Understanding and Sharing Intentions: The Origins of Cultural Cognition." *Behavioral and Brain Sciences* 28 (5) (Oct.): 675–691.

Tomasello, M., M. Carpenter, and U. Liszkowski. 2007. "A New Look at Infant Pointing." *Child Development* 78 (3) (May–June): 705–722.

Touretzky, D. S., and E. J. Tira-Thompson. 2005. "Tekkotsu: A Framework for Aibo Cognitive Robotics." Paper presented at the National Conference on Artificial Intelligence, Menlo Park, CA.

Traver, V. J., and A. Bernardino. 2010. "A Review of Log-Polar Imaging for Visual Perception in Robotics." *Robotics and Autonomous Systems* 58 (4): 378–398.

Trevarthen, C. 1975. "Growth of Visuomotor Coordination in Infants." *Journal of Human Movement Studies* 1: 57.

Triesch, J., C. Teuscher, G. O. Deák, and E. Carlson. 2006. "Gaze Following: Why (Not) Learn It?" *Developmental Science* 9 (2): 125–147.

Trivedi, D., C. D. Rahn, W. M. Kier, and I. D. Walker. 2008. "Soft Robotics: Biological Inspiration, State of the Art, and Future Research." *Applied Bionics and Biomechanics* 5 (3): 99–117.

Tuci, E., T. Ferrauto, A. Zeschel, G. Massera, and S. Nolfi. 2010. "An Experiment on the Evolution of Compositional Semantics and Behaviour Generalisation in Artificial Agents." Special issue on "Grounding Language in Action." *IEEE Transactions on Autonomous Mental Development* 3(2): 1–14.

Turing, A. M. 1950. "Computing Machinery and Intelligence." *Mind* 59 (236): 433–460.

Turkle, S., O. Dasté, C. Breazeal, and B. Scassellati. 2004. "Encounters with Kismet and Cog." Paper presented at the 2004 IEEE-RAS/RSJ International Conference on Humanoid Robots, Los Angeles, November.

Ude, A., and C. G. Atkeson. 2003. "Online Tracking and Mimicking of Human Movements by a Humanoid Robot." *Advanced Robotics* 17 (2): 165–178.

Ude, A., V. Wyart, L. H. Lin, and G. Cheng. 2005. "Distributed Visual Attention on a Humanoid Robot." Paper presented at the 5th IEEE-RAS International Conference on Humanoid Robots, New York.

Ugur, E., M. R. Dogar, M. Cakmak, and E. Sahin. 2007. "The Learning and Use of Traversability Affordance Using Range Images on a Mobile Robot." Paper presented at the 2007 IEEE International Conference on Robotics and Automation (ICRA), New York.

Valenza, E., F. Simion, V. M. Cassia, and C. Umilta. 1996. "Face Preference at Birth." *Journal of Experimental Psychology. Human Perception and Performance* 22 (4) (Aug.): 892–903.

van Leeuwen, L., A. Smitsman, and C. van Leeuwen. 1994. "Affordances, Perceptual Complexity, and the Development of Tool Use." *Journal of Experimental Psychology. Human Perception and Performance* 20 (1): 174–191.

van Sleuwen, B. E., A. C. Engelberts, M. M. Boere-Boonekamp, W. Kuis, T. W. J. Schulpen, and M. P. L'Hoir. 2007. "Swaddling: A Systematic Review." *Pediatrics* 120 (4): e1097–e1106.

van Wynsberghe, A. 2012. "Designing Robots with Care: Creating an Ethical Framework for the Future Design and Implementation of Care Robots." Ph.D. diss., University of Twente, Enschede, Netherlands.

Varela, F. J., E. T. Thompson, and E. Rosch. 1991. *The Embodied Mind: Cognitive Science and Human Experience*. Cambridge, MA: MIT Press.

Vaughan, R. 2008. "Massively Multi-Robot Simulation in Stage." *Swarm Intelligence* 2 (2–4): 189–208.

Vereijken, B., and K. Adolph. 1999. "Transitions in the Development of Locomotion." In *Non-Linear Developmental Processes*, ed. G. J. P. Savelsbergh, H. L. J. VanderMaas, and P. L. C. VanGeert, 137–149. Amsterdam: Elsevier.

Verguts, T., W. Fias, and M. Stevens. 2005. "A Model of Exact Small-Number Representation." *Psychonomic Bulletin & Review* 12 (1): 66–80.

Vernon, D. 2010. "Enaction as a Conceptual Framework for Developmental Cognitive Robotics." *Paladyn Journal of Behavioral Robotics* 1 (2): 89–98.

Vernon, D., G. Metta, and G. Sandini. 2007. "A Survey of Artificial Cognitive Systems: Implications for the Autonomous Development of Mental Capabilities in Computational Agents." *IEEE Transactions on Evolutionary Computation* 11 (2) (Apr.): 151–180.

Vernon, D., C. von Hofsten, and L. Fadiga. 2010. *A Roadmap for Cognitive Development in Humanoid Robots. Cognitive Systems Monographs (COSMOS)*. Vol. 11. Berlin: Springer-Verlag.

Verschure, P. F. M. J., T. Voegtlin, and R. J. Douglas. 2003. "Environmentally Mediated Synergy between Perception and Behaviour in Mobile Robots." *Nature* 425 (6958) (Oct.): 620–624.

Veruggio, G., and F. Operto. 2008. "64. Roboethics: Social and Ethical Implications of Robotics." In *Springer Handbook of Robotics*, ed. B. Siciliano and O. Khatib, 1499–1524. Berlin: Springer.

Vieira-Neto, H., and U. Nehmzow. 2007. "Real-Time Automated Visual Inspection Using Mobile Robots." *Journal of Intelligent & Robotic Systems* 49 (3): 293–307.

Vihman, M. M. 1996. *Phonological Development: The Origins of Language in the Child.* Oxford, UK: Blackwell Publishers.

Vinogradova, O. S. 1975. "The Hippocampus and the Orienting Reflex." In *Neuronal Mechanisms of the Orienting Reflex*, ed. E. N. Sokolov and O. S. Vinogradova, 128–154. Hillsdale, NJ: Erlbaum.

Vogt, P. 2000. "Bootstrapping Grounded Symbols by Minimal Autonomous Robots." *Evolution of Communication* 4 (1): 89–118.

Vollmer, A. L., K. Pitsch, K. S. Lohan, J. Fritsch, K. J. Rohlfing, and B. Wrede. 2010. "Developing Feedback: How Children of Different Age Contribute to a Tutoring Interaction with Adults." Paper presented at the IEEE 9th International Conference on Development and Learning (ICDL), Ann Arbor, MI, August 18–21.

von Hofsten, C. 2007. "Action in Development." *Developmental Science* 10 (1): 54–60.

von Hofsten, C. 1984. "Developmental Changes in the Organization of Prereaching Movements." *Developmental Psychology* 20 (3): 378–388.

von Hofsten, C. 1982. "Eye–Hand Coordination in the Newborn." *Developmental Psychology* 18 (3) (May): 450–461.

von Hofsten, C., and S. Fazel-Zandy. 1984. "Development of Visually Guided Hand Orientation in Reaching." *Journal of Experimental Child Psychology* 38 (2) (Oct.): 208–219.

von Hofsten, C., and L. Rönnqvist. 1993. "The Structuring of Neonatal Arm Movements." *Child Development* 64 (4) (Aug.): 1046–1057.

Vos, J. E., and K. A. Scheepstra. 1993. "Computer-Simulated Neural Networks—an Appropriate Model for Motor Development." *Early Human Development* 34 (1–2) (Sept.): 101–112.

Vygotsky, L. L. S. 1978. *Mind in Society: The Development of Higher Psychological Processes.* Cambridge, MA: Harvard University Press.

Wainer, J., K. Dautenhahn, B. Robins, and F. Amirabdollahian. 2010. "Collaborating with Kaspar: Using an Autonomous Humanoid Robot to Foster Cooperative Dyadic Play among Children with Autism." Paper presented at the Humanoid Robots (Humanoids), 10th IEEE-RAS International Conference, Nashville, TN, December 6–8.

Wainer, J., K. Dautenhahn, B. Robins, and F. Amirabdollahian. 2013. "A Pilot Study with a Novel Setup for Collaborative Play of the Humanoid Robot KASPAR with Children with Autism." *International Journal of Social Robotics* 6 (1): 45–65.

Wakeley, A., S. Rivera, and J. Langer. 2000. "Can Young Infants Add and Subtract?" *Child Development* 71 (6) (Nov.–Dec.): 1525–1534.

Wallach, W., and C. Allen. 2008. *Moral Machines: Teaching Robots Right from Wrong*. Oxford, UK: Oxford University Press.

Walters, M. L., K. Dautenhahn, R. Te Boekhorst, K. L. Koay, D. S. Syrdal, and C. L. Nehaniv. 2009. "An Empirical Framework for Human-Robot Proxemics." Paper presented at the New Frontiers in Human-Robot Interaction (symposium at the AISB2009 Convention), Heriot Watt University, Edinburgh, UK, April 8–9.

Walters, M. L., D. S. Syrdal, K. Dautenhahn, R. te Boekhorst, and K. L. Koay. 2008. "Avoiding the Uncanny Valley: Robot Appearance, Personality and Consistency of Behavior in an Attention-Seeking Home Scenario for a Robot Companion." *Autonomous Robots* 24 (2): 159–178.

Wang, H., T. R. Johnson, and J. Zhang. 2001. "The Mind's Views of Space." Paper presented at the 3rd International Conference of Cognitive Science, Tehran, Iran, May 10–12.

Warneken, F., F. Chen, and M. Tomasello. 2006. "Cooperative Activities in Young Children and Chimpanzees." *Child Development* 77 (3) (May–June): 640–663.

Warneken, F., and M. Tomasello. 2006. "Altruistic Helping in Human Infants and Young Chimpanzees." *Science* 311 (5765): 1301–1303.

Watanabe, A., M. Ogino, and M. Asada. 2007. "Mapping Facial Expression to Internal States Based on Intuitive Parenting." *Journal of Robotics and Mechatronics* 19 (3): 315–323.

Wauters, L. N., A. E. J. M. Tellings, W. H. J. van Bon, and A. W. van Haafren. 2003. "Mode of Acquisition of Word Meanings: The Viability of a Theoretical Construct." *Applied Psycholinguistics* 24 (3) (July): 385–406.

Wei, L., C. Jaramillo, and L. Yunyi. 2012. "Development of Mind Control System for Humanoid Robot through a Brain Computer Interface." Paper presented at the 2nd International Conference on Intelligent System Design and Engineering Application (ISDEA), Sanya, Hainan, January 6–7.

Weir, S., and R. Emanuel. 1976. *Using Logo to Catalyse Communication in an Autistic Child*. Edinburgh, UK: Department of Artificial Intelligence, University of Edinburgh.

Weng, J. 2004. "A Theory of Developmental Architecture." Paper presented at the 3rd International Conference on Development and Learning, La Jolla, CA, October 20–22.

Weng, J. Y., J. McClelland, A. Pentland, O. Sporns, I. Stockman, M. Sur, and E. Thelen. 2001. "Artificial Intelligence—Autonomous Mental Development by Robots and Animals." *Science* 291 (5504) (Jan.): 599–600.

Wentworth, N., M. M. Haith, and R. Hood. 2002. "Spatiotemporal Regularity and Interevent Contingencies as Information for Infants' Visual Expectations." *Infancy* 3 (3): 303–321.

Westermann, G., and E. R. Miranda. 2004. "A New Model of Sensorimotor Coupling in the Development of Speech." *Brain and Language* 82 (2): 393–400.

Wetherford, M. J., and L. B. Cohen. 1973. "Developmental Changes in Infant Visual Preferences for Novelty and Familiarity." *Child Development* 44 (3): 416–424.

Wheeler, D. S., A. H. Fagg, and R. A. Grupen. 2002. "Learning Prospective Pick and Place Behavior." Paper presented at the 2nd International Conference on Development and Learning (ICDL 2002), Massachusetts Institute of Technology, Cambridge, MA, June 12–15.

White, B. L., P. Castle, and R. Held. 1964. "Observations on the Development of Visually-Directed Reaching." *Child Development* 35: 349–364.

White, R. W. 1959. "Motivation Reconsidered: The Concept of Competence." *Psychological Review* 66: 297–333.

Wiemer-Hastings, K., J. Krug, and X. Xu. 2001. "Imagery, Context Availability, Contextual Constraint, and Abstractness." Paper presented at the 23rd Annual Conference of the Cognitive Science Society, Edinburgh, Scotland, August 1–4.

Wilson, M. 2002. "Six Views of Embodied Cognition." *Psychonomic Bulletin & Review* 9 (4) (Dec.): 625–636.

Witherington, D. C. 2005. "The Development of Prospective Grasping Control between 5 and 7 Months: A Longitudinal Study." *Infancy* 7 (2): 143–161.

Wolfe, J. M. 1994. "Guided Search 2.0: A Revised Model of Visual Search." *Psychonomic Bulletin & Review* 1 (2): 202–238.

Wolpert, D. M., and M. Kawato. 1998. "Multiple Paired Forward and Inverse Models for Motor Control." *Neural Networks* 11 (7–8) (Oct.–Nov.): 1317–1329.

Wood, L. J., K. Dautenhahn, A. Rainer, B. Robins, H. Lehmann, and D. S. Syrdal. 2013. "Robot-Mediated Interviews—How Effective Is a Humanoid Robot as a Tool for Interviewing Young Children?" *PLoS ONE* 8 (3) (Mar): e59448.

Wright, J. S., and J. Panksepp. 2012. "An Evolutionary Framework to Understand Foraging, Wanting, and Desire: The Neuropsychology of the Seeking System." *Neuropsychoanalysis: An Interdisciplinary Journal for Psychoanalysis and the Neurosciences* 14 (1): 5–39.

Wu, Q. D., C. J. Liu, J. Q. Zhang, and Q. J. Chen. 2009. "Survey of Locomotion Control of Legged Robots Inspired by Biological Concept." *Science in China Series F-Information Sciences* 52 (10): 1715–1729.

Wynn, K. 1998. "Psychological Foundations of Number: Numerical Competence in Human Infants." *Trends in Cognitive Sciences* 2 (8): 296–303.

Wynn, K. 1992. "Addition and Subtraction by Human Infants." *Nature* 358 (6389) (Aug.): 749–750.

Wynn, K. 1990. "Children's Understanding of Counting." *Cognition* 36 (2) (Aug.): 155–193.

Xu, F., and E. S. Spelke. 2000. "Large Number Discrimination in Six-Month-Old Infants." *Cognition* 74 (1): B1–B11.

Yamashita, Y., and J. Tani. 2008. "Emergence of Functional Hierarchy in a Multiple Timescale Neural Network Model: A Humanoid Robot Experiment." *PLoS Computational Biology* 4 (11): e1000220.

Yim, M., W. M. Shen, B. Salemi, D. Rus, M. Moll, H. Lipson, E. Klavins, and G. S. Chirikjian. 2007. "Modular Self-Reconfigurable Robot Systems—Challenges and Opportunities for the Future." *IEEE Robotics & Automation Magazine* 14 (1): 43–52.

Yoshikawa, Y., M. Asada, K. Hosoda, and J. Koga. 2003. "A Constructivist Approach to Infants' Vowel Acquisition through Mother–Infant Interaction." *Connection Science* 15 (4): 245–258.

Yu, C. 2005. "The Emergence of Links between Lexical Acquisition and Object Categorization: A Computational Study." *Connection Science* 17 (3–4) (Sept.–Dec.): 381–397.

Yucel, Z., A. A. Salah, C. Mericli, and T. Mericli. 2009. "Joint Visual Attention Modeling for Naturally Interacting Robotic Agents." Paper presented at the 24th International Symposium on Computer and Information Sciences (ISCIS), New York.

Yürüten, O., K. F. Uyanık, Y. Çalı⊠kan, A. K. Bozcuo⊠lu, E. ⊠ahin, and S. Kalkan. 2012. "Learning Adjectives and Nouns from Affordances on the iCub Humanoid Robot." Paper presented at the SAB-2012 Simulation of Adaptive Behavior Conference, Odense.

Zelazo, P. R., N. A. Zelazo, and S. Kolb. 1972. "'Walking' in the Newborn." *Science* 176 (4032): 314–315.

Zhang, X., and M. H. Lee. 2006. "Early Perceptual and Cognitive Development in Robot Vision." Paper presented at the 9th International Conference on Simulation of Adaptive Behavior (SAB 2006), Berlin, Germany.

Ziemke, T. 2003. "On the Role of Robot Simulations in Embodied Cognitive Science." *AISB Journal* 1 (4): 389–399.

Ziemke, T. 2001. "Are Robots Embodied?" Paper presented at the 1st International Workshop on Epigenetic Robotics: Modeling Cognitive Development in Robotic Systems, Lund, Sweden, September 17–18.

Zlatev, J., and C. Balkenius. 2001. "Introduction: Why 'Epigenetic Robotics'?: Paper presented at the 1st International Workshop on Epigenetic Robotics: Modeling Cognitive Development in Robotic Systems, Lund, Sweden, September 17–18.

Zöllner, R., T. Asfour, and R. Dillmann. 2004. "Programming by Demonstration: Dual-Arm Manipulation Tasks for Humanoid Robots." Paper presented at the IEEE/RSJ International Conference on Intelligent Robots and Systems, Sendai, Japan, September 28–October 2.

Index of Names

Index of Subjects